BUS

Radios
of the Baby Boom Era
1946 to 1960

VOLUME 4
Motorola to RCA Victor

PROMPT™
PUBLICATIONS

Imprint of Howard W. Sams & Company
Indianapolis, Indiana 46214-2012

For radio enthusiasts everywhere,
keeping the memories alive

Special thanks to Doug Heimstead, whose knowledge of and enthusiasm for old radios
was instrumental in creating this series. Thanks also to George Fathauer, Jr., of Antique
Electronic Supply, and John Terrey of Antique Radio Classified, for their helpful
suggestions.

PHOTOFACT® is a registered trademark of Howard W. Sams & Company

Cover Illustration: Ned Shaw Studios
Cover Production: Woodfield Hi-Tech Prepress
Database Design: Kevin Etter, Bill Skinner, Bill Smith WB9TUI, Barry Buchanan
Database Input: Kelly Jenkins, Deloris Bibb
Text: Brian McCaffrey, Wendy Ford
Technical Advisers: Bill Fink, Leon Lewis, Stan Scott, Dave Stitt, George Weliver
Editing: Bill Skinner, Ken Cobb
PHOTOFACT Research: Demetrice Bruno, Angelina Cooper, Lisa Gorman, Crystal
Johnson, Rose Smith
Production Assistance: Jim Beever, Lynne Clark, Alejandro Heyartz, Michele Lambert,
Jonathan Mangin
Printing: Ripon Community Printers, Ripon, Wisconsin

FIRST EDITION

ISBN 0-7906-1005-1

Library of Congress Catalog Number: 91-61691

Printed in the United States of America

CONTENTS

SAMS PHOTOFACT® PICTURE GUIDES

Radios of the Baby Boom Era is only one of a whole series of picture references to your favorite home entertainment devices of the years 1946 to 1960.

We have assembled thousands of photographs of old and collectible *televisions, car radios,* and *hi-fi components* from the PHOTOFACT files and bound them in handy book form.

Each book presents technical data, helpful indexes, and hints for restoring these devices to working order.

Look for Sams PHOTOFACT Picture Guides at your local authorized Sams distributor or your nearest bookstore.

INTRODUCTION

The baby boom years, 1946 to 1960...a time of bebop, rhythm-and-blues, and rock-and-roll, a time that forever transformed the views and values of America's youth...and radio played a central role. Radio makers produced thousands of different brands and models to satisfy eager consumers, from consoles to table models to portables.

Whether you're looking for one particular radio that sparks fond memories or are building a wide-ranging collection, this book can help you identify your radio and obtain the information needed to bring it back to operating condition.

The period covered by the *Radios of the Baby Boom Era* series produced profound changes in American culture, and radio was instrumental in bringing them about. From Frank Sinatra and the bobby soxers to Elvis and beyond, radio is intertwined with people's memories of the best years of the 1940s and 1950s. Radio is a key element of the nostalgia for this period, and the great variety of radios produced during those years are now becoming a nationwide focus for collectors.

Radio in the Baby Boom Years

Radio itself underwent a metamorphosis during the years after World War II, not because of dramatic changes in technology, but because of how people came to use it in their daily lives.

From its beginning, radio had wrought significant changes in our culture. Through the 1920s and 1930s it was a daily indulgence of millions, bringing sports, drama, comedy, news, and music to virtually every corner of the nation. Network broadcasting, with its standardized live programming, created a common thread which helped unify the nation.

When war came, involving the United States in a massive military effort that sealed its place as the world's leading economic and military power, radio played a key role in creating awareness of the situation. America learned of Japan's attack on Pearl Harbor through President Roosevelt's stirring "Day of Infamy" speech. Throughout World War II, radio was a vital communications and entertainment medium...a constant companion for civilians and soldiers alike, bringing news and entertainment which rallied the country to the cause and saw it through to victory.

The end of the war brought a new world order and an audience with different wants, needs, and values. This was a major crisis for radio, a crisis deepened by the emergence of television and its quick rise into the information and entertainment medium of choice. Radio was about to go through a profound change. Some even said that radio would die.

But the predictions of radio's death were greatly exaggerated. While TV became the focal point of entertainment for the family, radio became the companion to the individual. It provided a portability which TV couldn't match, with music and news as the predominant programming. Radio experienced a metamorphosis that reflected the new America...mobile, affluent, and commercialized.

Radio Manufacturers

Consumers were eager to buy radios, and manufacturers obliged by producing a wide assortment of models, some simple, some elegant, some whimsical, some disguised as other objects. Radio manufacturers came and went, others made fortunes. Some of the leading names were:

- *Admiral* Started during the depression on a $3,400 investment, it rose to be one of the largest corporations in America because it dared to be innovative.

- *Arvin* Which began by making automobile heaters and became one of a very few manufacturers to produce a metal cabinet radio.

- *Fada* Highly prized by collectors because of their distinctive cabinet designs by Catalin Corporation.

- *Philco* Rose from near-bankruptcy as a producer of batteries to become one of the top three names in radio.

- *RCA* Another of the "Big Three" in radio whose productivity in developing and manufacturing vacuum tubes was a major factor in developing the entire radio-electronics industry.

- *Zenith* The third of the "Big Three" whose involvement in radio spans the history of the industry from 1918 to the present.

Howard W. Sams and Radio

While manufacturers turned out thousands of radios, Howard W. Sams built a company to supply the people who repaired them--service technicians. In those days the average tube radio needed servicing six times a year, and technicians used Sams service data to make the job easier. That service data was called Sams PHOTOFACT, because it included a cabinet photo along with basic facts to help the repairer: a schematic, a list of vacuum tubes and other replaceable parts, power supply information, tuning ranges, adjustment and alignment guidelines, and troubleshooting techniques. Soon home handymen and hobbyists discovered PHOTOFACT and used them to repair their own radios at home.

The pictures and index data in this book are taken from those early PHOTOFACT sets, which are available to you today to help restore your old radio. The details of how to order a specific PHOTOFACT are found with the repair information at the back of this book.

Collecting Classic Radios

Collecting has been a distinctly American passion. It's a passion partly to possess things which have significant meaning to us, partly for financial gain, and partly for the thrill of the hunt. The amount of money invested or, indeed, the overall worth of the collection are strictly side issues. The real thrill of collecting is to possess a piece of the past. The real value of a collection is assessed in the satisfaction gained through the collecting process.

For years, collectors have focused on pre-World War II radios, some of which brought great emotional and financial rewards to those who pursued them. Unlike the pre-World War II sets, most radios of the baby boom era are still in relatively plentiful supply, creating the opportunity to get in on the ground floor and build an extremely diverse collection which will bring years of pleasure.

To get started in collecting, you can join a national club such as the Antique Wireless Association or the Antique Radio Club of America, or find a local antique radio club and start haunting their swap meets.

Subscribe to publications such as *Antique Radio Classified* and *Radio Age,* and get on the mailing lists for Antique Electronic Supply and for Vintage TV & Radio.

How to Use This Book

Radio Photos   The old radios themselves are the "stars" of this series, and you'll learn a lot just by browsing through the hundreds of pictures. If you know what a radio looks like but not the brand name or model number, perhaps you'll recognize it among these photos. With each picture, we furnish a description and basic technical information that helps in classifying the radio.

Indexes   The indexes provide a wealth of useful information cross-referenced to the pictured radios. If you know the model number of a radio, you can look it up to find out on which page it or a similar model is pictured, which tubes it uses, and which tubes would be suitable replacements. If you know the manufacturer of a radio but not the brand name, there's an index that summarizes all the manufacturers and brand names of the period.

Fix-Up Information   Because the purpose of this series is not only to list radios, but also to help in restoring them to operating condition, we provide a basic troubleshooting guide for working on tube radios. We also help you find sources for parts, collector's information, and radio restoration aids. There's a schematic of a typical 5-tube radio to help you understand the circuitry.

We hope you enjoy reading and using this book as much as we did bringing it to you. For those of us who appreciate what radio has accomplished over the years, it was a labor of love. Happy dialing!

THE RADIOS

From Motorola to RCA Victor, these pictures from the Sams PHOTOFACT library represent the radios that were America's constant companions during the baby boom years. From the living room to the kitchen and the bedroom to the beach, they provided continuous entertainment whether we were at work, at rest, or at play.

About the Brands in This Volume

Volume 4 shows 36 brand names, with pictures of more than 675 models and more than 1300 other models listed in the index. Some of the more noteworthy brands and models include:

- *Motorola*   Model 5A7 (p.9)   The radio with the flip-top box. It plays when you open the lid.

- *Packard-Bell*   Model 100 (p.143)   A throwback to an earlier time with its attractive Deco styling.

- *Philco*   Model 49-501 Transitone (p.181)   From Deco to Modern. Its left arched grill reminds us of a 1957 Plymouth.

- *Porto Baradio*   Model PB-520 (p.240)   The party machine of the 50s, complete with glassware and beverages.

- *RCA Victor*   Model 8BX5 (p.265)   Snakes alive! Hang this imitation snakeskin beauty from your shoulder for music on the move.

Names to look for in Volumes 1 through 3 and in Volumes 5 and 6 include Air King, Automatic, DeWald, Fada, General Electric, Remler, and Zenith.

The radio pictures are grouped together by brand name. Within brand names, they are organized roughly by historical sequence as determined by when their specific PHOTOFACT set was published.

TOP TEN NETWORK RADIO PROGRAMS OF 1946	TOP TEN NETWORK RADIO PROGRAMS OF 1957
1. Jack Benny	1. Our Miss Brooks
2. Fibber McGee and Molly	2. Edgar Bergen
3. Bob Hope	3. Two for the Money
4. The Charlie McCarthy Show	4. Dragnet
5. Fred Allen	5. News from NBC
6. Radio Theatre	6. Gene Autry
7. Amos 'n' Andy	7. The Great Gildersleeve
8. Walter Winchell	8. You Bet Your Life
9. Red Skelton	9. Gunsmoke
10. The Screen Guild Players	10. People Are Funny

TOP TEN DAYTIME RADIO PROGRAMS OF 1946	TOP TEN DAYTIME RADIO PROGRAMS OF 1957
1. When a Girl Marries	1. Wendy Warren
2. Young Widder Brown	2. Arthur Godfrey
3. Our Gal, Sunday	3. Helen Trent
4. Portia Faces Life	4. Guiding Light
5. Kate Smith Speaks	5. Young Dr. Malone
6. Ma Perkins	6. Our Gal Sunday
7. Breakfast in Hollywood	7. The Second Mrs. Burton
8. Aunt Jenny	8. The Road of Life
9. Right to Happiness	9. Aunt Jenny
10. The Romance of Helen Trent	10. My True Story

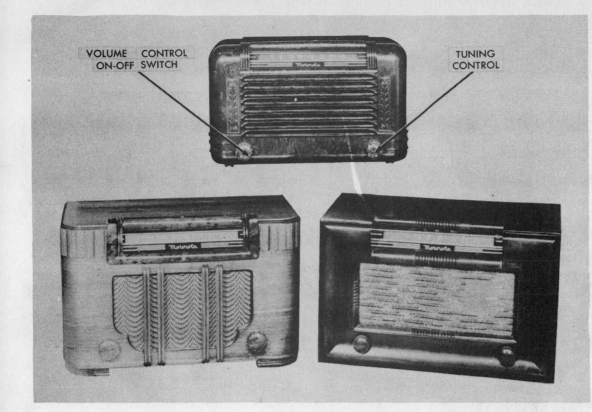

VOLUME CONTROL ON-OFF SWITCH

TUNING CONTROL

MOTOROLA

MODEL PICTURED
65X14A

AC/DC operated AM superheterodyne receiver with self-contained loop antenna

TUBES
6

POWER SUPPLY
117 volts AC/DC

TUNING RANGE
538-1720KC

MFR/SUPPLIER
Motorola Inc.

PHOTOFACT SET
4-8

PUBLISHED
1946

MOTOROLA

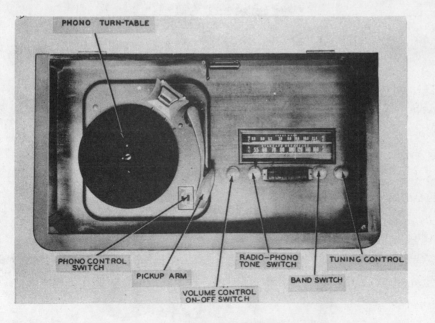

PHONO TURN-TABLE

PHONO CONTROL SWITCH

PICKUP ARM

VOLUME CONTROL ON-OFF SWITCH

RADIO-PHONO TONE SWITCH

BAND SWITCH

TUNING CONTROL

MODEL PICTURED
65F21

AC operated phono-radio combo two-band superheterodyne receiver with self-contained loop antenna

TUBES
6

POWER SUPPLY
117 volts AC

TUNING RANGE
535-1620KC,
5.6-12.2MC

MFR/SUPPLIER
Motorola Inc.

PHOTOFACT SET
4-12

PUBLISHED
1946

MOTOROLA

MODEL PICTURED
55F11

AC operated phono-radio combo AM superheterodyne receiver with self-contained loop antenna

TUBES
5

POWER SUPPLY
117 volts AC

TUNING RANGE
535-1620KC

MFR/SUPPLIER
Motorola Inc.

PHOTOFACT SET
4-14

PUBLISHED
1946

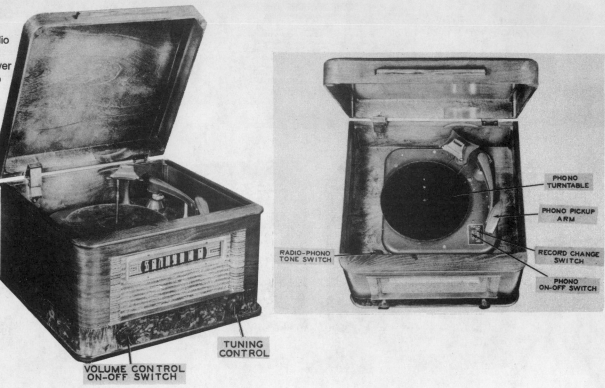

MOTOROLA

MODEL PICTURED
85K21

AC operated two-band superheterodyne receiver with self-contained loop antenna and automatic tuning

TUBES
8

POWER SUPPLY
117 volts AC

TUNING RANGE
535-1620KC,
5.6-12.2MC

MFR/SUPPLIER
Motorola Inc.

PHOTOFACT SET
5-3

PUBLISHED
1946

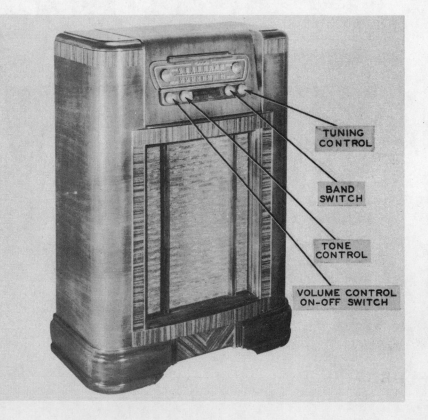

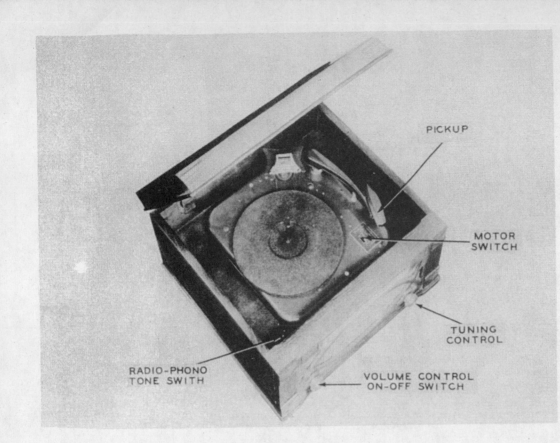

PICKUP

MOTOR
SWITCH

TUNING
CONTROL

RADIO-PHONO
TONE SWITH

VOLUME CONTROL
ON-OFF SWITCH

MOTOROLA

MODEL PICTURED
65F11

AC operated phono-radio
combo AM
superheterodyne receiver
with self-contained loop
antenna

TUBES
6

POWER SUPPLY
117 volts AC

TUNING RANGE
535-1620KC

MFR/SUPPLIER
Motorola Inc.

PHOTOFACT SET
6-19

PUBLISHED
1946

RADIO TONE
PHONO-RADIO
PHONO TONE
SWITCH

BAND
SWITCH

TUNING
CONTROL

VOLUME CONTROL
ON-OFF SWITCH

PHONO
TURNTABLE

PHONO
PICKUP

PHONO CONTROL
SWITCH

MOTOROLA

MODEL PICTURED
85F21

AC operated phono-radio
combo two-band
superheterodyne receiver

TUBES
8

POWER SUPPLY
117 volts AC

TUNING RANGE
535-1620KC,
5.6-12.2MC

MFR/SUPPLIER
Motorola Inc.

PHOTOFACT SET
6-20

PUBLISHED
1946

MOTOROLA

MODEL PICTURED
65L11

Three-power operated
portable AM
superheterodyne receiver
with self-contained loop
antenna

TUBES
6

POWER SUPPLY
117 volts AC/DC or two
4.5 volt batteries

TUNING RANGE
535-1600KC

MFR/SUPPLIER
Motorola Inc.

PHOTOFACT SET
8-22

PUBLISHED
1946

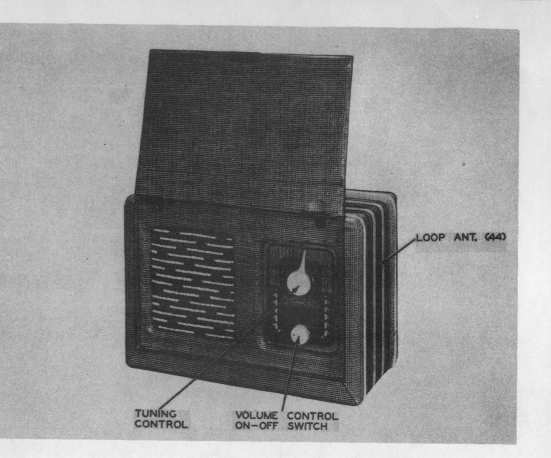

LOOP ANT. (44)

TUNING
CONTROL

VOLUME CONTROL
ON-OFF SWITCH

MOTOROLA

MODEL PICTURED
45B12

Battery operated AM
superheterodyne receiver

TUBES
4

POWER SUPPLY
1.5 volts A & 90 volts B
supply

TUNING RANGE
538-1720KC

MFR/SUPPLIER
Motorola Inc.

PHOTOFACT SET
9-23

PUBLISHED
1946

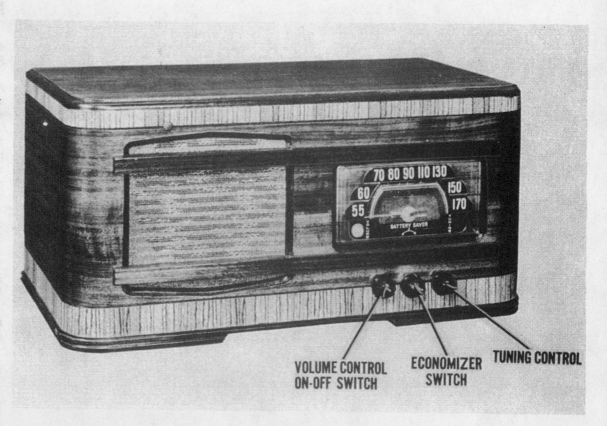

VOLUME CONTROL
ON-OFF SWITCH

ECONOMIZER
SWITCH

TUNING CONTROL

MOTOROLA

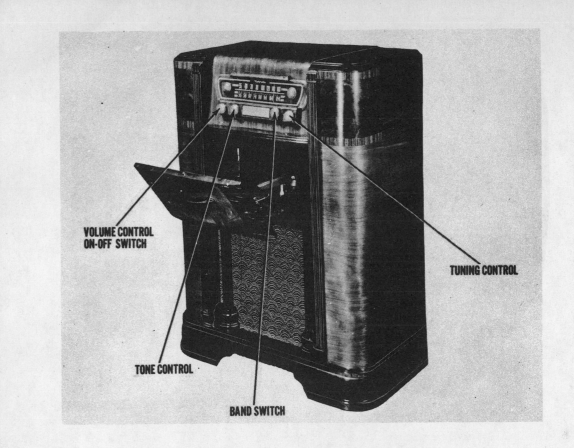

VOLUME CONTROL
ON-OFF SWITCH

TUNING CONTROL

TONE CONTROL

BAND SWITCH

MODEL PICTURED
75F21

AC operated phono-radio
combo AM
superheterodyne receiver
with loop antenna

TUBES
7

POWER SUPPLY
110-120 volts AC

TUNING RANGE
540-1620KC,
5.6-12.2MC

MFR/SUPPLIER
Motorola Inc.

PHOTOFACT SET
19-21

PUBLISHED
1947

MOTOROLA

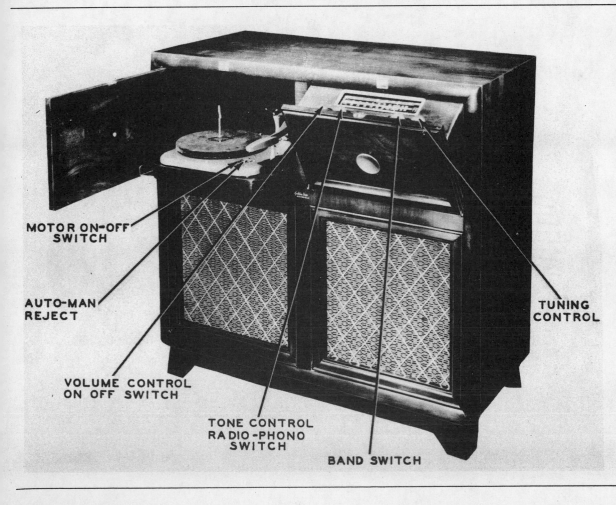

MOTOR ON-OFF
SWITCH

AUTO-MAN
REJECT

VOLUME CONTROL
ON OFF SWITCH

TONE CONTROL
RADIO-PHONO
SWITCH

BAND SWITCH

TUNING
CONTROL

MODEL PICTURED
95F31

AC operated phono-radio
combo FM-AM
superheterodyne receiver
with self-contained loop
antenna

TUBES
9

POWER SUPPLY
110-120 volts AC

TUNING RANGE
535-1620KC,
88-108MC,
5.6-12.2MC

MFR/SUPPLIER
Motorola Inc.

PHOTOFACT SET
19-22

PUBLISHED
1947

MOTOROLA

MODEL PICTURED
56X11

AC/DC operated AM
superheterodyne receiver
with loop antenna

TUBES
5

POWER SUPPLY
110-120 volts AC/DC

TUNING RANGE
538-1720KC

MFR/SUPPLIER
Motorola Inc.

PHOTOFACT SET
28-24

PUBLISHED
1947

VOLUME CONTROL
ON-OFF SWITCH   TUNING CONTROL

MOTOROLA

MODEL PICTURED
57X11, 57X12

AC/DC operated AM
superheterodyne receiver
with loop antenna

TUBES
5

POWER SUPPLY
110-120 volts AC/DC

TUNING RANGE
535-1620KC

MFR/SUPPLIER
Motorola Inc.

PHOTOFACT SET
28-25

PUBLISHED
1947

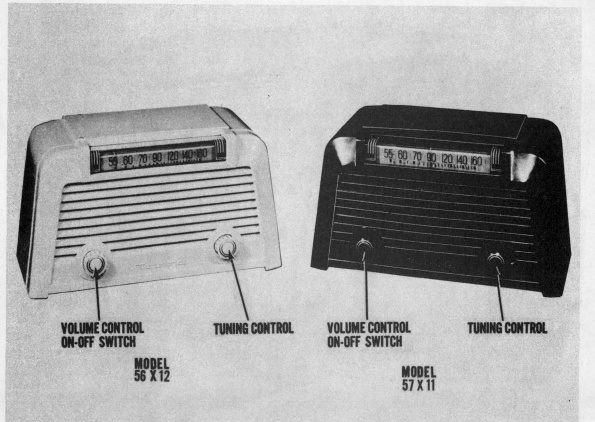

VOLUME CONTROL
ON-OFF SWITCH   TUNING CONTROL   VOLUME CONTROL
ON-OFF SWITCH   TUNING CONTROL

MODEL
56 X 12

MODEL
57 X 11

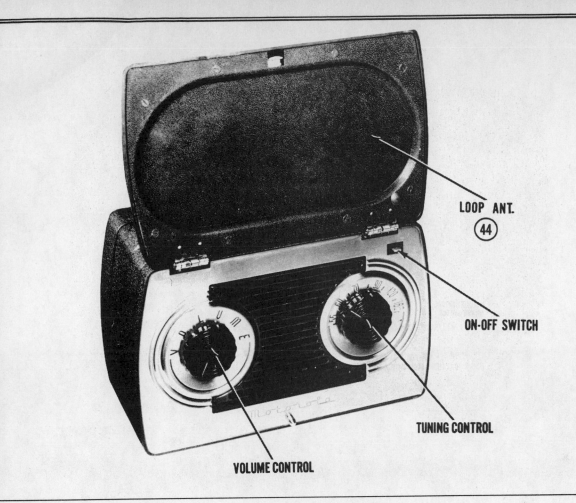

LOOP ANT.
(44)

ON-OFF SWITCH

TUNING CONTROL

VOLUME CONTROL

MOTOROLA

MODEL PICTURED
5A7

Three-power operated portable AM superheterodyne receiver with loop antenna

TUBES
4

POWER SUPPLY
110-120 volts AC/DC or 3 volts A & 67.5 volts B supply

TUNING RANGE
535-1620KC

MFR/SUPPLIER
Motorola Inc.

PHOTOFACT SET
29-16

PUBLISHED
1947

VOLUME CONTROL
ON-OFF SWITCH

BATTERY
SAVER
SWITCH

TUNING CONTROL

MOTOROLA

MODEL PICTURED
47B11

Battery operated AM superheterodyne receiver

TUBES
4

POWER SUPPLY
1.5 volts A & 90 volts B supply

TUNING RANGE
538-1720KC

MFR/SUPPLIER
Motorola Inc.

PHOTOFACT SET
29-17

PUBLISHED
1947

MOTOROLA

MODEL PICTURED
75F31

AC operated phono-radio
combo AM-FM-SW
superheterodyne receiver
with loop antenna

TUBES
7

POWER SUPPLY
110-120 volts AC

TUNING RANGE
535-1620KC,
88-108MC,
5.6-12.2MC

MFR/SUPPLIER
Motorola Inc.

PHOTOFACT SET
29-18

PUBLISHED
1947

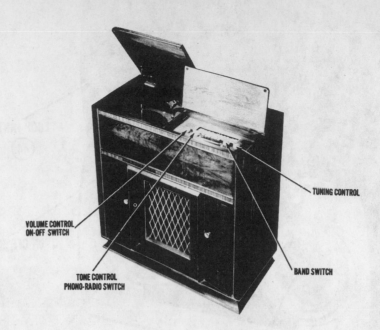

VOLUME CONTROL
ON-OFF SWITCH

TONE CONTROL
PHONO-RADIO SWITCH

TUNING CONTROL

BAND SWITCH

MOTOROLA

MODEL PICTURED
67X11, 67X12, 67X13

AC/DC operated AM
superheterodyne receiver
with loop antenna

TUBES
6

POWER SUPPLY
110-120 volts AC/DC

TUNING RANGE
535-1620KC

MFR/SUPPLIER
Motorola Inc.

PHOTOFACT SET
30-20

PUBLISHED
1947

MODEL 67X13

MODEL 67X12

MODEL 67X11

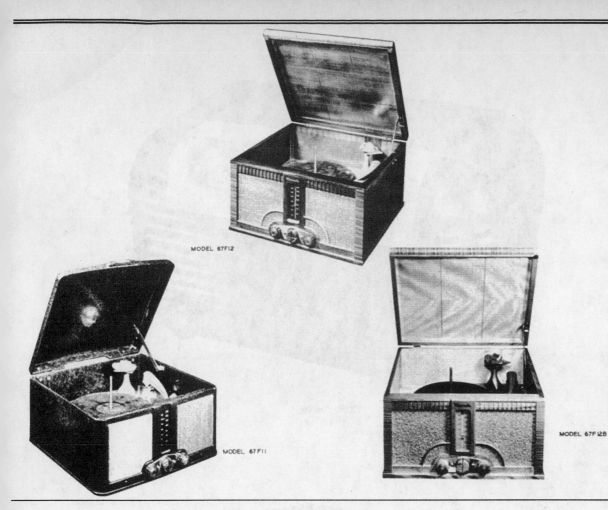

MODEL 67F12

MODEL 67F11

MODEL 67F12B

MOTOROLA

MODEL PICTURED
67F11,67F12, 67F12B

AC operated phono-radio combo AM superheterodyne receiver with loop antenna

TUBES
6

POWER SUPPLY
110-120 volts AC

TUNING RANGE
535-1620KC

MFR/SUPPLIER
Motorola Inc.

PHOTOFACT SET
31-20

PUBLISHED
1948

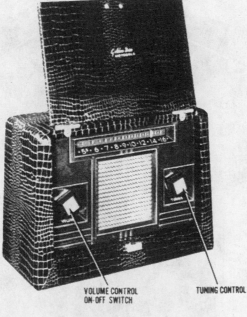

VOLUME CONTROL
ON-OFF SWITCH

TUNING CONTROL

MOTOROLA

MODEL PICTURED
67L11

Three-power operated portable AM superheterodyne receiver with loop antenna

TUBES
5

POWER SUPPLY
110-120 volts AC/DC or 9 volts A & 90 volts B supply

TUNING RANGE
535-1620KC

MFR/SUPPLIER
Motorola Inc.

PHOTOFACT SET
31-21

PUBLISHED
1948

MOTOROLA

MODEL PICTURED
67XM21

AC/DC operated AM-FM
superheterodyne receiver
with loop antenna

TUBES
5

POWER SUPPLY
105-125 volts AC/DC

TUNING RANGE
535-1620KC,
88-108MC

MFR/SUPPLIER
Motorola Inc.

PHOTOFACT SET
32-14

PUBLISHED
1948

TONE SWITCH

VOLUME CONTROL
ON-OFF SWITCH

TUNING CONTROL

BAND SWITCH

MOTOROLA

MODEL PICTURED
77FM21

AC operated phono-radio
combo AM-FM
superheterodyne receiver
with loop antenna

TUBES
6

POWER SUPPLY
110-120 volts AC

TUNING RANGE
535-1620KC,
88-108MC

MFR/SUPPLIER
Motorola Inc.

PHOTOFACT SET
33-13

PUBLISHED
1948

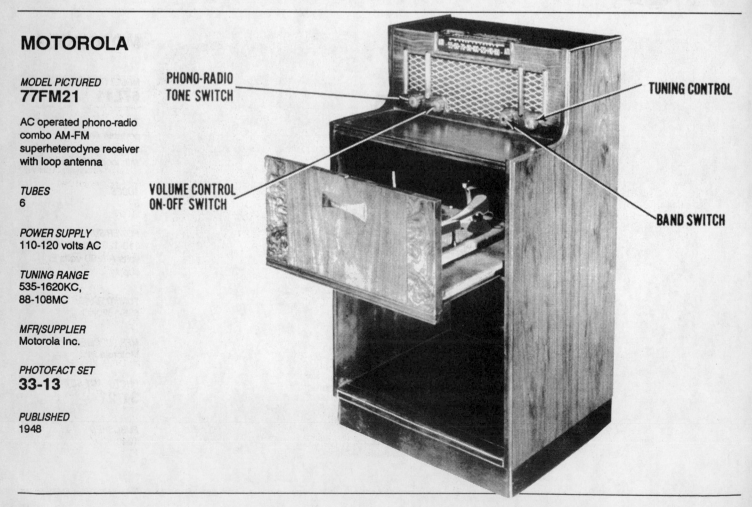

PHONO-RADIO
TONE SWITCH

TUNING CONTROL

VOLUME CONTROL
ON-OFF SWITCH

BAND SWITCH

MOTOROLA

MODEL PICTURED
107F31

AC operated phono-radio combo AM-FM-SW superheterodyne receiver with loop antenna

TUBES
10

POWER SUPPLY
110-120 volts AC

TUNING RANGE
535-1620KC,
88-108MC,
5.5-12.2MC

MFR/SUPPLIER
Motorola Inc.

PHOTOFACT SET
33-14

PUBLISHED
1948

TUNING CONTROL

BAND SWITCH

VOLUME CONTROL
ON-OFF SWITCH

TONE CONTROL
PHONO-RADIO SWITCH

MODEL 77XM22

MODEL 77XM22B

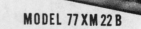

MODEL 77XM21

MOTOROLA

MODEL PICTURED
**77XM21,
77XM22,
77XM22B**

AC/DC operated AM-FM superheterodyne receiver with loop antenna

TUBES
6

POWER SUPPLY
105-125 volts AC/DC

TUNING RANGE
535-1620KC,
88-108MC

MFR/SUPPLIER
Motorola Inc.

PHOTOFACT SET
34-12

PUBLISHED
1948

MOTOROLA

MODEL PICTURED
58L11

Three-power operated
portable AM
superheterodyne receiver
with loop antenna

TUBES
4

POWER SUPPLY
110-120 volts AC/DC or 3
volts A & 67.5 volts B
supply

TUNING RANGE
540-1650KC

MFR/SUPPLIER
Motorola Inc.

PHOTOFACT SET
45-17

PUBLISHED
1948

VOLUME CONTROL

POWER
SWITCH

TUNING CONTROL

MOTOROLA

MODEL PICTURED
68L11

Three-power operated
portable AM
superheterodyne receiver
with loop antenna

TUBES
5

POWER SUPPLY
110-120 volts AC/DC or 9
volts A & 90 volts B
supply

TUNING RANGE
535-1620KC

MFR/SUPPLIER
Motorola Inc.

PHOTOFACT SET
45-18

PUBLISHED
1948

VOLUME CONTROL
ON-OFF SWITCH

TUNING CONTROL

A8

VOLUME CONTROL ON-OFF SWITCH

TUNING CONTROL

MOTOROLA

MODEL PICTURED
48L11

Battery operated portable AM superheterodyne receiver with loop antenna

TUBES
4

POWER SUPPLY
1.5 volts A & 67.5 volts B supply

TUNING RANGE
535-1620KC

MFR/SUPPLIER
Motorola Inc.

PHOTOFACT SET
47-13

PUBLISHED
1948

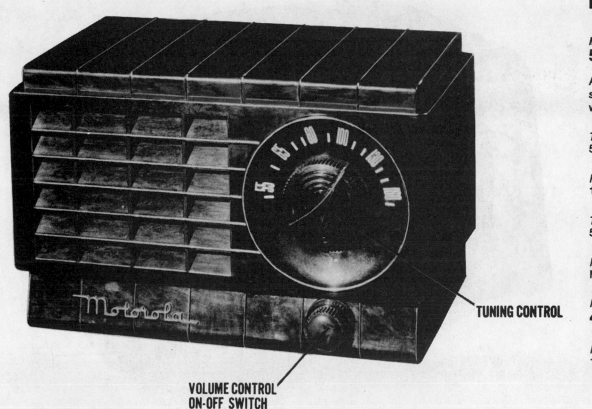

TUNING CONTROL

VOLUME CONTROL ON-OFF SWITCH

MOTOROLA

MODEL PICTURED
58R11

AC/DC operated AM superheterodyne receiver with loop antenna

TUBES
5

POWER SUPPLY
105-125 volts AC/DC

TUNING RANGE
535-1620KC

MFR/SUPPLIER
Motorola Inc.

PHOTOFACT SET
49-14

PUBLISHED
1948

15

MOTOROLA

MODEL PICTURED
58A11

AC/DC operated AM
superheterodyne receiver
with loop antenna

TUBES
5

POWER SUPPLY
105-125 volts AC

TUNING RANGE
535-1620KC

MFR/SUPPLIER
Motorola Inc.

PHOTOFACT SET
52-13

PUBLISHED
1948

VOLUME CONTROL
ON-OFF SWITCH

TUNING CONTROL

MOTOROLA

MODEL PICTURED
58X11

AC/DC operated AM
superheterodyne receiver
with loop antenna

TUBES
5

POWER SUPPLY
105-125 volts AC/DC

TUNING RANGE
535-1620KC

MFR/SUPPLIER
Motorola Inc.

PHOTOFACT SET
53-15

PUBLISHED
1949

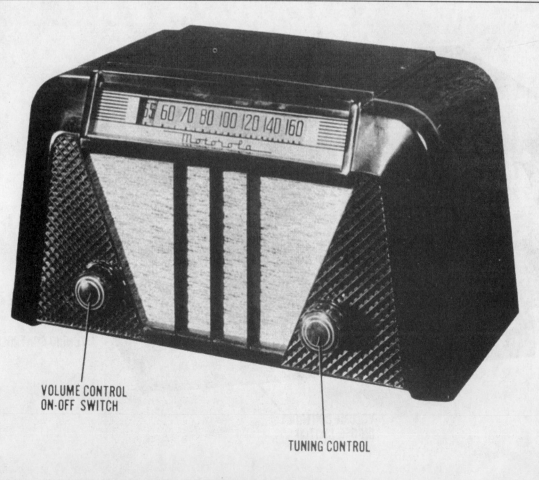

VOLUME CONTROL
ON-OFF SWITCH

TUNING CONTROL

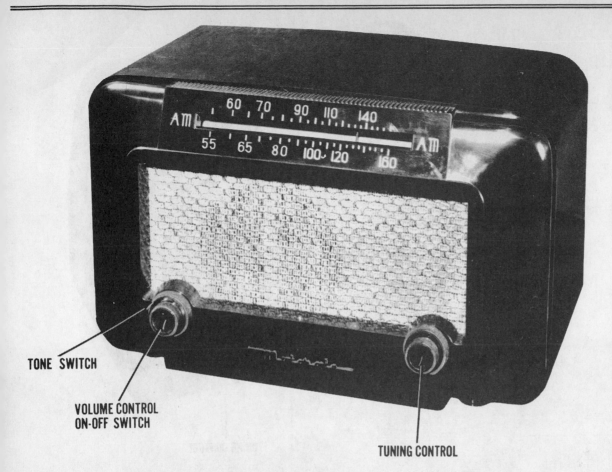

TONE SWITCH

VOLUME CONTROL
ON-OFF SWITCH

TUNING CONTROL

MOTOROLA

MODEL PICTURED
68T11

AC operated AM
superheterodyne receiver
with loop antenna

TUBES
5

POWER SUPPLY
105-125 volts AC

TUNING RANGE
540-1640KC

MFR/SUPPLIER
Motorola Inc.

PHOTOFACT SET
54-14

PUBLISHED
1949

VOLUME CONTROL
ON-OFF SWITCH

BAND SWITCH

TONE CONTROL
PHONO-RADIO SWITCH

TUNING CONTROL

MOTOROLA

MODEL PICTURED
88FM21

AC operated phono-radio
combo AM-FM
superheterodyne receiver
with loop antenna

TUBES
7

POWER SUPPLY
105-125 volts AC

TUNING RANGE
535-1620KC,
88-108MC

MFR/SUPPLIER
Motorola Inc.

PHOTOFACT SET
54-15

PUBLISHED
1949

MOTOROLA

MODEL PICTURED
68X11

AC/DC operated AM
superheterodyne receiver
with loop antenna

TUBES
6

POWER SUPPLY
110-120 volts AC/DC

TUNING RANGE
535-1620KC

MFR/SUPPLIER
Motorola Inc.

PHOTOFACT SET
56-16

PUBLISHED
1949

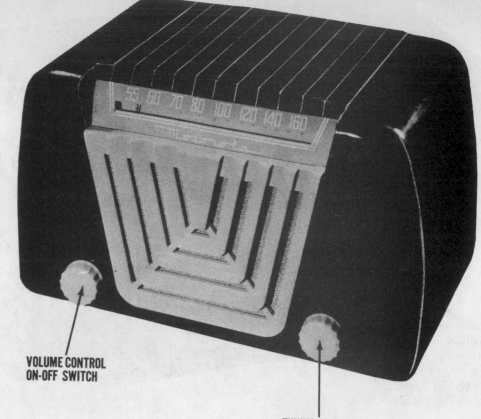

VOLUME CONTROL
ON-OFF SWITCH

TUNING CONTROL

MOTOROLA

MODEL PICTURED
78F11, 78F12

AC operated phono-radio
combo AM
superheterodyne receiver
with loop antenna

TUBES
6

POWER SUPPLY
105-125 volts AC

TUNING RANGE
535-1620KC

MFR/SUPPLIER
Motorola Inc.

PHOTOFACT SET
56-17

PUBLISHED
1949

TUNING CONTROL

PHONO-RADIO SWITCH

TONE CONTROL

VOLUME CONTROL
ON-OFF SWITCH

MODEL 78F12

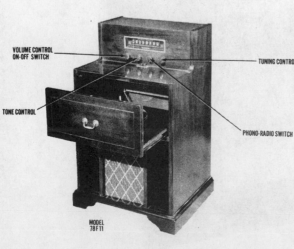

VOLUME CONTROL
ON-OFF SWITCH

TUNING CONTROL

TONE CONTROL

PHONO-RADIO SWITCH

MODEL
78F11

MODEL 68F12

MODEL 68F11

MODEL 68F14

MOTOROLA

MODEL PICTURED
68F11, 68F12, 68F14

AC operated phono-radio combo AM superheterodyne receiver with loop antenna

TUBES
6

POWER SUPPLY
105-125 volts AC

TUNING RANGE
535-1620KC

MFR/SUPPLIER
Motorola Inc.

PHOTOFACT SET
58-13

PUBLISHED
1949

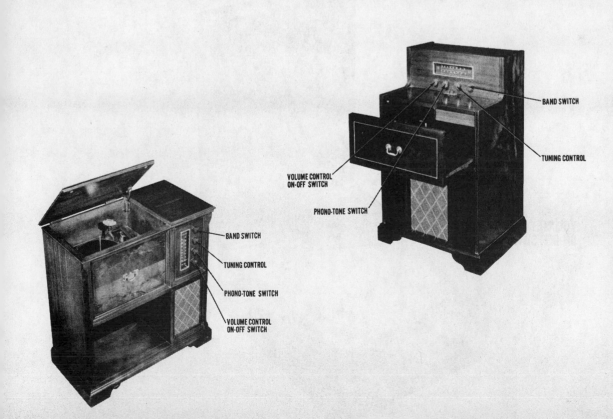

BAND SWITCH

TUNING CONTROL

VOLUME CONTROL ON-OFF SWITCH

PHONO-TONE SWITCH

BAND SWITCH

TUNING CONTROL

PHONO-TONE SWITCH

VOLUME CONTROL ON-OFF SWITCH

MOTOROLA

MODEL PICTURED
78FM21, 78FM22M

AC operated phono-radio combo AM-FM superheterodyne receiver with loop antenna

TUBES
6

POWER SUPPLY
105-125 volts AC

TUNING RANGE
540-1620KC, 88-108MC

MFR/SUPPLIER
Motorola Inc.

PHOTOFACT SET
59-13

PUBLISHED
1949

MOTOROLA

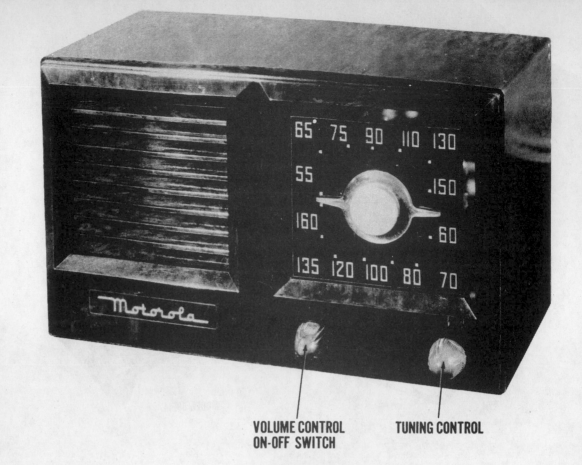

MODEL PICTURED
58G11

AC/DC operated AM
superheterodyne receiver
with loop antenna

TUBES
5

POWER SUPPLY
105-125 volts AC

TUNING RANGE
535-1620KC

MFR/SUPPLIER
Motorola Inc.

PHOTOFACT SET
64-8

PUBLISHED
1949

**VOLUME CONTROL
ON-OFF SWITCH**   **TUNING CONTROL**

MOTOROLA

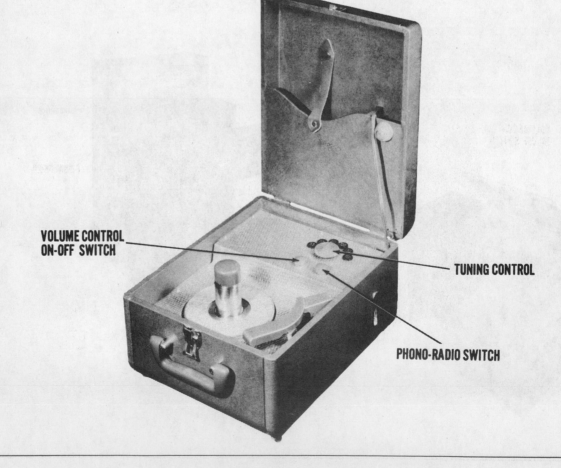

MODEL PICTURED
59F11

AC operated portable
phono-radio combo AM
superheterodyne receiver
with loop antenna

TUBES
5

POWER SUPPLY
110-120 volts AC

TUNING RANGE
535-1620KC

MFR/SUPPLIER
Motorola Inc.

PHOTOFACT SET
68-12

PUBLISHED
1949

**VOLUME CONTROL
ON-OFF SWITCH**   **TUNING CONTROL**

PHONO-RADIO SWITCH

MOTOROLA

MODEL PICTURED
58R11A

AC/DC operated AM
superheterodyne receiver
with loop antenna

TUBES
5

POWER SUPPLY
110-120 volts AC/DC

TUNING RANGE
535-1620KC

MFR/SUPPLIER
Motorola Inc.

PHOTOFACT SET
69-11

PUBLISHED
1949

TUNING CONTROL

**VOLUME CONTROL
ON-OFF SWITCH**

MOTOROLA

MODEL PICTURED
69L11

Three-power operated
portable AM
superheterodyne receiver
with loop antenna

TUBES
5

POWER SUPPLY
110-120 volts AC/DC or 9
volts A & 90 volts B
supply

TUNING RANGE
535-1620KC

MFR/SUPPLIER
Motorola Inc.

PHOTOFACT SET
76-15

PUBLISHED
1949

**VOLUME CONTROL
ON-OFF SWITCH**

TUNING CONTROL

A7

MOTOROLA

MODEL PICTURED
49L11Q

Battery operated portable
AM superheterodyne
receiver with loop
antenna

TUBES
4

POWER SUPPLY
1.5 volts A & 67.5 volts B
supply

TUNING RANGE
535-1620KC

MFR/SUPPLIER
Motorola Inc.

PHOTOFACT SET
77-7

PUBLISHED
1949

VOLUME CONTROL
ON-OFF SWITCH

TUNING CONTROL

MOTOROLA

MODEL PICTURED
59L12Q

Three-power operated
portable AM
superheterodyne receiver
with loop antenna

TUBES
4

POWER SUPPLY
110-120 volts AC/DC or 3
volts A & 67.5 volts B
supply

TUNING RANGE
535-1620KC

MFR/SUPPLIER
Motorola Inc.

PHOTOFACT SET
78-10

PUBLISHED
1949

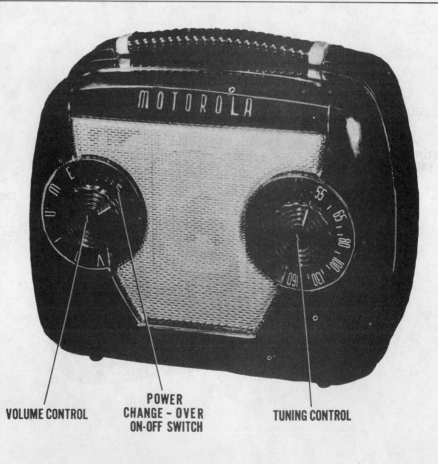

VOLUME CONTROL

POWER
CHANGE – OVER
ON-OFF SWITCH

TUNING CONTROL

TUNING CONTROL   VOLUME CONTROL
ON-OFF SWITCH

MOTOROLA

MODEL PICTURED
59R11

AC/DC operated AM
superheterodyne receiver
with loop antenna

TUBES
5

POWER SUPPLY
110-120 volts AC

TUNING RANGE
535-1620KC

MFR/SUPPLIER
Motorola Inc.

PHOTOFACT SET
79-10

PUBLISHED
1949

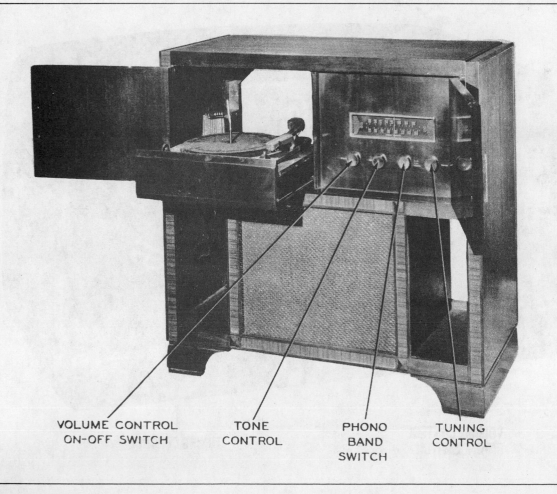

VOLUME CONTROL   TONE   PHONO   TUNING
ON-OFF SWITCH   CONTROL   BAND   CONTROL
                                    SWITCH

MOTOROLA

MODEL PICTURED
99FM21R

AC operated phono-radio
combo AM-FM
superheterodyne receiver
with loop antenna

TUBES
9

POWER SUPPLY
110-120 volts AC

TUNING RANGE
540-1620KC,
88-108MC

MFR/SUPPLIER
Motorola Inc.

PHOTOFACT SET
80-10

PUBLISHED
1949

MOTOROLA

MODEL PICTURED
59X11

AC/DC operated AM
superheterodyne receiver
with loop antenna

TUBES
5

POWER SUPPLY
110-120 volts AC/DC

TUNING RANGE
535-1620KC

MFR/SUPPLIER
Motorola Inc.

PHOTOFACT SET
81-11

PUBLISHED
1950

**VOLUME CONTROL
ON-OFF SWITCH**

TUNING CONTROL

MOTOROLA

MODEL PICTURED
69X11

AC/DC operated AM
superheterodyne receiver
with loop antenna

TUBES
6

POWER SUPPLY
110-120 volts AC/DC

TUNING RANGE
535-1620KC

MFR/SUPPLIER
Motorola Inc.

PHOTOFACT SET
82-9

PUBLISHED
1950

**VOLUME CONTROL
ON-OFF SWITCH**

TUNING CONTROL

MOTOROLA

MODEL PICTURED
79XM21

AC/DC operated AM-FM
superheterodyne receiver
with loop antenna

TUBES
6

POWER SUPPLY
110-120 volts AC

TUNING RANGE
535-1620KC,
88-108MC

MFR/SUPPLIER
Motorola Inc.

PHOTOFACT SET
85-9

PUBLISHED
1950

VOLUME CONTROL
ON-OFF SWITCH

TONE SWITCH

BAND SWITCH

TUNING CONTROL

MOTOROLA

TONE CONTROL
RADIO-PHONO
SWITCH

BAND SWITCH

VOLUME CONTROL
ON-OFF SWITCH

TUNING
CONTROL

MODEL PICTURED
79FM21R

AC operated phono-radio
combo AM-FM
superheterodyne receiver
with loop antenna

TUBES
6

POWER SUPPLY
110-120 volts AC

TUNING RANGE
535-1620KC,
88-108MC

MFR/SUPPLIER
Motorola Inc.

PHOTOFACT SET
88-7

PUBLISHED
1950

MOTOROLA

MODEL PICTURED
59H11U

AC/DC operated AM
superheterodyne receiver
with loop antenna

TUBES
5

POWER SUPPLY
110-120 volts AC/DC

TUNING RANGE
535-1620KC

MFR/SUPPLIER
Motorola Inc.

PHOTOFACT SET
97-9

PUBLISHED
1950

VOLUME CONTROL
ON-OFF SWITCH

TUNING CONTROL

MOTOROLA

MODEL PICTURED
59X21U

AC/DC operated AM
superheterodyne receiver
with loop antenna

TUBES
5

POWER SUPPLY
110-120 volts AC/DC

TUNING RANGE
535-1620KC,
5.85-18.1MC

MFR/SUPPLIER
Motorola Inc.

PHOTOFACT SET
98-6

PUBLISHED
1950

VOLUME CONTROL
ON-OFF SWITCH

BAND SWITCH

TUNING CONTROL

MOTOROLA

MODEL PICTURED
5L1, 5J1

Three-power operated portable AM superheterodyne receiver with loop antenna

TUBES
4

POWER SUPPLY
110-120 volts AC/DC or 3 volts A & 67.5 volts B supply

TUNING RANGE
535-1620KC

MFR/SUPPLIER
Motorola Inc.

PHOTOFACT SET
100-7

PUBLISHED
1950

MODEL 5LI   MODEL 5JI

MOTOROLA

MODEL PICTURED
5M1

Three-power operated portable AM superheterodyne receiver with loop antenna

TUBES
4

POWER SUPPLY
110-120 volts AC/DC or 3 volts A & 67.5 volts B supply

TUNING RANGE
535-1620KC

MFR/SUPPLIER
Motorola Inc.

PHOTOFACT SET
101-7

PUBLISHED
1950

ON-OFF SWITCH

VOLUME CONTROL

TUNING CONTROL

MOTOROLA

MODEL PICTURED
6L1

Three-power operated
portable AM
superheterodyne receiver
with loop antenna

TUBES
5

POWER SUPPLY
110-120 volts AC/DC or 9
volts A & 90 volts B
supply

TUNING RANGE
535-1620KC

MFR/SUPPLIER
Motorola Inc.

PHOTOFACT SET
102-7

PUBLISHED
1950

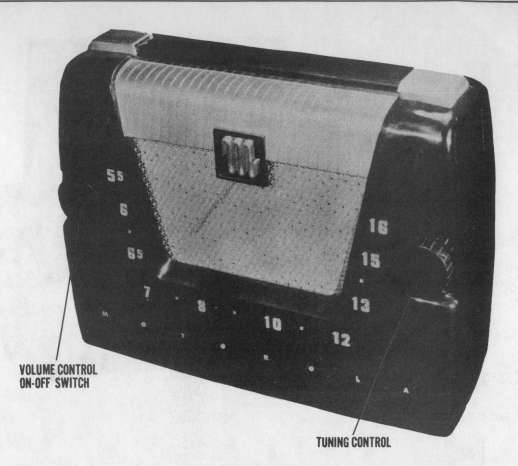

VOLUME CONTROL
ON-OFF SWITCH

TUNING CONTROL

MOTOROLA

MODEL PICTURED
6X11U

AC/DC operated AM
superheterodyne receiver
with loop antenna

TUBES
6

POWER SUPPLY
110-120 volts AC/DC

TUNING RANGE
535-1620KC

MFR/SUPPLIER
Motorola Inc.

PHOTOFACT SET
112-5

PUBLISHED
1950

VOLUME CONTROL
ON-OFF SWITCH

TUNING CONTROL

MOTOROLA

MODEL PICTURED
7F11

AC operated phono-radio combo AM superheterodyne receiver with loop antenna

TUBES
7

POWER SUPPLY
110-120 volts AC

TUNING RANGE
535-1620KC

MFR/SUPPLIER
Motorola Inc.

PHOTOFACT SET
113-5

PUBLISHED
1950

VOLUME CONTROL
ON-OFF SWITCH

RADIO-PHONO SWITCH

TONE CONTROL

TUNING CONTROL

MOTOROLA

MODEL PICTURED
5X11U

AC/DC operated AM superheterodyne receiver with loop antenna

TUBES
5

POWER SUPPLY
110-120 volts AC/DC

TUNING RANGE
535-1620KC

MFR/SUPPLIER
Motorola Inc.

PHOTOFACT SET
114-7

PUBLISHED
1950

VOLUME CONTROL
ON-OFF SWITCH

TUNING
CONTROL

MOTOROLA

MODEL PICTURED
9FM21

AC operated phono-radio
combo AM-FM
superheterodyne receiver
with loop antenna

TUBES
9

POWER SUPPLY
110-120 volts AC

TUNING RANGE
535-1620KC,
88-108MC

MFR/SUPPLIER
Motorola Inc.

PHOTOFACT SET
114-8

PUBLISHED
1950

TONE CONTROL AM-FM TUNING
PHONO SWITCH CONTROL

VOLUME CONTROL
ON-OFF SWITCH

MOTOROLA

MODEL PICTURED
5R11U

AC/DC operated AM
superheterodyne receiver
with loop antenna

TUBES
5

POWER SUPPLY
110-120 volts AC/DC

TUNING RANGE
535-1620KC

MFR/SUPPLIER
Motorola Inc.

PHOTOFACT SET
115-6

PUBLISHED
1950

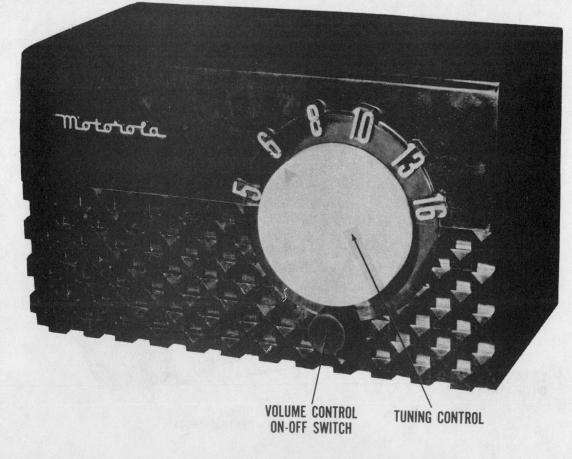

VOLUME CONTROL   TUNING CONTROL
ON-OFF SWITCH

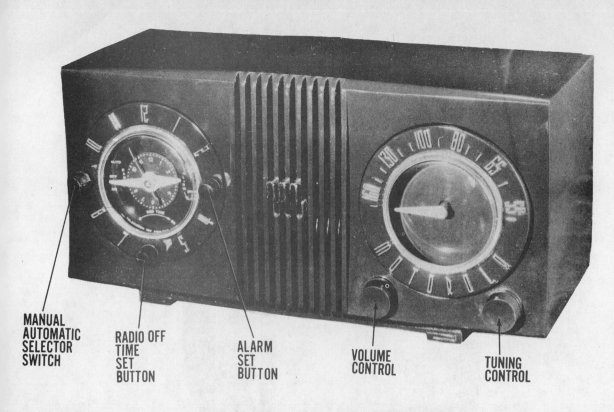

MOTOROLA

MODEL PICTURED
5C1

AC operated AM
superheterodyne receiver
with electric clock

TUBES
5

POWER SUPPLY
110-120 volts AC

TUNING RANGE
535-1620KC

MFR/SUPPLIER
Motorola Inc.

PHOTOFACT SET
116-9

PUBLISHED
1950

MANUAL
AUTOMATIC
SELECTOR
SWITCH

RADIO OFF
TIME
SET
BUTTON

ALARM
SET
BUTTON

VOLUME
CONTROL

TUNING
CONTROL

MOTOROLA

MODEL PICTURED
5H11U

AC/DC operated AM
superheterodyne receiver
with loop antenna

TUBES
5

POWER SUPPLY
110-120 volts AC/DC

TUNING RANGE
535-1620KC

MFR/SUPPLIER
Motorola Inc.

PHOTOFACT SET
117-9

PUBLISHED
1950

VOLUME CONTROL
ON-OFF SWITCH

TUNING CONTROL

MOTOROLA

MODEL PICTURED
6F11

AC operated phono-radio combo AM superheterodyne receiver with loop antenna

TUBES
6

POWER SUPPLY
110-120 volts AC

TUNING RANGE
535-1620KC

MFR/SUPPLIER
Motorola Inc.

PHOTOFACT SET
117-10

PUBLISHED
1950

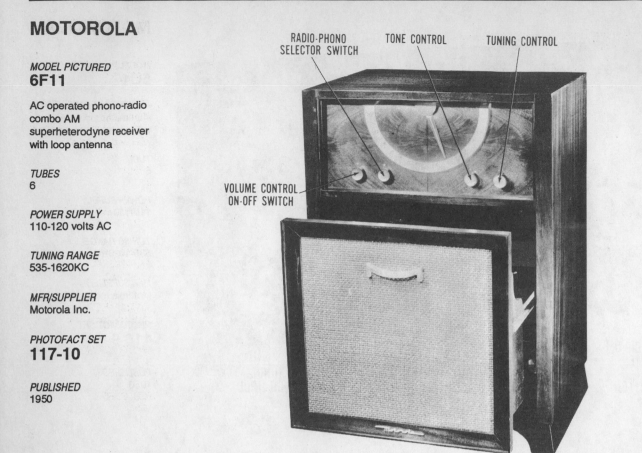

RADIO-PHONO SELECTOR SWITCH   TONE CONTROL   TUNING CONTROL

VOLUME CONTROL ON-OFF SWITCH

MOTOROLA

MODEL PICTURED
5X21U

AC/DC operated two-band superheterodyne receiver with loop antenna

TUBES
5

POWER SUPPLY
110-120 volts AC/DC

TUNING RANGE
535-1620KC,
5.85-18.1MC

MFR/SUPPLIER
Motorola Inc.

PHOTOFACT SET
120-9

PUBLISHED
1951

VOLUME CONTROL ON-OFF SWITCH

TUNING CONTROL

BAND SELECTOR SWITCH

VOLUME CONTROL
ON-OFF SWITCH   RADIO-PHONO SWITCH
TONE CONTROL   BAND
SELECTOR SWITCH   TUNING CONTROL

MOTOROLA

MODEL PICTURED
8FM21

AC operated phono-radio combo AM-FM superheterodyne receiver with loop antenna

TUBES
8

POWER SUPPLY
110-120 volts AC

TUNING RANGE
535-1620KC,
88-108MC

MFR/SUPPLIER
Motorola Inc.

PHOTOFACT SET
121-9

PUBLISHED
1951

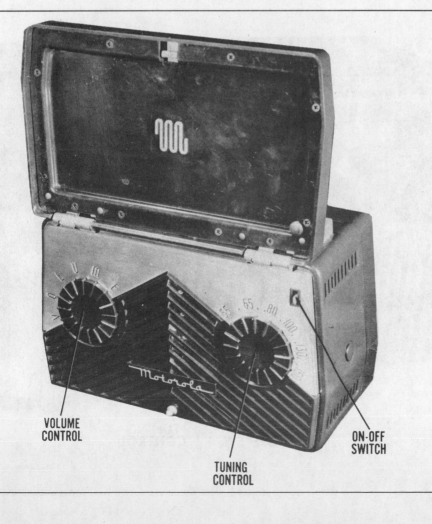

VOLUME
CONTROL   TUNING
CONTROL   ON-OFF
SWITCH

MOTOROLA

MODEL PICTURED
51M1U

Three-power operated portable AM superheterodyne receiver with loop antenna

TUBES
4

POWER SUPPLY
110-120 volts AC/DC or 3 volts A & 67.5 volts B supply

TUNING RANGE
535-1620KC

MFR/SUPPLIER
Motorola Inc.

PHOTOFACT SET
149-8

PUBLISHED
1951

MOTOROLA

MODEL PICTURED
62X11U

AC/DC operated AM
superheterodyne receiver

TUBES
6

POWER SUPPLY
110-120 volts AC/DC

TUNING RANGE
535-1620KC

MFR/SUPPLIER
Motorola Inc.

PHOTOFACT SET
175-14

PUBLISHED
1952

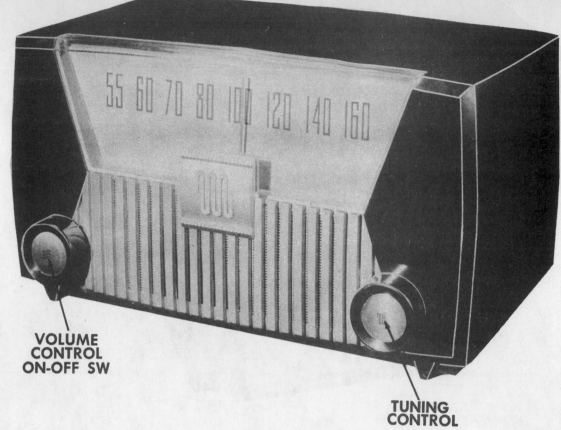

VOLUME
CONTROL
ON-OFF SW

TUNING
CONTROL

MOTOROLA

MODEL PICTURED
52H11U

AC/DC operated AM
superheterodyne receiver
with loop antenna

TUBES
5

POWER SUPPLY
110-120 volts AC/DC

TUNING RANGE
535-1620KC

MFR/SUPPLIER
Motorola Inc.

PHOTOFACT SET
176-6

PUBLISHED
1952

VOLUME
CONTROL
ON-OFF SW.

TUNING
CONTROL

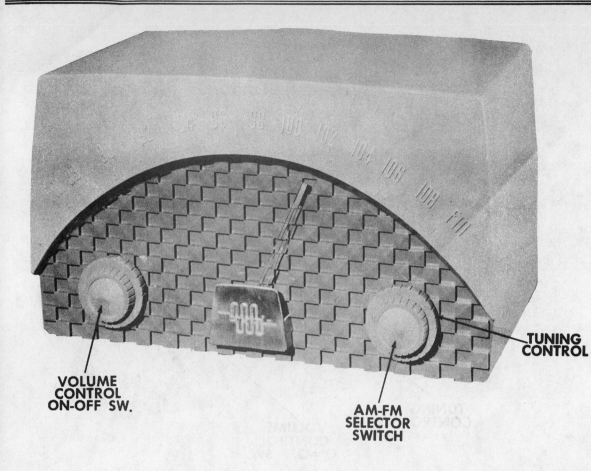

MOTOROLA

MODEL PICTURED
72XM21

AC/DC operated AM-FM
superheterodyne receiver

TUBES
6

POWER SUPPLY
110-120 volts AC/DC

TUNING RANGE
535-1620KC,
88-108MC

MFR/SUPPLIER
Motorola Inc.

PHOTOFACT SET
176-7

PUBLISHED
1952

VOLUME
CONTROL
ON-OFF SW.

AM-FM
SELECTOR
SWITCH

TUNING
CONTROL

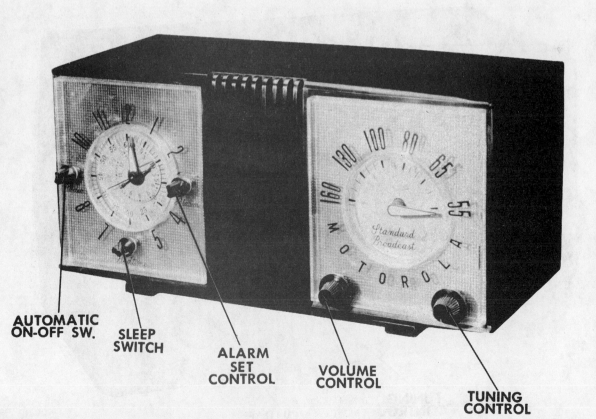

MOTOROLA

MODEL PICTURED
52C6

AC operated AM
superheterodyne receiver
with electric clock

TUBES
5

POWER SUPPLY
110-120 volts AC

TUNING RANGE
535-1620KC

MFR/SUPPLIER
Motorola Inc.

PHOTOFACT SET
177-10

PUBLISHED
1952

AUTOMATIC
ON-OFF SW.

SLEEP
SWITCH

ALARM
SET
CONTROL

VOLUME
CONTROL

TUNING
CONTROL

MOTOROLA

MODEL PICTURED
52R12U

AC/DC operated AM
superheterodyne receiver

TUBES
5

POWER SUPPLY
110-120 volts AC/DC

TUNING RANGE
535-1620KC

MFR/SUPPLIER
Motorola Inc.

PHOTOFACT SET
177-11

PUBLISHED
1952

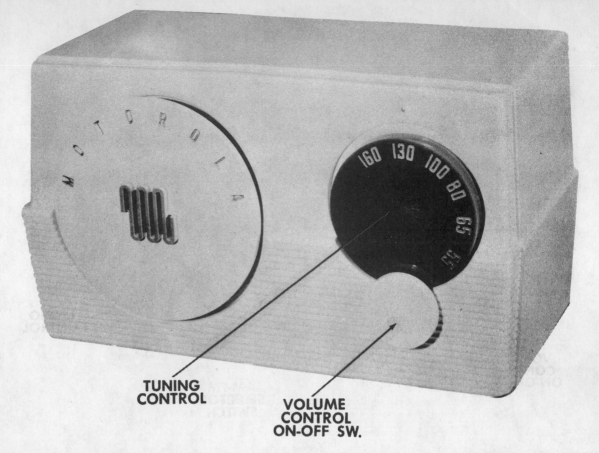

TUNING
CONTROL

VOLUME
CONTROL
ON-OFF SW.

MOTOROLA

MODEL PICTURED
52R12A

AC/DC operated AM
superheterodyne receiver

TUBES
5

POWER SUPPLY
110-120 volts AC/DC

TUNING RANGE
535-1620KC

MFR/SUPPLIER
Motorola Inc.

PHOTOFACT SET
178-7

PUBLISHED
1952

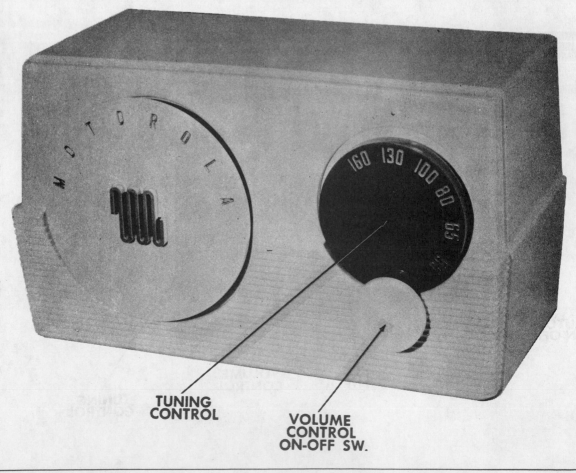

TUNING
CONTROL

VOLUME
CONTROL
ON-OFF SW.

VOLUME CONTROL ON-OFF SW

TUNING CONTROL

MOTOROLA

MODEL PICTURED
62L1U

Three-power operated portable AM superheterodyne receiver

TUBES
5

POWER SUPPLY
110-120 volts AC/DC or 9 volts A & 90 volts B supply

TUNING RANGE
535-1620KC

MFR/SUPPLIER
Motorola Inc.

PHOTOFACT SET
183-10

PUBLISHED
1952

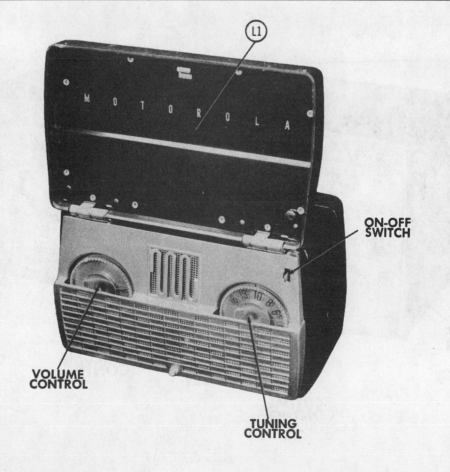

L1

ON-OFF SWITCH

VOLUME CONTROL

TUNING CONTROL

MOTOROLA

MODEL PICTURED
52M1U

Three-power operated portable AM superheterodyne receiver

TUBES
4

POWER SUPPLY
110-120 volts AC/DC or 3 volts A & 67.5 volts B supply

TUNING RANGE
535-1620KC

MFR/SUPPLIER
Motorola Inc.

PHOTOFACT SET
188-10

PUBLISHED
1952

MOTOROLA

MODEL PICTURED
52R14

AC/DC operated AM
superheterodyne receiver

TUBES
5

POWER SUPPLY
110-120 volts AC/DC

TUNING RANGE
535-1620KC

MFR/SUPPLIER
Motorola Inc.

PHOTOFACT SET
188-11

PUBLISHED
1952

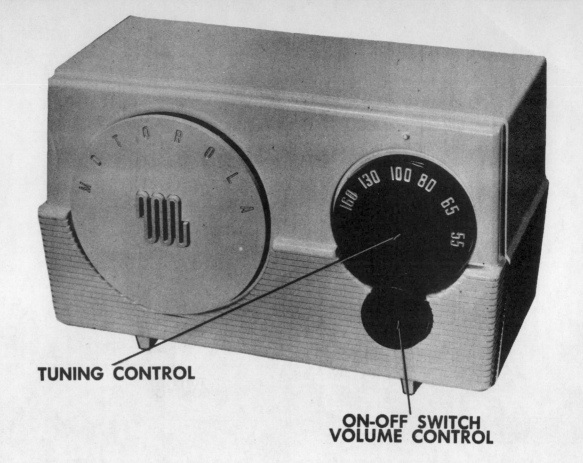

TUNING CONTROL

ON-OFF SWITCH
VOLUME CONTROL

MOTOROLA

MODEL PICTURED
62C1

AC operated AM
superheterodyne receiver
with electric clock

TUBES
6

POWER SUPPLY
110-120 volts AC/DC

TUNING RANGE
535-1620KC

MFR/SUPPLIER
Motorola Inc.

PHOTOFACT SET
189-12

PUBLISHED
1952

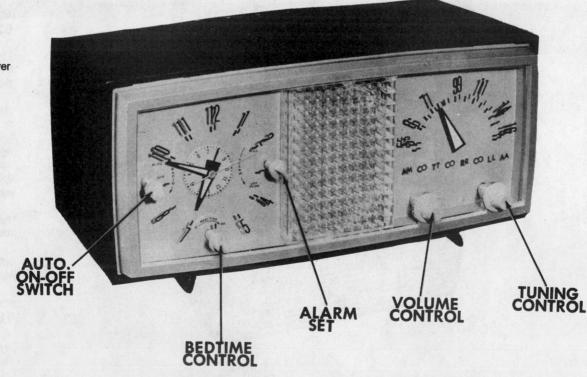

AUTO.
ON-OFF
SWITCH

BEDTIME
CONTROL

ALARM
SET

VOLUME
CONTROL

TUNING
CONTROL

TUNING CONTROL

VOLUME CONTROL
ON-OFF SWITCH

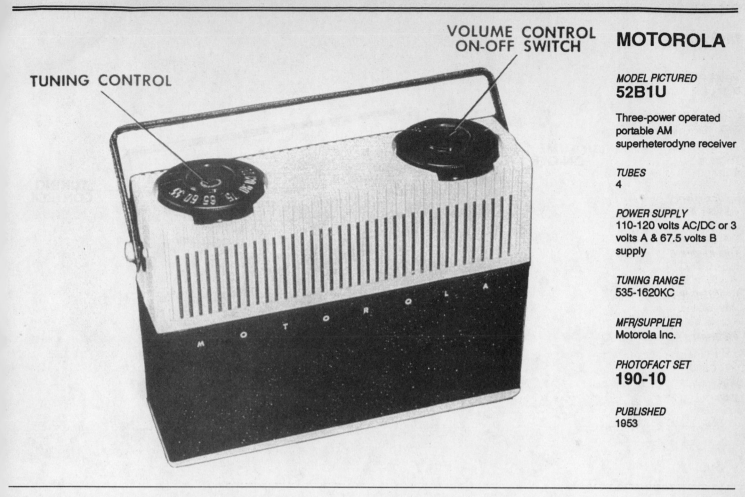

MOTOROLA

MODEL PICTURED
52B1U

Three-power operated
portable AM
superheterodyne receiver

TUBES
4

POWER SUPPLY
110-120 volts AC/DC or 3
volts A & 67.5 volts B
supply

TUNING RANGE
535-1620KC

MFR/SUPPLIER
Motorola Inc.

PHOTOFACT SET
190-10

PUBLISHED
1953

VOLUME CONTROL
ON-OFF SWITCH

TUNING
CONTROL

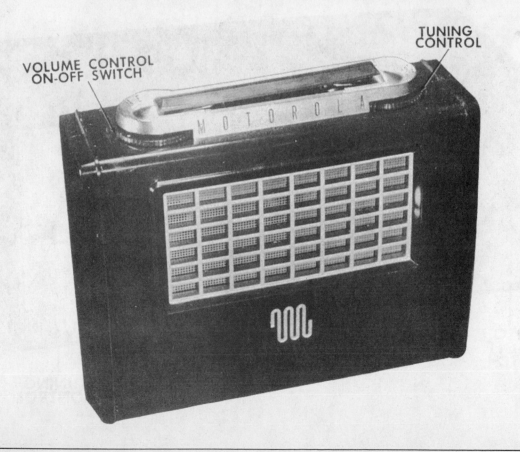

MOTOROLA

MODEL PICTURED
52L1

Three-power operated
portable AM
superheterodyne receiver

TUBES
4

POWER SUPPLY
110-120 volts AC/DC or 3
volts A & 67.5 volts B
supply

TUNING RANGE
535-1620KC

MFR/SUPPLIER
Motorola Inc.

PHOTOFACT SET
190-11

PUBLISHED
1953

MOTOROLA

MODEL PICTURED
42B1

Battery operated AM
superheterodyne receiver

TUBES
4

POWER SUPPLY
1.5 volts A & 67.5 volts B
supply

TUNING RANGE
535-1620KC

MFR/SUPPLIER
Motorola Inc.

PHOTOFACT SET
191-14

PUBLISHED
1953

VOLUME CONTROL
ON-OFF SWITCH

TUNING
CONTROL

MOTOROLA

MODEL PICTURED
52C1

AC operated AM
superheterodyne receiver
with electric clock

TUBES
5

POWER SUPPLY
110-120 volts AC

TUNING RANGE
535-1620KC

MFR/SUPPLIER
Motorola Inc.

PHOTOFACT SET
191-15

PUBLISHED
1953

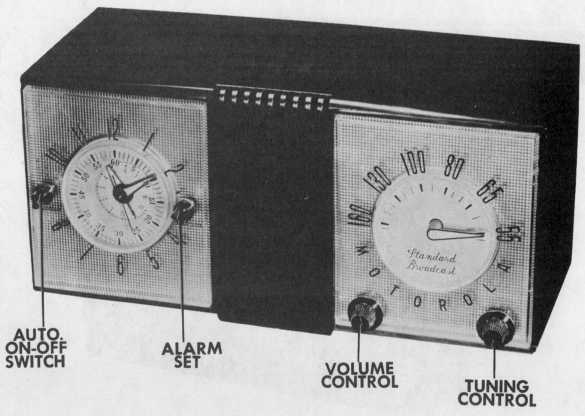

AUTO.
ON-OFF
SWITCH

ALARM
SET

VOLUME
CONTROL

TUNING
CONTROL

MOTOROLA

MODEL PICTURED
62CW1

AC operated AM
superheterodyne receiver
with electric clock

TUBES
6

POWER SUPPLY
110-120 volts AC

TUNING RANGE
535-1620KC

MFR/SUPPLIER
Motorola Inc.

PHOTOFACT SET
196-7

PUBLISHED
1953

ON-OFF
SWITCH
VOLUME
CONTROL

TONE
CONTROL

TUNING
CONTROL

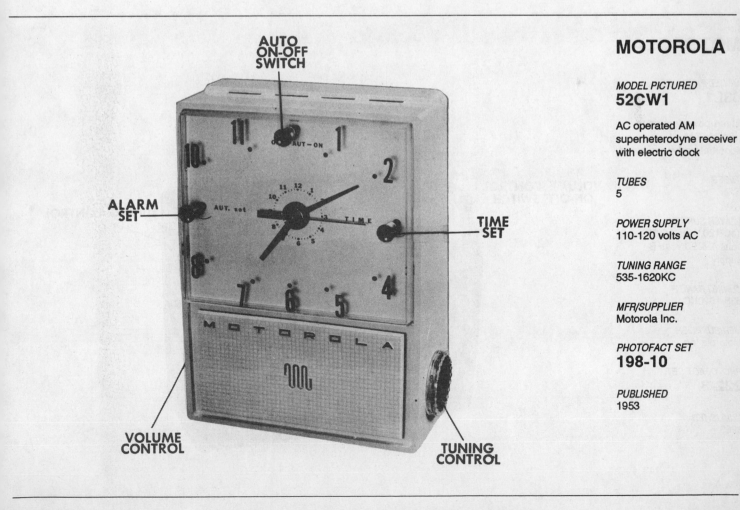

MOTOROLA

MODEL PICTURED
52CW1

AC operated AM
superheterodyne receiver
with electric clock

TUBES
5

POWER SUPPLY
110-120 volts AC

TUNING RANGE
535-1620KC

MFR/SUPPLIER
Motorola Inc.

PHOTOFACT SET
198-10

PUBLISHED
1953

AUTO
ON-OFF
SWITCH

ALARM
SET

TIME
SET

VOLUME
CONTROL

TUNING
CONTROL

MOTOROLA

53LC1

Three-power operated
portable AM
superheterodyne receiver
with spring wound clock

TUBES
4

POWER SUPPLY
110-120 volts AC/DC or 3
volts A & 67.5 volts B
supply

TUNING RANGE
535-1620KC

MFR/SUPPLIER
Motorola Inc.

PHOTOFACT SET
217-10

PUBLISHED
1953

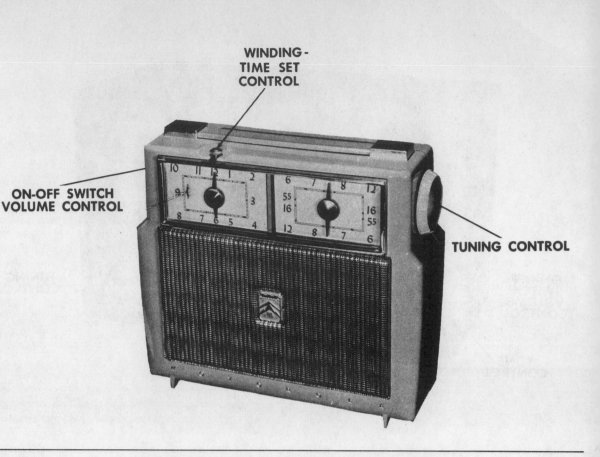

WINDING -
TIME SET
CONTROL

ON-OFF SWITCH
VOLUME CONTROL

TUNING CONTROL

MOTOROLA

MODEL PICTURED
63L1

Three-power operated
portable AM
superheterodyne receiver

TUBES
5

POWER SUPPLY
110-120 volts AC/DC or 9
volts A & 90 volts B
supply

TUNING RANGE
535-1620KC

MFR/SUPPLIER
Motorola Inc.

PHOTOFACT SET
222-8

PUBLISHED
1953

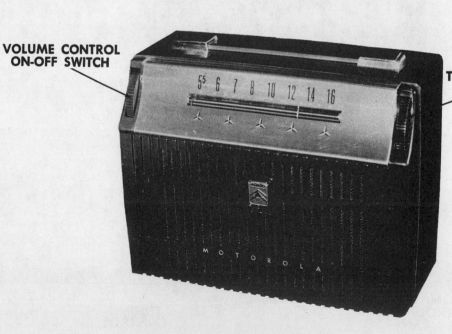

VOLUME CONTROL
ON-OFF SWITCH

TUNING CONTROL

MOTOROLA

MODEL PICTURED
62X21

AC/DC operated
two-band
superheterodyne receiver

TUBES
6

POWER SUPPLY
110-120 volts AC/DC

TUNING RANGE
535-1620KC,
6-16MC

MFR/SUPPLIER
Motorola Inc.

PHOTOFACT SET
228-12

PUBLISHED
1954

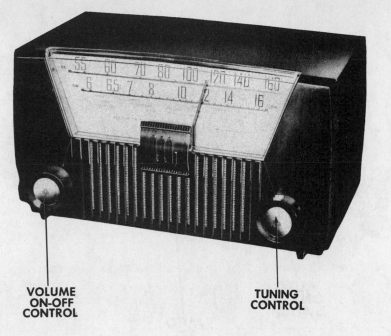

**VOLUME
ON-OFF
CONTROL**

**TUNING
CONTROL**

MOTOROLA

MODEL PICTURED
53F2

AC operated phono-radio
combo AM
superheterodyne receiver

TUBES
5

POWER SUPPLY
110-120 volts AC

TUNING RANGE
535-1620KC

MFR/SUPPLIER
Motorola Inc.

PHOTOFACT SET
234-9

PUBLISHED
1954

**TUNING
CONTROL**

**TONE
CONTROL
RADIO-
PHONO
SWITCH**

**VOLUME
CONTROL
ON-OFF
SWITCH**

MOTOROLA

MODEL PICTURED
53C6

AC operated AM
superheterodyne receiver
with electric clock

TUBES
5

POWER SUPPLY
110-120 volts AC

TUNING RANGE
535-1620KC

MFR/SUPPLIER
Motorola Inc.

PHOTOFACT SET
235-7

PUBLISHED
1954

VOLUME CONTROL

AUTO. ON-OFF SWITCH

AUTO SET

TUNING CONTROL

MOTOROLA

MODEL PICTURED
53C1

AC operated AM
superheterodyne receiver
with electric clock

TUBES
5

POWER SUPPLY
110-120 volts AC

TUNING RANGE
535-1620KC

MFR/SUPPLIER
Motorola Inc.

PHOTOFACT SET
236-7

PUBLISHED
1954

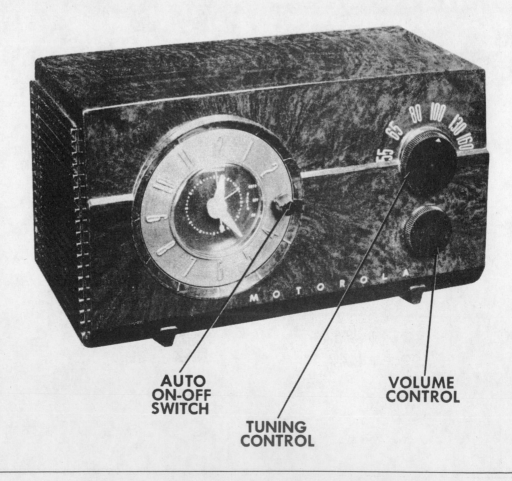

AUTO ON-OFF SWITCH

TUNING CONTROL

VOLUME CONTROL

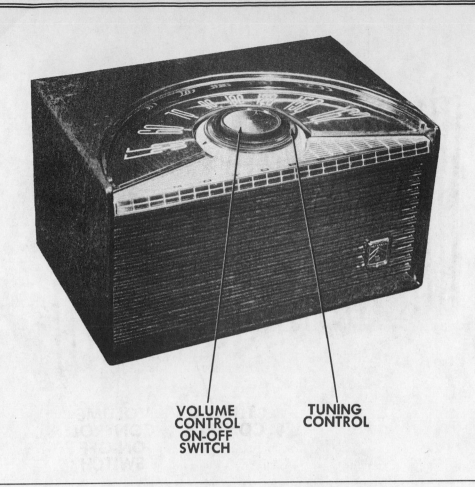

MOTOROLA

MODEL PICTURED
53X1

AC/DC operated AM
superheterodyne receiver

TUBES
5

POWER SUPPLY
110-120 volts AC/DC

TUNING RANGE
535-1620KC

MFR/SUPPLIER
Motorola Inc.

PHOTOFACT SET
236-8

PUBLISHED
1954

VOLUME
CONTROL
ON-OFF
SWITCH

TUNING
CONTROL

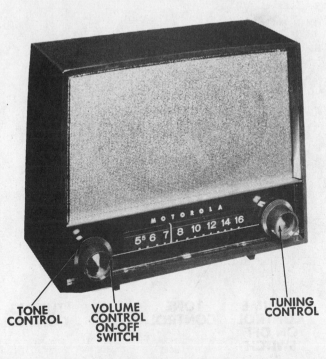

MOTOROLA

MODEL PICTURED
63X1

AC/DC operated AM
superheterodyne receiver

TUBES
6

POWER SUPPLY
110-120 volts AC/DC

TUNING RANGE
535-1620KC

MFR/SUPPLIER
Motorola Inc.

PHOTOFACT SET
238-9

PUBLISHED
1954

TONE
CONTROL

VOLUME
CONTROL
ON-OFF
SWITCH

TUNING
CONTROL

MOTOROLA

MODEL PICTURED
53R1

AC/DC operated AM
superheterodyne receiver

TUBES
5

POWER SUPPLY
110-120 volts AC/DC

TUNING RANGE
535-1620KC

MFR/SUPPLIER
Motorola Inc.

PHOTOFACT SET
247-8

PUBLISHED
1954

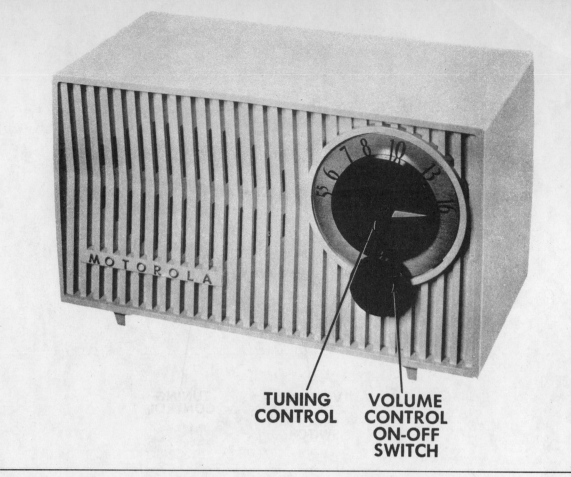

**TUNING
CONTROL**   **VOLUME
CONTROL
ON-OFF
SWITCH**

MOTOROLA

MODEL PICTURED
63X21

AC/DC operated
two-band
superheterodyne receiver

TUBES
6

POWER SUPPLY
110-120 volts AC/DC

TUNING RANGE
535-1620KC,
6-16MC

MFR/SUPPLIER
Motorola Inc.

PHOTOFACT SET
249-11

PUBLISHED
1954

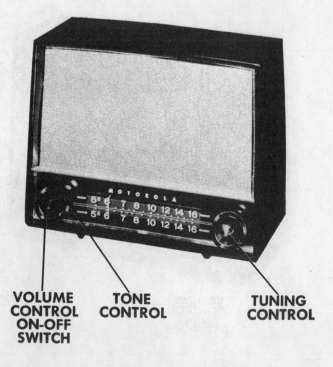

**VOLUME
CONTROL
ON-OFF
SWITCH**   **TONE
CONTROL**   **TUNING
CONTROL**

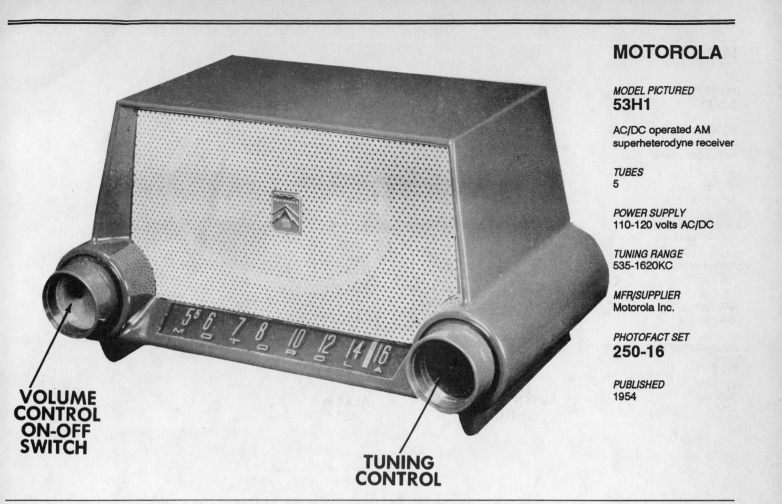

MOTOROLA

MODEL PICTURED
53H1

AC/DC operated AM
superheterodyne receiver

TUBES
5

POWER SUPPLY
110-120 volts AC/DC

TUNING RANGE
535-1620KC

MFR/SUPPLIER
Motorola Inc.

PHOTOFACT SET
250-16

PUBLISHED
1954

**VOLUME
CONTROL
ON-OFF
SWITCH**

**TUNING
CONTROL**

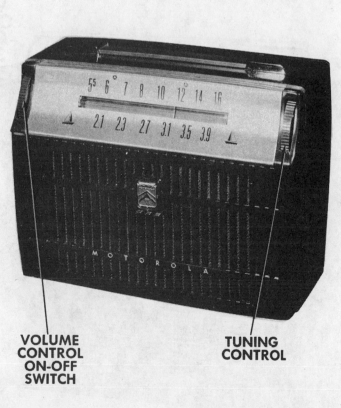

MOTOROLA

MODEL PICTURED
63LSS

Three-power operated
portable two-band
superheterodyne receiver

TUBES
5

POWER SUPPLY
110-120 volts AC/DC or 9
volts A & 90 volts B
supply

TUNING RANGE
535-1620KC,
2-4MC

MFR/SUPPLIER
Motorola Inc.

PHOTOFACT SET
251-13

PUBLISHED
1954

**VOLUME
CONTROL
ON-OFF
SWITCH**

**TUNING
CONTROL**

MOTOROLA

MODEL PICTURED
53D1

AC operated AM
superheterodyne receiver
with electric clock

TUBES
5

POWER SUPPLY
110-120 volts AC

TUNING RANGE
535-1620KC

MFR/SUPPLIER
Motorola Inc.

PHOTOFACT SET
253-9

PUBLISHED
1954

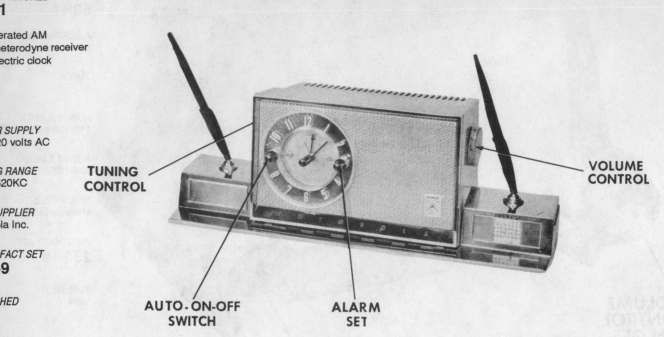

TUNING
CONTROL

VOLUME
CONTROL

AUTO-ON-OFF
SWITCH

ALARM
SET

MOTOROLA

MODEL PICTURED
54L1

Three-power operated
portable AM
superheterodyne receiver

TUBES
4

POWER SUPPLY
110-120 volts AC/DC or 3
volts A & 67.5 volts B
supply

TUNING RANGE
535-1620KC

MFR/SUPPLIER
Motorola Inc.

PHOTOFACT SET
266-9

PUBLISHED
1955

VOLUME
CONTROL
ON-OFF
SWITCH

TUNING

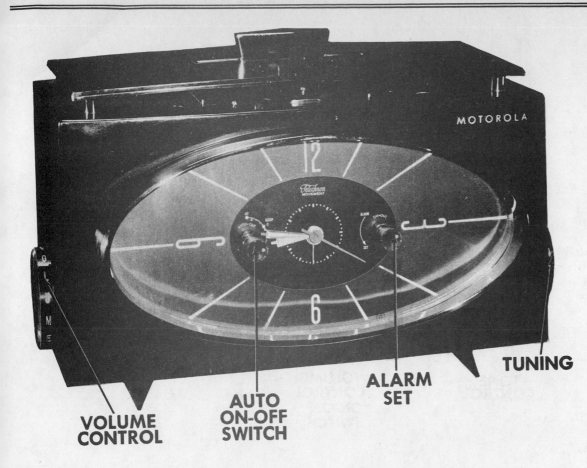

VOLUME CONTROL

AUTO ON-OFF SWITCH

ALARM SET

TUNING

MOTOROLA

MODEL PICTURED
63C1

AC operated AM
superheterodyne receiver
with electric clock

TUBES
6

POWER SUPPLY
110-120 volts AC

TUNING RANGE
535-1620KC

MFR/SUPPLIER
Motorola Inc.

PHOTOFACT SET
266-10

PUBLISHED
1955

TUNING

VOLUME CONTROL ON-OFF SWITCH

MOTOROLA

MODEL PICTURED
53R1A

AC/DC operated AM
superheterodyne receiver

TUBES
5

POWER SUPPLY
110-120 volts AC/DC

TUNING RANGE
535-1620KC

MFR/SUPPLIER
Motorola Inc.

PHOTOFACT SET
273-8

PUBLISHED
1955

MOTOROLA

MODEL PICTURED
64X1

AC/DC operated AM
superheterodyne receiver

TUBES
6

POWER SUPPLY
110-120 volts AC/DC

TUNING RANGE
535-1620KC

MFR/SUPPLIER
Motorola Inc.

PHOTOFACT SET
277-9

PUBLISHED
1955

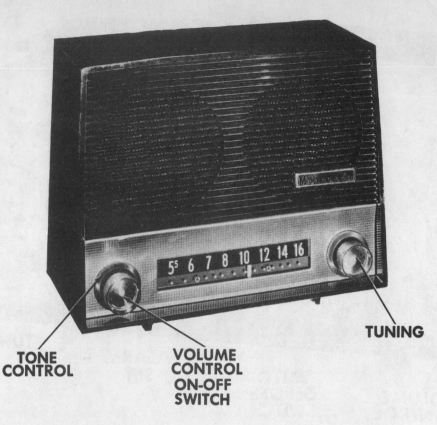

**TONE
CONTROL**

**VOLUME
CONTROL
ON-OFF
SWITCH**

TUNING

MOTOROLA

MODEL PICTURED
55C1

AC operated AM
superheterodyne receiver
with electric clock

TUBES
5

POWER SUPPLY
110-120 volts AC

TUNING RANGE
535-1620KC

MFR/SUPPLIER
Motorola Inc.

PHOTOFACT SET
280-7

PUBLISHED
1955

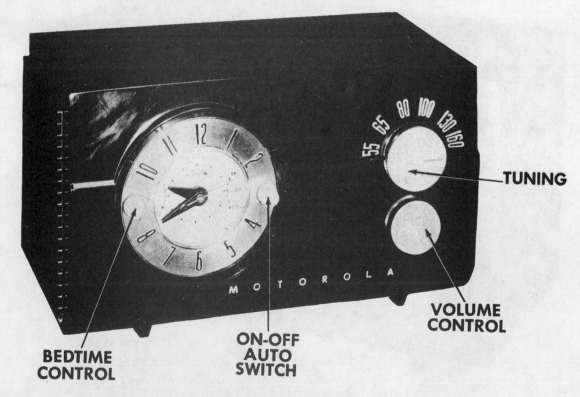

TUNING

**VOLUME
CONTROL**

**BEDTIME
CONTROL**

**ON-OFF
AUTO
SWITCH**

TUNING

VOLUME
CONTROL
ON-OFF
SWITCH

MOTOROLA

MODEL PICTURED
54X1

AC/DC operated AM
superheterodyne receiver

TUBES
5

POWER SUPPLY
110-120 volts AC/DC

TUNING RANGE
535-1620KC

MFR/SUPPLIER
Motorola Inc.

PHOTOFACT SET
282-9

PUBLISHED
1955

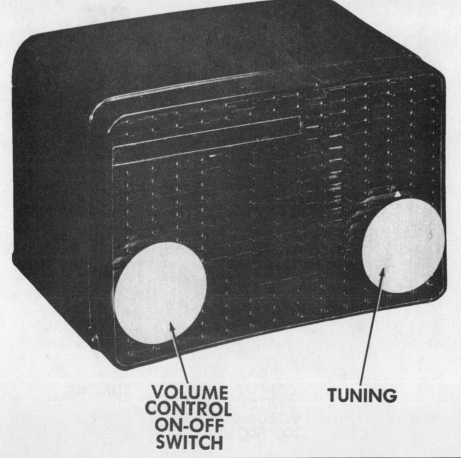

VOLUME
CONTROL
ON-OFF
SWITCH

TUNING

MOTOROLA

MODEL PICTURED
55A1

AC/DC operated AM
superheterodyne receiver

TUBES
5

POWER SUPPLY
110-120 volts AC/DC

TUNING RANGE
535-1620KC

MFR/SUPPLIER
Motorola Inc.

PHOTOFACT SET
299-5

PUBLISHED
1955

MOTOROLA

MODEL PICTURED
55J1

Three-power operated
portable AM
superheterodyne receiver

TUBES
4

POWER SUPPLY
110-120 volts AC/DC or 3
volts A & 90 volts B
supply

TUNING RANGE
535-1620KC

MFR/SUPPLIER
Motorola Inc.

PHOTOFACT SET
301-7

PUBLISHED
1955

**ON-OFF
SWITCH
VOLUME
CONTROL**　　　　**TUNING**

MOTOROLA

MODEL PICTURED
55B1

Three-power operated
portable AM
superheterodyne receiver

TUBES
4

POWER SUPPLY
110-120 volts AC/DC or
7.5 volts A & 90 volts B
supply

TUNING RANGE
535-1620KC

MFR/SUPPLIER
Motorola Inc.

PHOTOFACT SET
303-6

PUBLISHED
1956

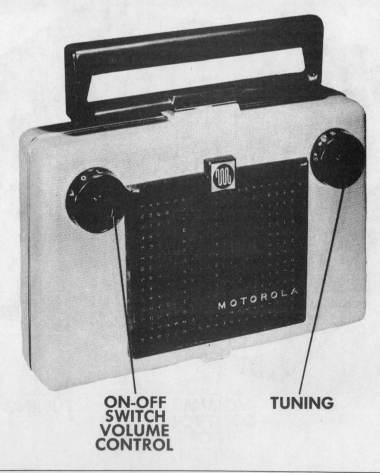

**ON-OFF
SWITCH
VOLUME
CONTROL**　　　　**TUNING**

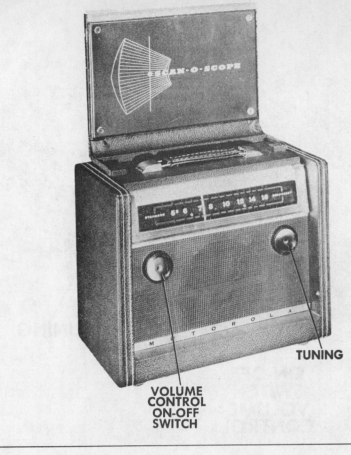

TUNING

VOLUME
CONTROL
ON-OFF
SWITCH

MOTOROLA

MODEL PICTURED
65L1

Three-power operated
portable AM
superheterodyne receiver

TUBES
5

POWER SUPPLY
110-120 volts AC/DC or 9
volts A & 90 volts B
supply

TUNING RANGE
535-1620KC

MFR/SUPPLIER
Motorola Inc.

PHOTOFACT SET
305-13

PUBLISHED
1956

TUNING

VOLUME
CONTROL
ON-OFF
SWITCH

MOTOROLA

MODEL PICTURED
45P1

Battery operated portable
AM superheterodyne
receiver

TUBES
4

POWER SUPPLY
1.5 volts A & 45 volts B
supply

TUNING RANGE
535-1620KC

MFR/SUPPLIER
Motorola Inc.

PHOTOFACT SET
308-8

PUBLISHED
1956

MOTOROLA

MODEL PICTURED
56W1

AC/DC operated AM
superheterodyne receiver

TUBES
5

POWER SUPPLY
110-120 volts AC/DC

TUNING RANGE
530-1620KC

MFR/SUPPLIER
Motorola Inc.

PHOTOFACT SET
316-11

PUBLISHED
1956

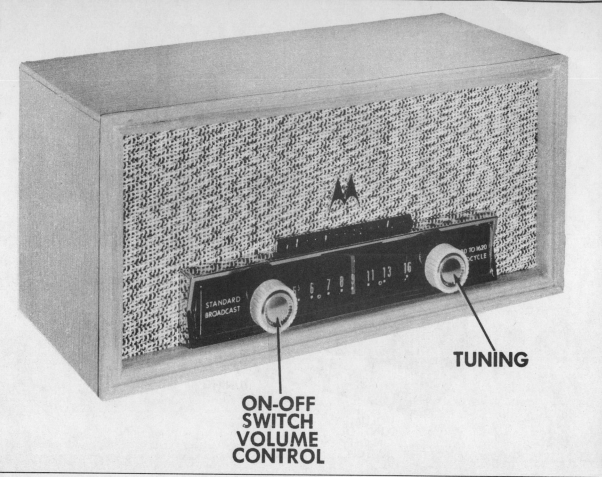

TUNING

**ON-OFF
SWITCH
VOLUME
CONTROL**

MOTOROLA

MODEL PICTURED
56CE1

AC operated AM
superheterodyne receiver
with electric clock

TUBES
5

POWER SUPPLY
110-120 volts AC

TUNING RANGE
535-1620KC

MFR/SUPPLIER
Motorola Inc.

PHOTOFACT SET
317-10

PUBLISHED
1956

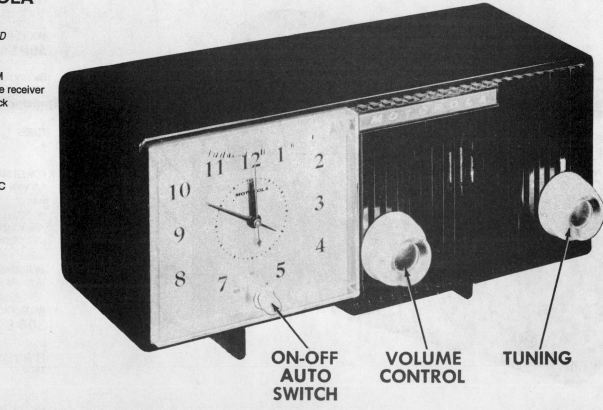

**ON-OFF
AUTO
SWITCH**

**VOLUME
CONTROL**

TUNING

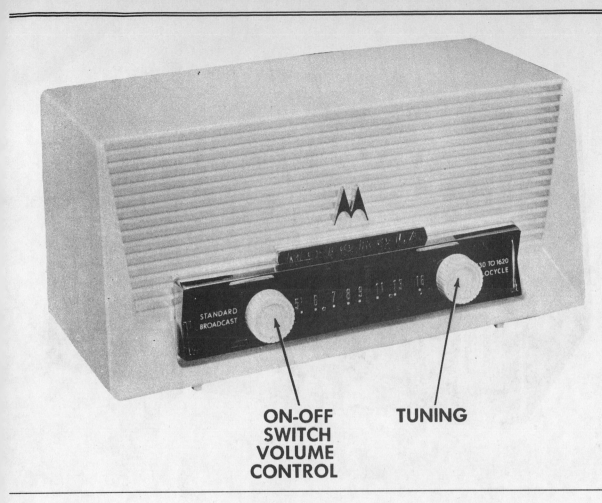

ON-OFF
SWITCH
VOLUME
CONTROL

TUNING

MOTOROLA

MODEL PICTURED
56X1

AC/DC operated AM
superheterodyne receiver

TUBES
5

POWER SUPPLY
110-120 volts AC/DC

TUNING RANGE
530-1620KC

MFR/SUPPLIER
Motorola Inc.

PHOTOFACT SET
318-8

PUBLISHED
1956

VOLUME
CONTROL

TUNING

ALARM
SET

SLEEP
SWITCH

ON-OFF
AUTO
SWITCH

MOTOROLA

MODEL PICTURED
56CC1

AC operated AM
superheterodyne receiver
with electric clock

TUBES
5

POWER SUPPLY
110-120 volts AC

TUNING RANGE
535-1620KC

MFR/SUPPLIER
Motorola Inc.

PHOTOFACT SET
319-8

PUBLISHED
1956

MOTOROLA

MODEL PICTURED
56R1

AC/DC operated AM
superheterodyne receiver

TUBES
5

POWER SUPPLY
110-120 volts AC/DC

TUNING RANGE
535-1620KC

MFR/SUPPLIER
Motorola Inc.

PHOTOFACT SET
320-10

PUBLISHED
1956

MOTOROLA

MODEL PICTURED
66X1

AC/DC operated AM
superheterodyne receiver

TUBES
6

POWER SUPPLY
110-120 volts AC/DC

TUNING RANGE
530-1620KC

MFR/SUPPLIER
Motorola Inc.

PHOTOFACT SET
321-8

PUBLISHED
1956

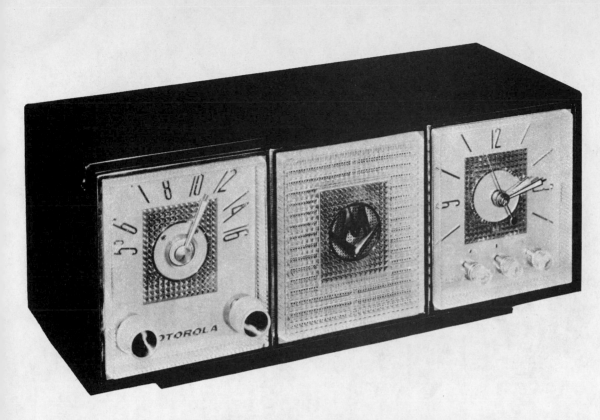

MOTOROLA

MODEL PICTURED
56CJ1

AC operated AM
superheterodyne receiver
with electric clock

TUBES
5

POWER SUPPLY
110-120 volts AC

TUNING RANGE
535-1620KC

MFR/SUPPLIER
Motorola Inc.

PHOTOFACT SET
322-9

PUBLISHED
1956

MOTOROLA

MODEL PICTURED
56RF1

AC operated AM
superheterodyne receiver
with three-speed record
player

TUBES
5

POWER SUPPLY
110-120 volts AC

TUNING RANGE
535-1620KC

MFR/SUPPLIER
Motorola Inc.

PHOTOFACT SET
324-10

PUBLISHED
1956

MOTOROLA

MODEL PICTURED
66C1

AC operated AM
superheterodyne receiver
with electric clock

TUBES
6

POWER SUPPLY
110-120 volts AC

TUNING RANGE
535-1620KC

MFR/SUPPLIER
Motorola Inc.

PHOTOFACT SET
325-8

PUBLISHED
1956

MOTOROLA

MODEL PICTURED
66L1

Three-power operated
portable AM
superheterodyne receiver

TUBES
5

POWER SUPPLY
110-120 volts AC/DC or 9
volts A & 90 volts B
supply

TUNING RANGE
535-1620KC

MFR/SUPPLIER
Motorola Inc.

PHOTOFACT SET
338-7

PUBLISHED
1956

MOTOROLA

MODEL PICTURED
56T1

Battery operated portable transistorized AM superheterodyne receiver

TUBES
5 transistor

POWER SUPPLY
9 volt battery

TUNING RANGE
530-1620KC

MFR/SUPPLIER
Motorola Inc.

PHOTOFACT SET
339-12

PUBLISHED
1956

MOTOROLA

MODEL PICTURED
56B1A

Three-power operated portable AM receiver

TUBES
4

POWER SUPPLY
110-120 volts AC/DC or 7.5 volts A & 90 volts B supply

TUNING RANGE
532-1620KC

MFR/SUPPLIER
Motorola Inc.

PHOTOFACT SET
341-9

PUBLISHED
1956

OTOROLA

MODEL PICTURED
57R1

AC/DC operated AM
receiver

TUBES
5

POWER SUPPLY
110-120 volts AC/DC

TUNING RANGE
535-1620KC

MFR/SUPPLIER
Motorola Inc.

PHOTOFACT SET
347-9

PUBLISHED
1957

MOTOROLA

MODEL PICTURED
57X1

AC/DC operated AM
receiver

TUBES
5

POWER SUPPLY
110-120 volts AC/DC

TUNING RANGE
530-1620KC

MFR/SUPPLIER
Motorola Inc.

PHOTOFACT SET
351-13

PUBLISHED
1957

MOTOROLA

MODEL PICTURED
57RF1

AC operated AM receiver
with four-speed auto
record changer

TUBES
5

POWER SUPPLY
110-120 volts AC

TUNING RANGE
535-1620KC

MFR/SUPPLIER
Motorola Inc.

PHOTOFACT SET
352-10

PUBLISHED
1957

MOTOROLA

MODEL PICTURED
67X1

AC/DC operated AM
receiver

TUBES
6

POWER SUPPLY
110-120 volts AC/DC

TUNING RANGE
535-1620KC

MFR/SUPPLIER
Motorola Inc.

PHOTOFACT SET
352-11

PUBLISHED
1957

MOTOROLA

MODEL PICTURED
57CE

AC operated AM receiver
with electric clock

TUBES
5

POWER SUPPLY
110-120 volts AC

TUNING RANGE
535-1620KC

MFR/SUPPLIER
Motorola Inc.

PHOTOFACT SET
353-8

PUBLISHED
1957

MOTOROLA

MODEL PICTURED
57W1

AC/DC operated AM
receiver

TUBES
5

POWER SUPPLY
110-120 volts AC/DC

TUNING RANGE
535-1620KC

MFR/SUPPLIER
Motorola Inc.

PHOTOFACT SET
353-9

PUBLISHED
1957

MOTOROLA

MODEL PICTURED
57A1

AC/DC operated AM
receiver

TUBES
5

POWER SUPPLY
110-120 volts AC/DC

TUNING RANGE
535-1620KC

MFR/SUPPLIER
Motorola Inc.

PHOTOFACT SET
355-7

PUBLISHED
1957

MOTOROLA

MODEL PICTURED
67C1

AC operated AM receiver
with electric clock

TUBES
6

POWER SUPPLY
110-120 volts AC

TUNING RANGE
530-1620KC

MFR/SUPPLIER
Motorola Inc.

PHOTOFACT SET
357-6

PUBLISHED
1957

MOTOROLA

MODEL PICTURED
57H1

AC/DC operated AM
receiver

TUBES
5

POWER SUPPLY
110-120 volts AC/DC

TUNING RANGE
535-1620KC

MFR/SUPPLIER
Motorola Inc.

PHOTOFACT SET
358-8

PUBLISHED
1957

MOTOROLA

MODEL PICTURED
57CD1

AC operated AM receiver
with electric clock

TUBES
5

POWER SUPPLY
110-120 volts AC

TUNING RANGE
535-1620KC

MFR/SUPPLIER
Motorola Inc.

PHOTOFACT SET
359-9

PUBLISHED
1957

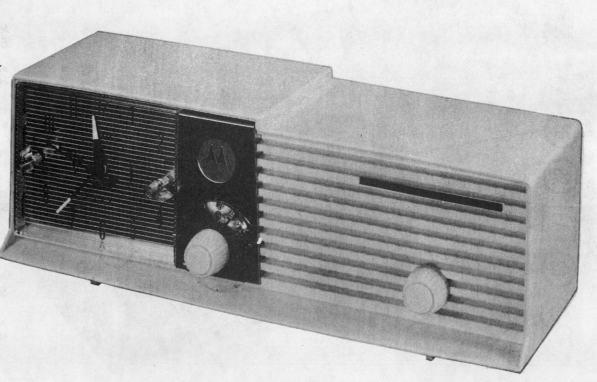

MOTOROLA

MODEL PICTURED
76T1

Battery operated portable
AM transistorized
receiver

TUBES
7 transistor

POWER SUPPLY
9 volts DC

TUNING RANGE
530-1620KC

MFR/SUPPLIER
Motorola Inc.

PHOTOFACT SET
360-7

PUBLISHED
1957

MOTOROLA

MODEL PICTURED
57CS1

AC operated AM receiver
with electric clock

TUBES
5

POWER SUPPLY
110-120 volts AC

TUNING RANGE
535-1620KC

MFR/SUPPLIER
Motorola Inc.

PHOTOFACT SET
361-7

PUBLISHED
1957

MOTOROLA

MODEL PICTURED
5P31A

Three-power operated
portable AM receiver

TUBES
4

POWER SUPPLY
110-120 volts AC/DC or
7.5 volts A & 90 volts B
supply

TUNING RANGE
535-1620KC

MFR/SUPPLIER
Motorola Inc.

PHOTOFACT SET
363-14

PUBLISHED
1957

MOTOROLA

MODEL PICTURED
66T1

Battery operated portable
AM transistorized
receiver

TUBES
6 transistor

POWER SUPPLY
6 volts DC

TUNING RANGE
530-1620KC

MFR/SUPPLIER
Motorola Inc.

PHOTOFACT SET
366-8

PUBLISHED
1957

MOTOROLA

MODEL PICTURED
10T28B

AC operated AM-FM
receiver

TUBES
10

POWER SUPPLY
110-120 volts AC

TUNING RANGE
540-1600KC,
88-108MC

MFR/SUPPLIER
Motorola Inc.

PHOTOFACT SET
402-11

PUBLISHED
1958

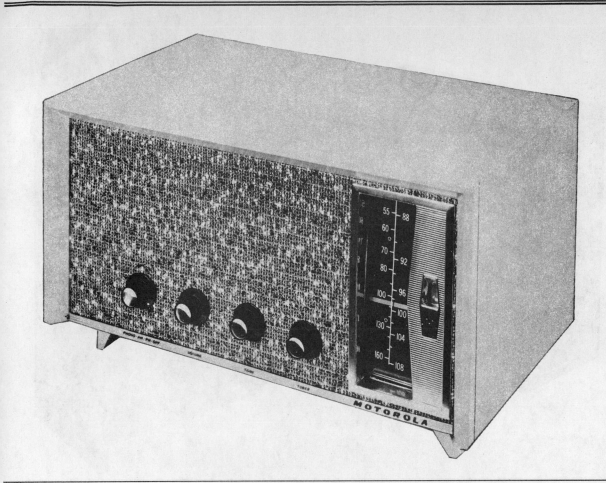

MOTOROLA

MODEL PICTURED
5T24GW-1

AC/DC operated AM
receiver

TUBES
5

POWER SUPPLY
110-120 volts AC/DC

TUNING RANGE
535-1620KC

MFR/SUPPLIER
Motorola Inc.

PHOTOFACT SET
407-14

PUBLISHED
1958

MOTOROLA

MODEL PICTURED
15KT25MC-1

AC operated AM-FM
receiver with four-speed
auto record changer

TUBES
15

POWER SUPPLY
110-120 volts AC

TUNING RANGE
540-1600KC,
88-108MC

MFR/SUPPLIER
Motorola Inc.

PHOTOFACT SET
408-12

PUBLISHED
1958

MOTOROLA

MODEL PICTURED
5R23G

AC operated AM receiver
with four-speed auto
record changer

TUBES
5

POWER SUPPLY
110-120 volts AC

TUNING RANGE
535-1620KC

MFR/SUPPLIER
Motorola Inc.

PHOTOFACT SET
409-13

PUBLISHED
1958

MOTOROLA

MODEL PICTURED
5C22M

AC operated AM receiver
with electric clock

TUBES
5

POWER SUPPLY
110-120 volts AC

TUNING RANGE
532-1620KC

MFR/SUPPLIER
Motorola Inc.

PHOTOFACT SET
410-11

PUBLISHED
1958

MOTOROLA

MODEL PICTURED
6X39A-2

Battery operated
two-band portable
transistorized receiver

TUBES
6 transistor

POWER SUPPLY
6 volts DC

TUNING RANGE
535-1620KC,
200-420KC

MFR/SUPPLIER
Motorola Inc.

PHOTOFACT SET
419-10

PUBLISHED
1958

MOTOROLA

MODEL PICTURED
5P21N

Three-power operated
portable AM receiver

TUBES
4

POWER SUPPLY
110-120 volts AC/DC or
7.5 volts A & 90 volts B
supply

TUNING RANGE
532-1620KC

MFR/SUPPLIER
Motorola Inc.

PHOTOFACT SET
431-15

PUBLISHED
1959

MOTOROLA

MODEL PICTURED
5T11M

AC/DC operated AM
receiver

TUBES
5

POWER SUPPLY
110-120 volts AC/DC

TUNING RANGE
535-1620KC

MFR/SUPPLIER
Motorola Inc.

PHOTOFACT SET
444-13

PUBLISHED
1959

MOTOROLA

MODEL PICTURED
6X28N

Battery operated portable transistorized AM receiver

TUBES
6 transistor

POWER SUPPLY
6 volts DC

TUNING RANGE
535-1620KC

MFR/SUPPLIER
Motorola Inc.

PHOTOFACT SET
444-14

PUBLISHED
1959

MOTOROLA

MODEL PICTURED
13KT15M

AC operated FM-AM receiver with four-speed auto record changer

TUBES
13

POWER SUPPLY
110-120 volts AC

TUNING RANGE
535-1620KC,
88-108MC

MFR/SUPPLIER
Motorola Inc.

PHOTOFACT SET
451-10

PUBLISHED
1959

MOTOROLA

MODEL PICTURED
5C11E

AC operated AM receiver
with electric clock

TUBES
5

POWER SUPPLY
110-120 volts AC

TUNING RANGE
535-1620KC

MFR/SUPPLIER
Motorola Inc.

PHOTOFACT SET
454-13

PUBLISHED
1959

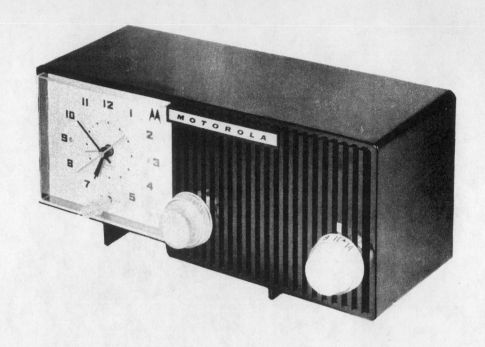

MOTOROLA

MODEL PICTURED
6T15S

AC/DC operated AM
receiver

TUBES
6

POWER SUPPLY
110-120 volts AC/DC

TUNING RANGE
532-1620KC

MFR/SUPPLIER
Motorola Inc.

PHOTOFACT SET
456-15

PUBLISHED
1959

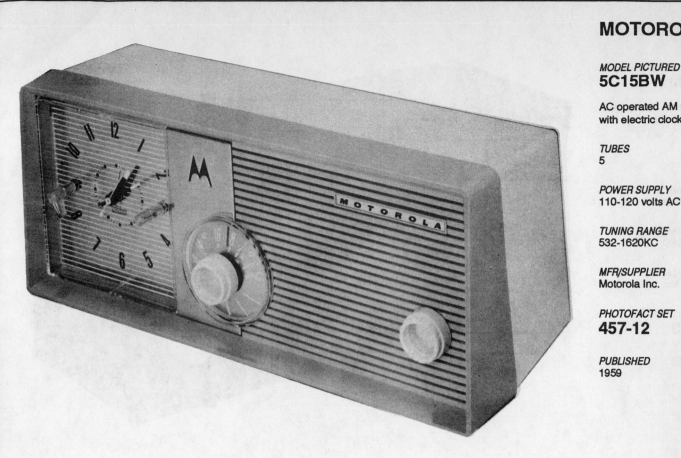

MOTOROLA

MODEL PICTURED
5C15BW

AC operated AM receiver
with electric clock

TUBES
5

POWER SUPPLY
110-120 volts AC

TUNING RANGE
532-1620KC

MFR/SUPPLIER
Motorola Inc.

PHOTOFACT SET
457-12

PUBLISHED
1959

MOTOROLA

MODEL PICTURED
8K26E

Battery operated portable
transistorized AM
receiver

TUBES
8 transistor

POWER SUPPLY
6 volts DC

TUNING RANGE
535-1620KC

MFR/SUPPLIER
Motorola Inc.

PHOTOFACT SET
459-8

PUBLISHED
1959

MOTOROLA

MODEL PICTURED
5C13M

AC operated AM receiver
with electric clock

TUBES
5

POWER SUPPLY
110-120 volts AC

TUNING RANGE
532-1620KC

MFR/SUPPLIER
Motorola Inc.

PHOTOFACT SET
460-13

PUBLISHED
1959

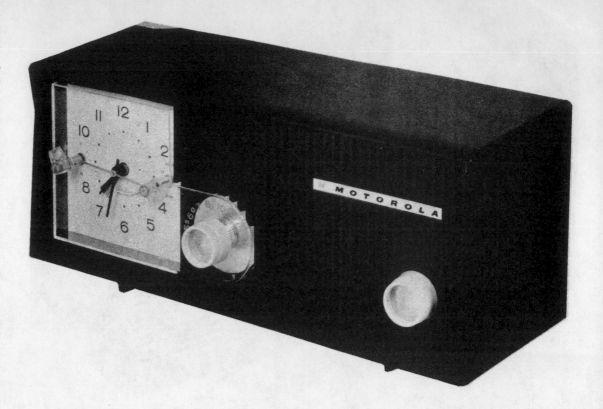

MOTOROLA

MODEL PICTURED
5T13P

AC/DC operated AM
receiver

TUBES
5

POWER SUPPLY
110-120 volts AC/DC

TUNING RANGE
535-1620KC

MFR/SUPPLIER
Motorola Inc.

PHOTOFACT SET
462-11

PUBLISHED
1959

MOTOROLA

MODEL PICTURED
7X24S

Battery operated portable
transistorized AM
receiver

TUBES
7 transistor

POWER SUPPLY
6 volts DC

TUNING RANGE
535-1620KC

MFR/SUPPLIER
Motorola Inc.

PHOTOFACT SET
462-12

PUBLISHED
1959

MOTOROLA

MODEL PICTURED
7X25P

Battery operated portable
transistorized AM
receiver

TUBES
7 transistor

POWER SUPPLY
6 volts DC

TUNING RANGE
535-1620KC

MFR/SUPPLIER
Motorola Inc.

PHOTOFACT SET
467-10

PUBLISHED
1959

MOTOROLA

MODEL PICTURED
L14E

Battery operated portable
transistorized AM
receiver

TUBES
8 transistor

POWER SUPPLY
9 volts DC

TUNING RANGE
535-1620KC

MFR/SUPPLIER
Motorola Inc.

PHOTOFACT SET
470-16

PUBLISHED
1960

MOTOROLA

MODEL PICTURED
L12N

Battery operated portable
transistorized AM
receiver

TUBES
6 transistor

POWER SUPPLY
9 volts DC

TUNING RANGE
535-1620KC

MFR/SUPPLIER
Motorola Inc.

PHOTOFACT SET
471-9

PUBLISHED
1960

MOTOROLA

MODEL PICTURED
10KT12M

AC operated FM-AM
receiver with four-speed
auto record changer

TUBES
10

POWER SUPPLY
110-120 volts AC

TUNING RANGE
535-1620KC,
88-108MC

MFR/SUPPLIER
Motorola Inc.

PHOTOFACT SET
471-10

PUBLISHED
1960

MOTOROLA

MODEL PICTURED
SK25MC

AC operated FM-AM
receiver with four-speed
auto stereo record
changer

TUBES
19

POWER SUPPLY
110-120 volts AC

TUNING RANGE
535-1620KC,
88-108MC

MFR/SUPPLIER
Motorola Inc.

PHOTOFACT SET
472-6

PUBLISHED
1960

MOTOROLA

MODEL PICTURED
X12A

Battery operated portable
transistorized AM
receiver

TUBES
6 transistor

POWER SUPPLY
6 volts DC

TUNING RANGE
535-1620KC

MFR/SUPPLIER
Motorola Inc.

PHOTOFACT SET
472-7

PUBLISHED
1960

MOTOROLA

MODEL PICTURED
X11E

Battery operated portable
transistorized AM
receiver

TUBES
6 transistor

POWER SUPPLY
9 volts DC

TUNING RANGE
532-1620KC

MFR/SUPPLIER
Motorola Inc.

PHOTOFACT SET
473-9

PUBLISHED
1960

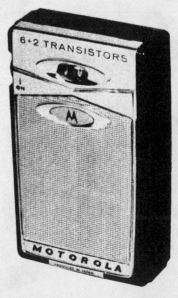

MOTOROLA

MODEL PICTURED
X11B

Battery operated
transistorized AM
portable receiver

TUBES
6 transistor

POWER SUPPLY
9 volts DC

TUNING RANGE
532-1620KC

MFR/SUPPLIER
Motorola Inc.

PHOTOFACT SET
481-9

PUBLISHED
1960

MOTOROLA

MODEL PICTURED
X12A-1

Battery operated
transistorized portable
AM receiver

TUBES
6 transistor

POWER SUPPLY
6 volts DC

TUNING RANGE
535-1620KC

MFR/SUPPLIER
Motorola Inc.

PHOTOFACT SET
486-18

PUBLISHED
1960

MOTOROLA

MODEL PICTURED
SK32W

AC operated FM-AM
receiver, stereo preamp,
and stereo amplifier with
auto record changer

TUBES
10

POWER SUPPLY
110-120 volts AC

TUNING RANGE
535-1620KC,
88-108MC

MFR/SUPPLIER
Motorola Inc.

PHOTOFACT SET
490-10

PUBLISHED
1960

MOTOROLA

MODEL PICTURED
A1W

AC/DC operated AM
receiver

TUBES
5

POWER SUPPLY
110-120 volts AC/DC

TUNING RANGE
535-1620KC

MFR/SUPPLIER
Motorola Inc.

PHOTOFACT SET
491-9

PUBLISHED
1960

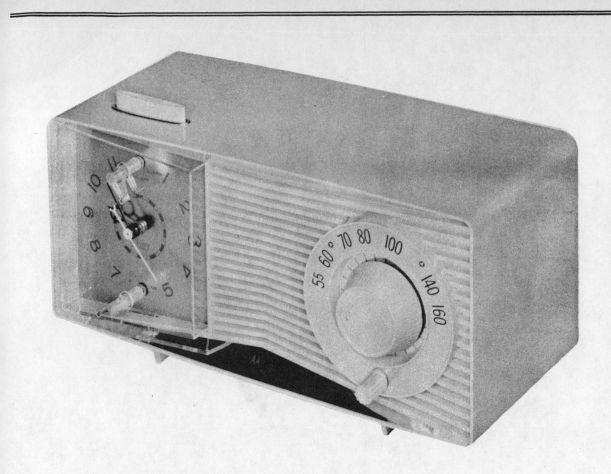

MOTOROLA

MODEL PICTURED
C3S-1

AC operated AM receiver

TUBES
5

POWER SUPPLY
110-120 volts AC

TUNING RANGE
535-1620KC

MFR/SUPPLIER
Motorola Inc.

PHOTOFACT SET
492-13

PUBLISHED
1960

MOTOROLA

MODEL PICTURED
C4B

AC operated AM receiver
with electric clock

TUBES
5

POWER SUPPLY
110-120 volts AC

TUNING RANGE
535-1620KC

MFR/SUPPLIER
Motorola Inc.

PHOTOFACT SET
494-11

PUBLISHED
1960

MUNTZ

804

AC operated AM receiver
with four-speed automatic
record changer

TUBES
6

POWER SUPPLY
110-120 volts AC

TUNING RANGE
535-1620KC

MFR/SUPPLIER
Muntz TV

PHOTOFACT SET
421-11

PUBLISHED
1958

MUNTZ

MODEL PICTURED
R-10

AC/DC operated AM
receiver

TUBES
4

POWER SUPPLY
110-120 volts AC/DC

TUNING RANGE
535-1600KC

MFR/SUPPLIER
Muntz TV

PHOTOFACT SET
428-9

PUBLISHED
1959

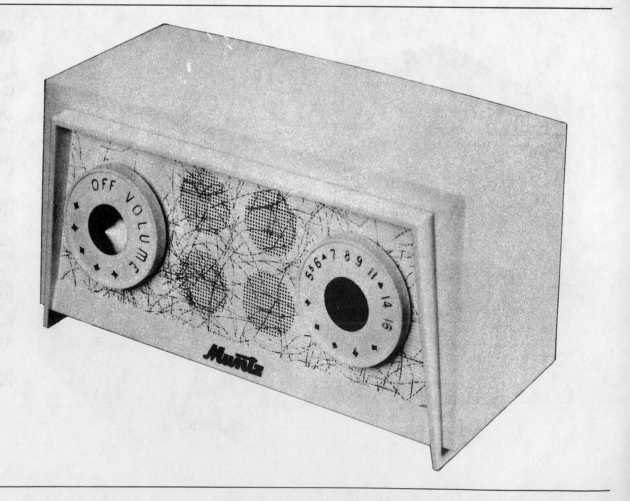

MUNTZ

MODEL PICTURED
806A

AC operated AM-FM receiver with four-speed auto record changer

TUBES
11

POWER SUPPLY
110-120 volts AC

TUNING RANGE
540-1650KC,
88-108MC

MFR/SUPPLIER
Muntz TV

PHOTOFACT SET
434-8

PUBLISHED
1959

MUNTZ

MODEL PICTURED
R-12

AC operated AM receiver with electric clock

TUBES
4

POWER SUPPLY
110-120 volts AC

TUNING RANGE
535-1620KC

MFR/SUPPLIER
Muntz TV

PHOTOFACT SET
456-16

PUBLISHED
1959

MUNTZ

MODEL PICTURED
R-13

AC operated AM receiver
with electric clock

TUBES
5

POWER SUPPLY
110-120 volts AC

TUNING RANGE
535-1620KC

MFR/SUPPLIER
Muntz TV

PHOTOFACT SET
457-13

PUBLISHED
1959

NANOLA
(NANAO)

MODEL PICTURED
6TP-106

Battery operated
transistorized portable
AM receiver

TUBES
6 transistor

POWER SUPPLY
9 volts DC

TUNING RANGE
535-1640KC

MFR/SUPPLIER
Leopold Sales Corp.

PHOTOFACT SET
476-8

PUBLISHED
1960

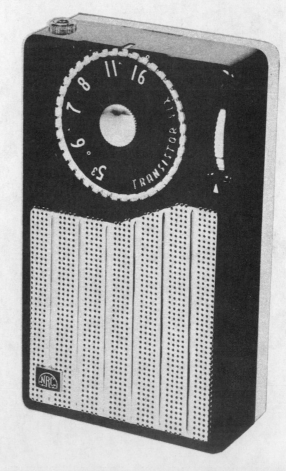

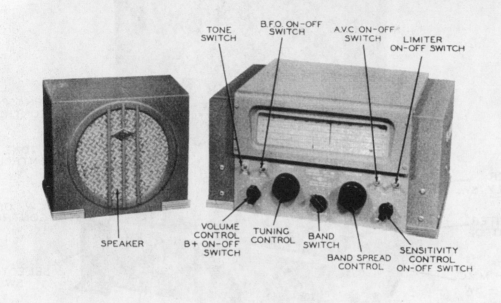

TONE SWITCH
B.F.O. ON-OFF SWITCH
A.V.C. ON-OFF SWITCH
LIMITER ON-OFF SWITCH

SPEAKER

VOLUME CONTROL B+ ON-OFF SWITCH
TUNING CONTROL
BAND SWITCH
BAND SPREAD CONTROL
SENSITIVITY CONTROL ON-OFF SWITCH

NATIONAL

MODEL PICTURED
NC-46

AC/DC operated
four-band
superheterodyne
communication receiver

TUBES
10

POWER SUPPLY
110-130 volts AC

TUNING RANGE
540-1600KC,
1.55-4.6MC,
4.4-12.0MC,
11.5-30MC

MFR/SUPPLIER
National Co.

PHOTOFACT SET
9-26

PUBLISHED
1946

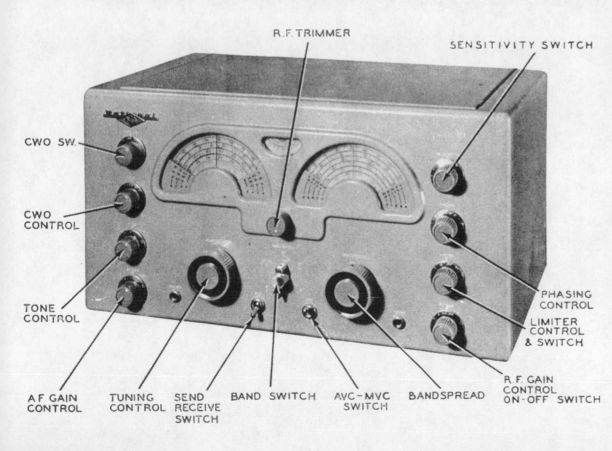

R.F. TRIMMER
SENSITIVITY SWITCH

CWO SW.
CWO CONTROL
TONE CONTROL

A F GAIN CONTROL
TUNING CONTROL
SEND RECEIVE SWITCH
BAND SWITCH
AVC-MVC SWITCH
BANDSPREAD
R.F. GAIN CONTROL ON-OFF SWITCH

PHASING CONTROL
LIMITER CONTROL & SWITCH

NATIONAL

MODEL PICTURED
NC-173T

AC operated multi-band
superheterodyne
commercial type receiver

TUBES
13

POWER SUPPLY
110-120 or 220-240
volts AC

TUNING RANGE
5 bands

MFR/SUPPLIER
National Co.

PHOTOFACT SET
40-13

PUBLISHED
1948

NATIONAL

MODEL PICTURED
NC-2-40DR

AC operated multi-band
superheterodyne
communication receiver

TUBES
12

POWER SUPPLY
110-120 or 220-240
volts AC

TUNING RANGE
10 bands

MFR/SUPPLIER
National Co.

PHOTOFACT SET
41-16

PUBLISHED
1948

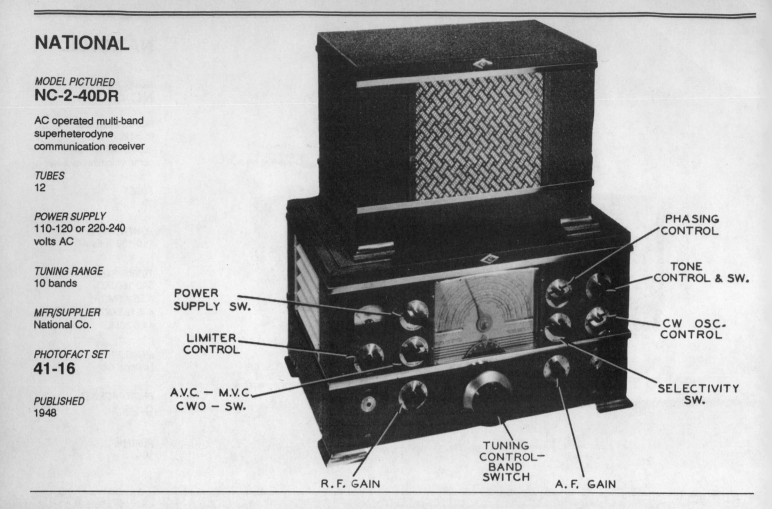

PHASING CONTROL

TONE CONTROL & SW.

POWER SUPPLY SW.

CW OSC. CONTROL

LIMITER CONTROL

A.V.C. — M.V.C. CWO — SW.

SELECTIVITY SW.

R.F. GAIN

TUNING CONTROL-BAND SWITCH

A.F. GAIN

NATIONAL

MODEL PICTURED
NC-33

AC/DC operated
multi-band commercial
type superheterodyne
receiver

TUBES
6

POWER SUPPLY
105-130 volts AC/DC

TUNING RANGE
4 bands

MFR/SUPPLIER
National Co.

PHOTOFACT SET
47-14

PUBLISHED
1948

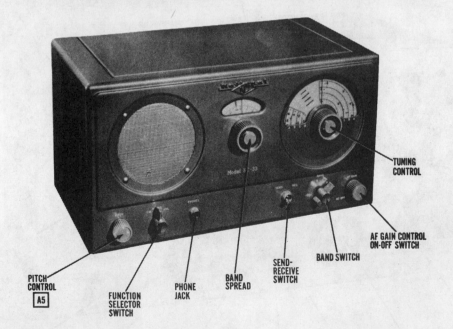

TUNING CONTROL

AF GAIN CONTROL ON-OFF SWITCH

BAND SWITCH

SEND-RECEIVE SWITCH

BAND SPREAD

PHONE JACK

FUNCTION SELECTOR SWITCH

PITCH CONTROL
A5

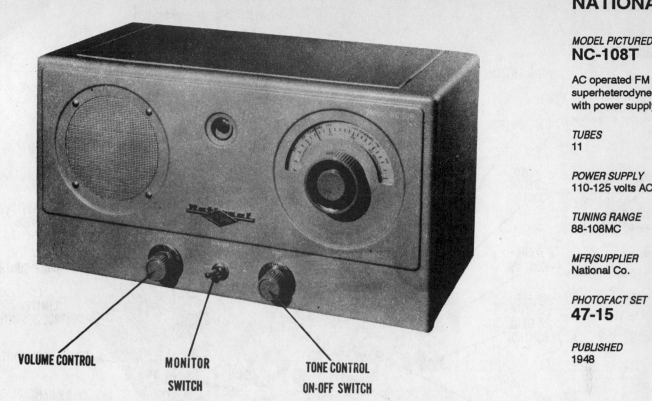

NATIONAL

MODEL PICTURED
NC-108T

AC operated FM superheterodyne receiver with power supply

TUBES
11

POWER SUPPLY
110-125 volts AC

TUNING RANGE
88-108MC

MFR/SUPPLIER
National Co.

PHOTOFACT SET
47-15

PUBLISHED
1948

VOLUME CONTROL

MONITOR SWITCH

TONE CONTROL ON-OFF SWITCH

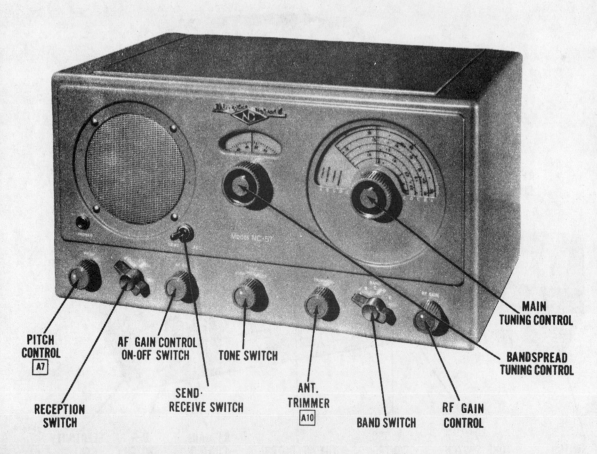

NATIONAL

MODEL PICTURED
NC-57

AC/Battery operated multi-band superheterodyne commercial type receiver

TUBES
9

POWER SUPPLY
105-130 volts AC or 6.3 volts & 250 volts DC

TUNING RANGE
5 bands

MFR/SUPPLIER
National Co.

PHOTOFACT SET
48-14

PUBLISHED
1948

PITCH CONTROL A7

AF GAIN CONTROL ON-OFF SWITCH

TONE SWITCH

MAIN TUNING CONTROL

BANDSPREAD TUNING CONTROL

RECEPTION SWITCH

SEND-RECEIVE SWITCH

ANT. TRIMMER A10

BAND SWITCH

RF GAIN CONTROL

NATIONAL

MODEL PICTURED
NC-183R

AC/Battery operated
commercial type
multi-band
superheterodyne receiver

TUBES
16

POWER SUPPLY
110-120 or 220-240 volts
AC or 6 volts A &
135-250 volts B supply

TUNING RANGE
5 bands

MFR/SUPPLIER
National Co.

PHOTOFACT SET
49-15

PUBLISHED
1948

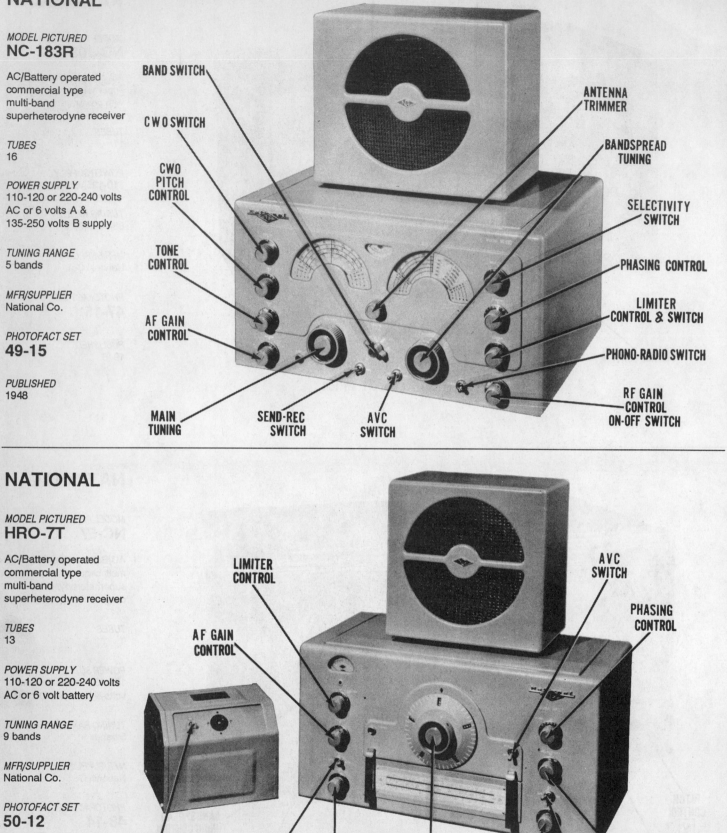

BAND SWITCH

CWO SWITCH

CWO PITCH CONTROL

TONE CONTROL

AF GAIN CONTROL

MAIN TUNING

SEND-REC SWITCH

AVC SWITCH

ANTENNA TRIMMER

BANDSPREAD TUNING

SELECTIVITY SWITCH

PHASING CONTROL

LIMITER CONTROL & SWITCH

PHONO-RADIO SWITCH

RF GAIN CONTROL ON-OFF SWITCH

NATIONAL

MODEL PICTURED
HRO-7T

AC/Battery operated
commercial type
multi-band
superheterodyne receiver

TUBES
13

POWER SUPPLY
110-120 or 220-240 volts
AC or 6 volt battery

TUNING RANGE
9 bands

MFR/SUPPLIER
National Co.

PHOTOFACT SET
50-12

PUBLISHED
1948

LIMITER CONTROL

AF GAIN CONTROL

AVC SWITCH

PHASING CONTROL

ON-OFF SWITCH

TONE SWITCH

CW OSC. CONTROL

TUNING CONTROL

RF GAIN CONTROL

B+ SWITCH

SELECTIVITY SWITCH

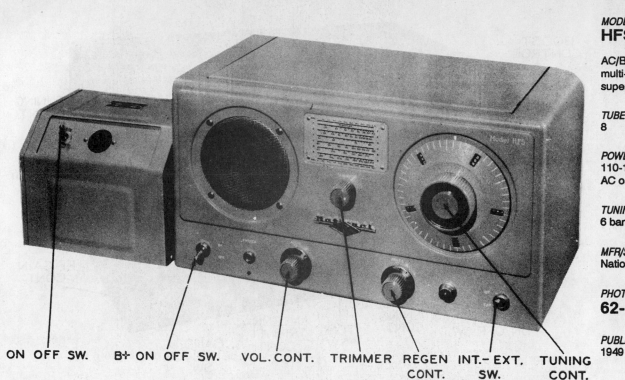

ON OFF SW.   B+ ON OFF SW.   VOL. CONT.   TRIMMER   REGEN CONT.   INT.- EXT. SW.   TUNING CONT.

MODEL PICTURED
HFS

AC/Battery operated multi-band superheterodyne receiver

TUBES
8

POWER SUPPLY
110-120 or 220-240 volts AC or 6 volt battery

TUNING RANGE
6 bands

MFR/SUPPLIER
National Co.

PHOTOFACT SET
62-14

PUBLISHED
1949

NATIONAL

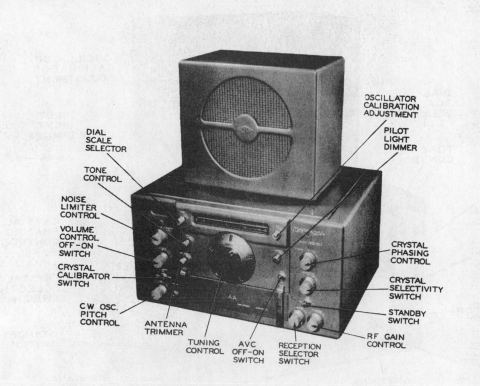

DIAL SCALE SELECTOR

TONE CONTROL

NOISE LIMITER CONTROL

VOLUME CONTROL OFF-ON SWITCH

CRYSTAL CALIBRATOR SWITCH

CW OSC. PITCH CONTROL

ANTENNA TRIMMER

TUNING CONTROL

AVC OFF-ON SWITCH

RECEPTION SELECTOR SWITCH

OSCILLATOR CALIBRATION ADJUSTMENT

PILOT LIGHT DIMMER

CRYSTAL PHASING CONTROL

CRYSTAL SELECTIVITY SWITCH

STANDBY SWITCH

RF GAIN CONTROL

MODEL PICTURED
HRO-50

AC operated multi-band communication receiver

TUBES
17

POWER SUPPLY
110-120 or 220-240 volts AC

TUNING RANGE
12 bands

MFR/SUPPLIER
National Co.

PHOTOFACT SET
112-7

PUBLISHED
1950

NATIONAL

MODEL PICTURED
NC-125

AC operated multi-band superheterodyne communication type receiver

TUBES
11

POWER SUPPLY
110-120 volts AC

TUNING RANGE
5 bands

MFR/SUPPLIER
National Co.

PHOTOFACT SET
139-10

PUBLISHED
1951

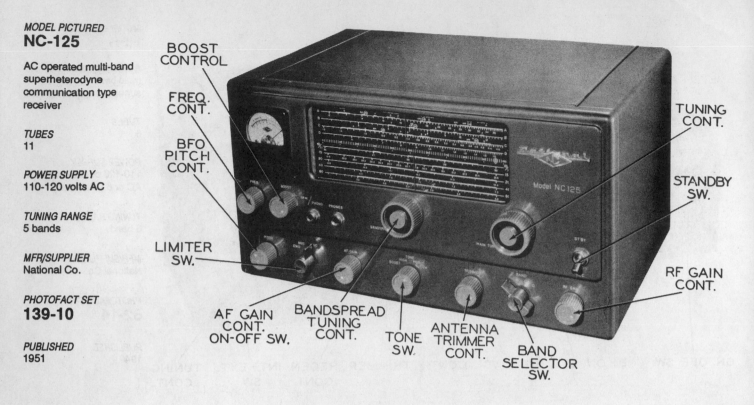

NATIONAL

MODEL PICTURED
HRO-50R1

AC operated multi-band communication receiver

TUBES
16

POWER SUPPLY
110-120 volts AC

TUNING RANGE
12 bands

MFR/SUPPLIER
National Co.

PHOTOFACT SET
169-11

PUBLISHED
1952

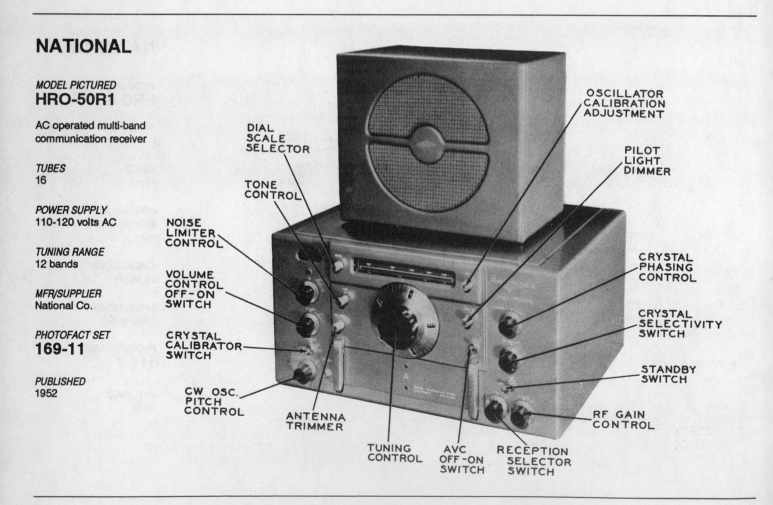

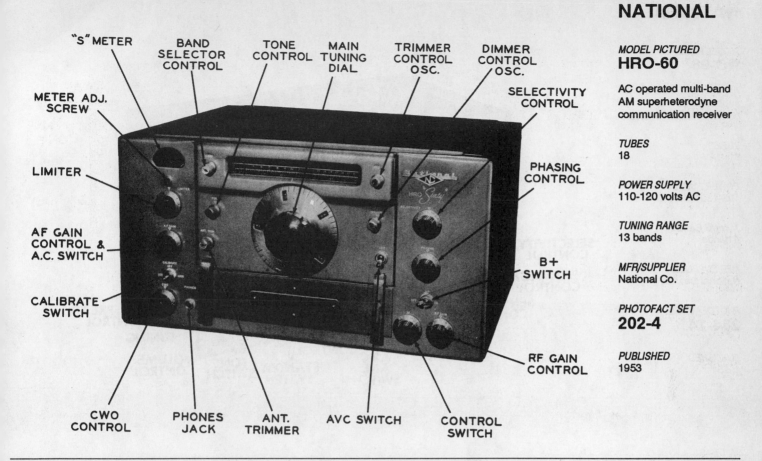

"S" METER

METER ADJ.
SCREW

LIMITER

AF GAIN
CONTROL &
A.C. SWITCH

CALIBRATE
SWITCH

BAND
SELECTOR
CONTROL

TONE
CONTROL

MAIN
TUNING
DIAL

TRIMMER
CONTROL
OSC.

DIMMER
CONTROL
OSC.

SELECTIVITY
CONTROL

PHASING
CONTROL

B+
SWITCH

RF GAIN
CONTROL

CWO
CONTROL

PHONES
JACK

ANT.
TRIMMER

AVC SWITCH

CONTROL
SWITCH

NATIONAL

MODEL PICTURED
HRO-60

AC operated multi-band
AM superheterodyne
communication receiver

TUBES
18

POWER SUPPLY
110-120 volts AC

TUNING RANGE
13 bands

MFR/SUPPLIER
National Co.

PHOTOFACT SET
202-4

PUBLISHED
1953

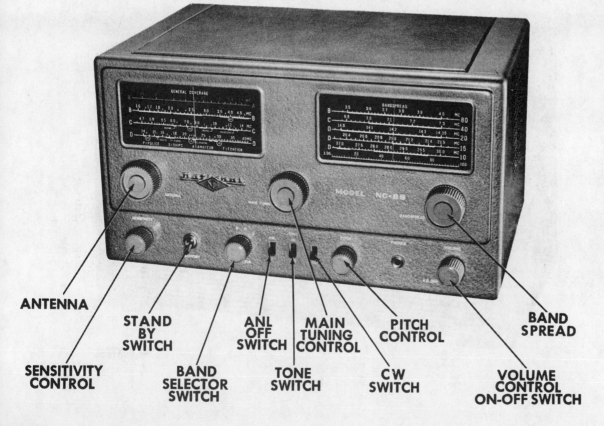

ANTENNA

STAND
BY
SWITCH

SENSITIVITY
CONTROL

BAND
SELECTOR
SWITCH

ANL
OFF
SWITCH

TONE
SWITCH

MAIN
TUNING
CONTROL

CW
SWITCH

PITCH
CONTROL

BAND
SPREAD

VOLUME
CONTROL
ON-OFF SWITCH

NATIONAL

MODEL PICTURED
NC-88

AC operated multi-band
superheterodyne
communication receiver

TUBES
9

POWER SUPPLY
105/130 volts AC

TUNING RANGE
4 bands

MFR/SUPPLIER
National Co.

PHOTOFACT SET
233-7

PUBLISHED
1954

NATIONAL

MODEL PICTURED
NC-98

AC operated multi-band
superheterodyne
communication receiver

TUBES
9

POWER SUPPLY
105-130 volts AC

TUNING RANGE
4 bands

MFR/SUPPLIER
National Co.

PHOTOFACT SET
264-14

PUBLISHED
1955

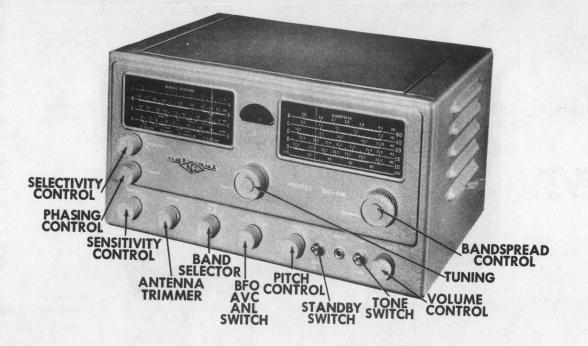

SELECTIVITY
CONTROL

PHASING
CONTROL

SENSITIVITY
CONTROL

ANTENNA
TRIMMER

BAND
SELECTOR

BFO
AVC
ANL
SWITCH

PITCH
CONTROL

STANDBY
SWITCH

TONE
SWITCH

VOLUME
CONTROL

TUNING

BANDSPREAD
CONTROL

NATIONAL UNION

MODEL PICTURED
571

AC/DC operated AM
superheterodyne receiver
with self-contained loop
antenna

TUBES
5

POWER SUPPLY
105-125 volts AC/DC

TUNING RANGE
540-1600KC

MFR/SUPPLIER
National Union Radio
Corp.

PHOTOFACT SET
17-22

PUBLISHED
1947

VOLUME CONTROL
ON-OFF SWITCH

TUNING CONTROL

VOLUME
CONTROL
ON-OFF
SWITCH

TUNING
CONTROL

ELECTRIC-BATTERY
SWITCH

NATIONAL UNION

MODEL PICTURED
G-613

Three-power operated portable AM superheterodyne receiver with loop antenna

TUBES
6

POWER SUPPLY
105-120 volts AC or 9 volts A & 90 volts B supply

TUNING RANGE
540-1650KC

MFR/SUPPLIER
National Union Radio Corp.

PHOTOFACT SET
19-23

PUBLISHED
1947

NEC

MODEL PICTURED
NT-61

Battery operated transistorized portable AM receiver

TUBES
6 transistor

POWER SUPPLY
9 volts DC

TUNING RANGE
535-1650KC

MFR/SUPPLIER
Kanematsu

PHOTOFACT SET
497-13

PUBLISHED
1960

93

NEC

MODEL PICTURED
NT-620

Battery operated
transistorized portable
AM receiver

TUBES
6 transistor

POWER SUPPLY
9 volts DC

TUNING RANGE
520-1700KC

MFR/SUPPLIER
Kanematsu

PHOTOFACT SET
498-13

PUBLISHED
1960

NORELCO

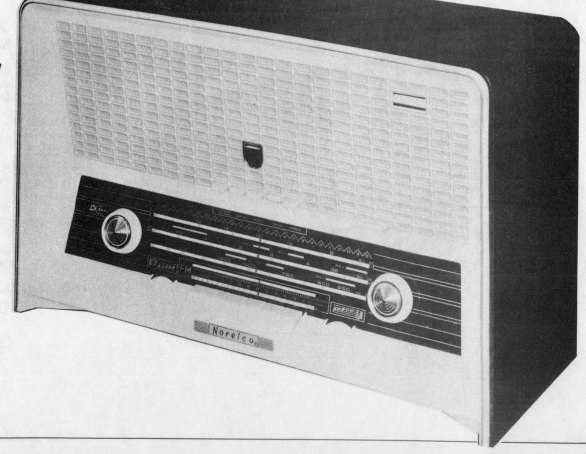

MODEL PICTURED
B5X88A

AC operated FM-AM-SW
receiver

TUBES
9

POWER SUPPLY
90/110/127/145/190/220
volts AC

TUNING RANGE
517-1622KC,
88-108MC,
7.3-15.8MC,
3.2-7.5MC

MFR/SUPPLIER
North American Philips
Co. Inc.

PHOTOFACT SET
501-11

PUBLISHED
1960

NORELCO

MODEL PICTURED
L2X97T

Battery operated
transistorized AM
receiver with clock

TUBES
7 transistor

POWER SUPPLY
6 volts DC

TUNING RANGE
512-1630KC

MFR/SUPPLIER
North American Philips
Co. Inc.

PHOTOFACT SET
503-14

PUBLISHED
1960

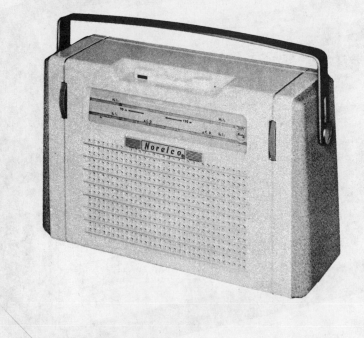

NORELCO

MODEL PICTURED
L3X86T

Battery operated
transistorized portable
AM receiver

TUBES
7 transistor

POWER SUPPLY
9 volts DC

TUNING RANGE
512-1635KC

MFR/SUPPLIER
North American Philips
Co. Inc.

PHOTOFACT SET
504-17

PUBLISHED
1960

NORELCO

MODEL PICTURED
L3X88T

Battery operated
transistorized portable
AM-SW receiver

TUBES
7 transistor

POWER SUPPLY
9 volts DC

TUNING RANGE
512-1635KC,
4.65-12.2MC

MFR/SUPPLIER
North American Philips
Co. Inc.

PHOTOFACT SET
506-16

PUBLISHED
1960

NORELCO

MODEL PICTURED
L4X95T

Battery operated
transistorized portable
AM-SW receiver

TUBES
7 transistor

POWER SUPPLY
9 volts DC

TUNING RANGE
517-1622KC,
11.6-22MC,
4.65-9.9MC,
1.6-3.9MC

MFR/SUPPLIER
North American Philips
Co. Inc.

PHOTOFACT SET
507-12

PUBLISHED
1960

NORELCO

MODEL PICTURED
B3X88U/71

AC/DC operated
FM-AM-SW receiver

TUBES
7

POWER SUPPLY
110/127/220 volts AC/DC

TUNING RANGE
517-1612KC,
88-108MC,
8.8-18.1MC,
3.1-7.5MC

MFR/SUPPLIER
North American Philips
Co. Inc.

PHOTOFACT SET
508-15

PUBLISHED
1960

NORELCO

MODEL PICTURED
L1X75T/64R

Battery operated
transistorized portable
AM receiver

TUBES
7 transistor

POWER SUPPLY
6 volts DC

TUNING RANGE
517-1622KC

MFR/SUPPLIER
North American Philips
Co. Inc.

PHOTOFACT SET
510-13

PUBLISHED
1960

NORELCO

MODEL PICTURED
B2X98A/70R

AC operated FM-AM-SW
receiver

TUBES
6

POWER SUPPLY
110/127/220 volts AC

TUNING RANGE
517-1622KC,
88-108MC,
6-18.3MC

MFR/SUPPLIER
North American Philips
Co. Inc.

PHOTOFACT SET
511-10

PUBLISHED
1960

OLSON

MODEL PICTURED
RA-315

Battery operated
transistorized portable
AM receiver

TUBES
4 transistor

POWER SUPPLY
9 volts DC

TUNING RANGE
535-1620KC

MFR/SUPPLIER
Olson Radio Corp.

PHOTOFACT SET
483-11

PUBLISHED
1960

OLSON

MODEL PICTURED
RA-323

AC/DC operated AM receiver

TUBES
5

POWER SUPPLY
110-120 volts AC

TUNING RANGE
530-1620KC

MFR/SUPPLIER
Olson Radio Corp.

PHOTOFACT SET
486-19

PUBLISHED
1960

OLYMPIC

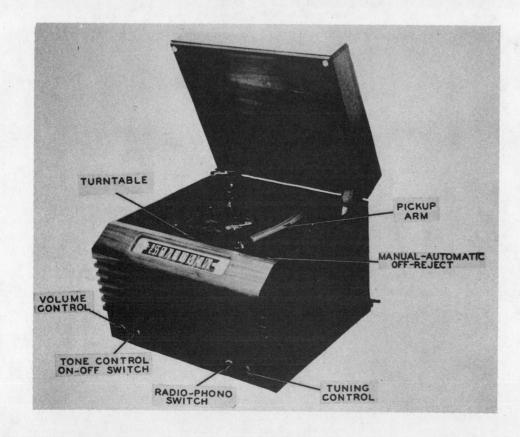

TURNTABLE

PICKUP ARM

MANUAL-AUTOMATIC OFF-REJECT

VOLUME CONTROL

TONE CONTROL ON-OFF SWITCH

RADIO-PHONO SWITCH

TUNING CONTROL

MODEL PICTURED
6-617

AC operated phono-radio combo AM superheterodyne receiver with self-contained loop antenna

TUBES
6

POWER SUPPLY
105-125 volts AC

TUNING RANGE
530-1700KC

MFR/SUPPLIER
Olympic Radio & Television

PHOTOFACT SET
4-7

PUBLISHED
1946

OLYMPIC

MODEL PICTURED
6-502P

AC/DC operated
superheterodyne receiver
with self-contained loop
antenna

TUBES
5

POWER SUPPLY
105-125 volts AC/DC

TUNING RANGE
535-1700KC

MFR/SUPPLIER
Olympic Radio &
Television

PHOTOFACT SET
4-10

PUBLISHED
1946

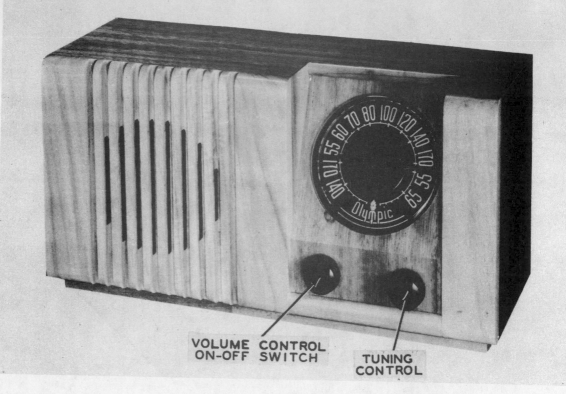

VOLUME CONTROL
ON-OFF SWITCH

TUNING
CONTROL

OLYMPIC

MODEL PICTURED
6-606

AC/DC operated portable
AM superheterodyne
receiver with
self-contained loop
antenna

TUBES
6

POWER SUPPLY
105-125 volts AC/DC or 9
volts A & 90 volts B
supply

TUNING RANGE
530-1700KC

MFR/SUPPLIER
Olympic Radio &
Television

PHOTOFACT SET
4-36

PUBLISHED
1946

TUNING
CONTROL

VOLUME CONTROL
ON-OFF SWITCH

OLYMPIC

MODEL PICTURED
6-601W

AC operated two-band superheterodyne receiver with self-contained loop antenna

TUBES
6

POWER SUPPLY
105-125 volts AC

TUNING RANGE
535-1700KC,
5.7-18.4MC

MFR/SUPPLIER
Olympic Radio & Television

PHOTOFACT SET
8-24

PUBLISHED
1946

OLYMPIC

MODEL PICTURED
6-604W

AC/DC operated two-band superheterodyne receiver with loop antenna

TUBES
6

POWER SUPPLY
110/150/220 volts AC

TUNING RANGE
530-1700KC,
5.7-18.4MC

MFR/SUPPLIER
Olympic Radio & Television

PHOTOFACT SET
22-21

PUBLISHED
1947

OLYMPIC

MODEL PICTURED
7-724

AC operated phono-radio
combo two-band
superheterodyne receiver
with loop antenna

TUBES
8

POWER SUPPLY
105-125 volts AC

TUNING RANGE
535-1700KC,
5.7-18.4MC

MFR/SUPPLIER
Olympic Radio &
Television

PHOTOFACT SET
29-19

PUBLISHED
1947

VOLUME CONTROL   TONE CONTROL ON-OFF SWITCH   PHONO-BAND SWITCH   TUNING CONTROL

OLYMPIC

MODEL PICTURED
7-526

Three-power operated
portable AM
superheterodyne receiver
with loop antenna

TUBES
5

POWER SUPPLY
105-125 volts AC/DC or 9
volts A & 90 volts B
supply

TUNING RANGE
535-1650KC

MFR/SUPPLIER
Olympic Radio &
Television

PHOTOFACT SET
30-21

PUBLISHED
1947

VOLUME CONTROL ON-OFF SWITCH   TUNING CONTROL

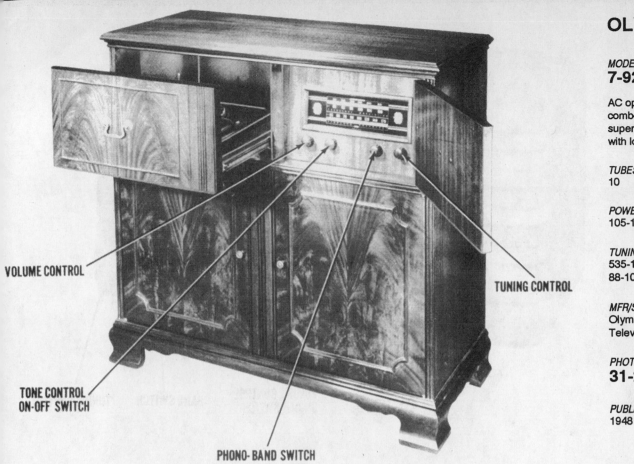

VOLUME CONTROL

TUNING CONTROL

TONE CONTROL
ON-OFF SWITCH

PHONO- BAND SWITCH

OLYMPIC

MODEL PICTURED
7-925

AC operated phono-radio
combo AM-FM
superheterodyne receiver
with loop antenna

TUBES
10

POWER SUPPLY
105-125 volts AC

TUNING RANGE
535-1700KC,
88-108MC

MFR/SUPPLIER
Olympic Radio &
Television

PHOTOFACT SET
31-22

PUBLISHED
1948

OLYMPIC

MODEL PICTURED
7-532W

AC/DC operated AM-FM
superheterodyne receiver
with loop antenna

TUBES
6

POWER SUPPLY
105-125 volts AC/DC

TUNING RANGE
530-1700KC,
88-108MC

MFR/SUPPLIER
Olympic Radio &
Television

PHOTOFACT SET
32-15

PUBLISHED
1948

VOLUME CONTROL

TONE CONTROL
ON-OFF SWITCH

BAND SWITCH   TUNING CONTROL

OLYMPIC

MODEL PICTURED
7-435V

AC/DC operated
two-band
superheterodyne receiver
with loop antenna

TUBES
5

POWER SUPPLY
105-125 volts AC/DC

TUNING RANGE
540-1650KC,
4.8-16.4MC

MFR/SUPPLIER
Olympic Radio &
Television

PHOTOFACT SET
34-13

PUBLISHED
1948

VOLUME CONTROL
ON-OFF SWITCH   BAND SWITCH   TUNING CONTROL

OLYMPIC

MODEL PICTURED
7-622

AC operated phono-radio
combo AM
superheterodyne receiver
with loop antenna

TUBES
6

POWER SUPPLY
105-125 volts AC

TUNING RANGE
537-1700KC

MFR/SUPPLIER
Olympic Radio &
Television

PHOTOFACT SET
34-14

PUBLISHED
1948

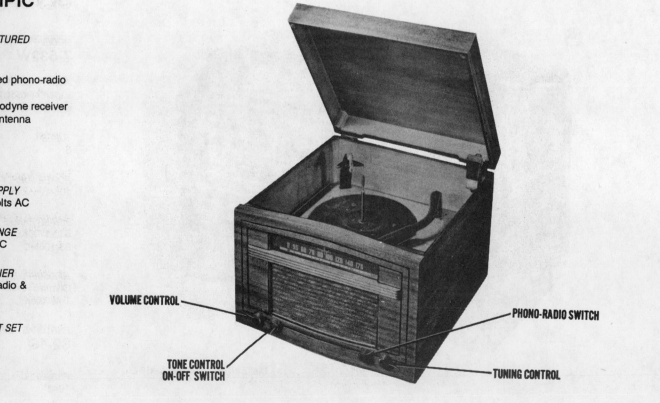

VOLUME CONTROL

PHONO-RADIO SWITCH

TONE CONTROL
ON-OFF SWITCH

TUNING CONTROL

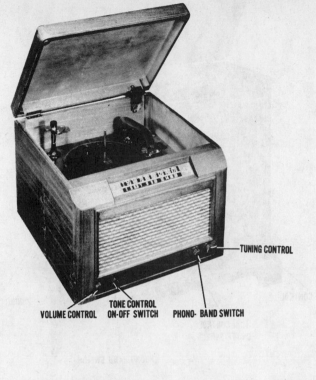

VOLUME CONTROL   TONE CONTROL   PHONO- BAND SWITCH
                   ON-OFF SWITCH

TUNING CONTROL

OLYMPIC

MODEL PICTURED
8-618

AC operated phono-radio
combo two-band
superheterodyne receiver
with loop antenna

TUBES
6

POWER SUPPLY
110-120 volts AC

TUNING RANGE
535-1700KC,
5.7-18.4MC

MFR/SUPPLIER
Olympic Radio &
Television

PHOTOFACT SET
35-16

PUBLISHED
1948

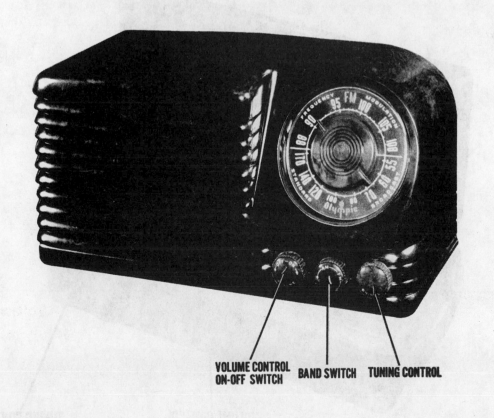

VOLUME CONTROL   BAND SWITCH   TUNING CONTROL
ON-OFF SWITCH

OLYMPIC

MODEL PICTURED
7-537

AC/DC operated AM-FM
superheterodyne receiver
with loop antenna

TUBES
6

POWER SUPPLY
105-125 volts AC/DC

TUNING RANGE
540-1730KC,
88-108MC

MFR/SUPPLIER
Olympic Radio &
Television

PHOTOFACT SET
37-13

PUBLISHED
1948

OLYMPIC

MODEL PICTURED
8-934

AC operated phono-radio
combo AM-FM
superheterodyne receiver
with loop antenna

TUBES
10

POWER SUPPLY
105-125 volts AC

TUNING RANGE
535-1700KC,
88-108MC

MFR/SUPPLIER
Olympic Radio &
Television

PHOTOFACT SET
45-19

PUBLISHED
1948

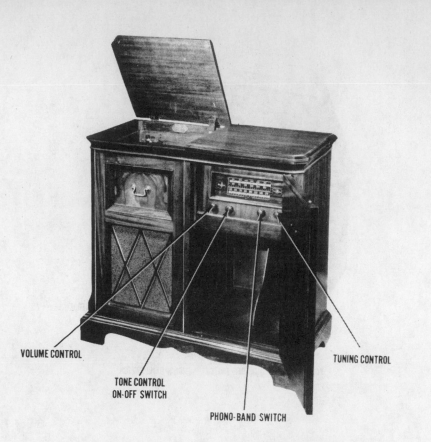

VOLUME CONTROL

TONE CONTROL
ON-OFF SWITCH

PHONO-BAND SWITCH

TUNING CONTROL

OLYMPIC

MODEL PICTURED
8-451

Battery operated portable
AM superheterodyne
receiver with loop
antenna

TUBES
4

POWER SUPPLY
1.5 volts A & 67.5 volts B
supply

TUNING RANGE
540-1600KC

MFR/SUPPLIER
Olympic Radio &
Television

PHOTOFACT SET
48-15

PUBLISHED
1948

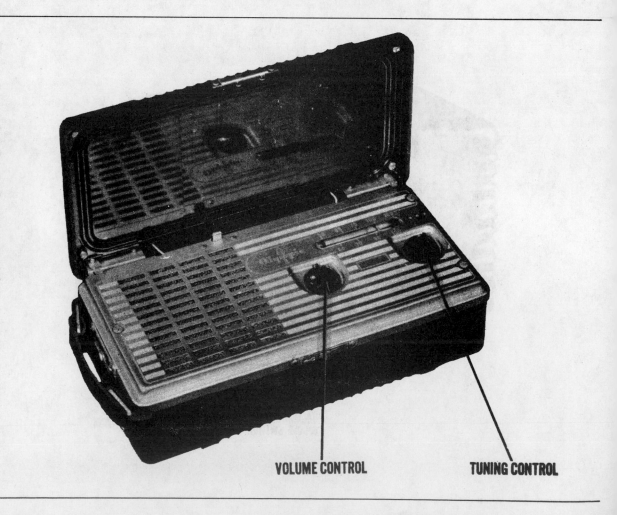

VOLUME CONTROL

TUNING CONTROL

OLYMPIC

**VOLUME CONTROL
ON-OFF SWITCH**

TUNING CONTROL

MODEL PICTURED
7-421W

AC/DC operated AM
superheterodyne receiver
with loop antenna

TUBES
5

POWER SUPPLY
105-125 volts AC/DC

TUNING RANGE
535-1700KC

MFR/SUPPLIER
Olympic Radio &
Television

PHOTOFACT SET
57-13

PUBLISHED
1949

OLYMPIC

VOLUME CONTROL **TONE CONTROL
ON-OFF SWITCH** **BAND SWITCH** **TUNING CONTROL**

MODEL PICTURED
8-533W

AC/DC operated AM-FM
superheterodyne receiver
with loop antenna

TUBES
6

POWER SUPPLY
105-125 volts AC/DC

TUNING RANGE
530-1700KC,
88-108MC

MFR/SUPPLIER
Olympic Radio &
Television

PHOTOFACT SET
57-14

PUBLISHED
1949

OLYMPIC

MODEL PICTURED
51-421W

AC/DC operated AM
superheterodyne receiver
with loop antenna

TUBES
5

POWER SUPPLY
105-125 volts AC/DC

TUNING RANGE
535-1620KC

MFR/SUPPLIER
Olympic Radio &
Television

PHOTOFACT SET
151-9

PUBLISHED
1951

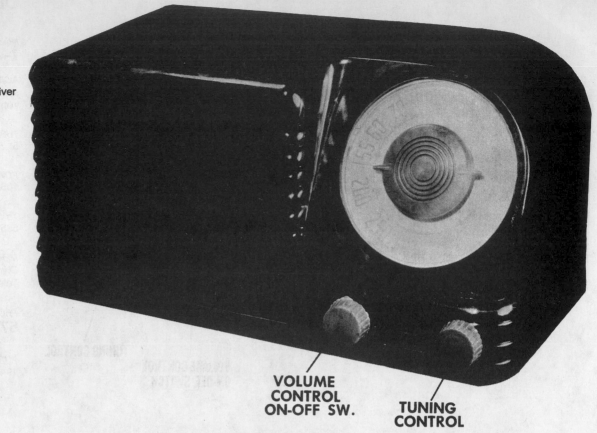

VOLUME
CONTROL
ON-OFF SW.

TUNING
CONTROL

OLYMPIC

MODEL PICTURED
9-435V

AC/DC operated
two-band
superheterodyne receiver
with loop antenna

TUBES
5

POWER SUPPLY
105-125 volts AC/DC

TUNING RANGE
540-1610KC,
4.7-16.1MC

MFR/SUPPLIER
Olympic Radio &
Television

PHOTOFACT SET
152-11

PUBLISHED
1951

VOLUME
CONTROL
ON-OFF SW

BC–SW
SELECTOR
SWITCH

TUNING
CONTROL

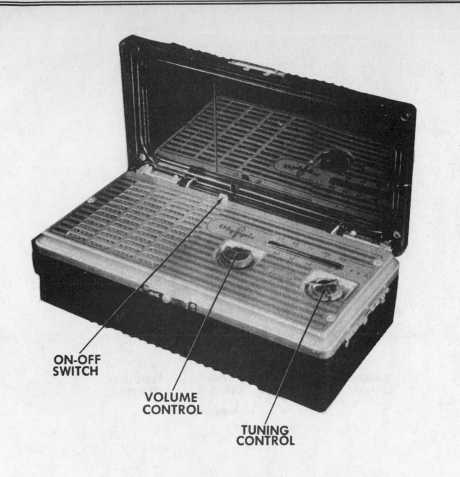

ON-OFF
SWITCH

VOLUME
CONTROL

TUNING
CONTROL

OLYMPIC

MODEL PICTURED
489

Battery operated portable
AM superheterodyne
receiver with loop
antenna

TUBES
4

POWER SUPPLY
1.5 volts A & 67.5 volts B
supply

TUNING RANGE
540-1600KC

MFR/SUPPLIER
Olympic Radio &
Television

PHOTOFACT SET
154-9

PUBLISHED
1951

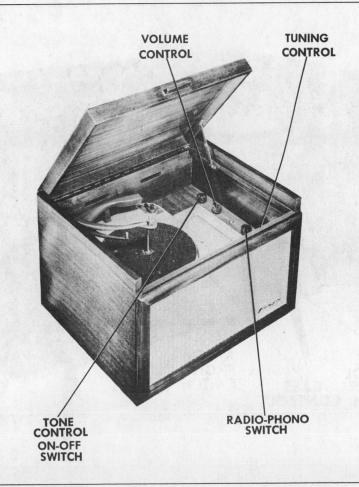

VOLUME
CONTROL

TUNING
CONTROL

TONE
CONTROL
ON-OFF
SWITCH

RADIO-PHONO
SWITCH

OLYMPIC

MODEL PICTURED
HF500

AC operated phono-radio
combo AM
superheterodyne receiver
with three-speed auto
record changer

TUBES
6

POWER SUPPLY
105-125 volts AC

TUNING RANGE
540-1620KC

MFR/SUPPLIER
Olympic Radio &
Television

PHOTOFACT SET
256-11

PUBLISHED
1954

OLYMPIC

MODEL PICTURED
572B

AC operated phono-radio
combo AM-FM
superheterodyne receiver
with three-speed auto
record changer

TUBES
10

POWER SUPPLY
105-120 volts AC

TUNING RANGE
535-1700KC,
88-108MC

MFR/SUPPLIER
Olympic Radio &
Television

PHOTOFACT SET
257-11

PUBLISHED
1954

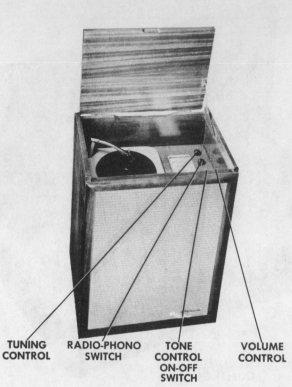

TUNING
CONTROL

RADIO-PHONO
SWITCH

TONE
CONTROL
ON-OFF
SWITCH

VOLUME
CONTROL

OLYMPIC

MODEL PICTURED
505

AC operated phono-radio
combo AM-FM
superheterodyne receiver

TUBES
10

POWER SUPPLY
110-120 volts AC

TUNING RANGE
535-1700KC,
88-108MC

MFR/SUPPLIER
Olympic Radio &
Television

PHOTOFACT SET
259-10

PUBLISHED
1954

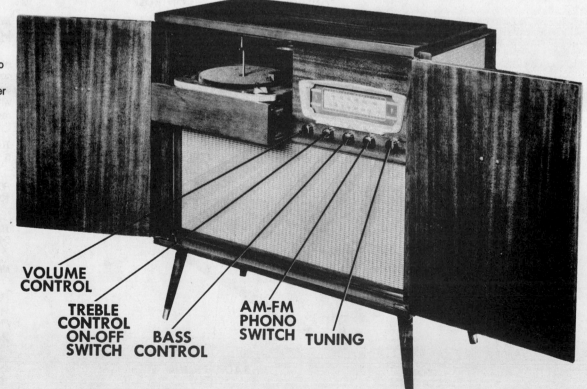

VOLUME
CONTROL

TREBLE
CONTROL
ON-OFF
SWITCH

BASS
CONTROL

AM-FM
PHONO
SWITCH

TUNING

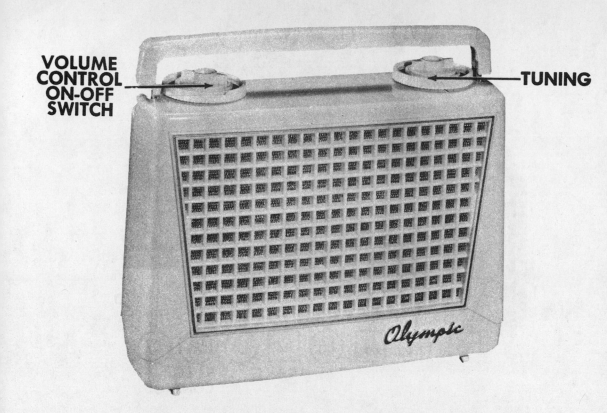

VOLUME CONTROL ON-OFF SWITCH

TUNING

OLYMPIC

MODEL PICTURED
445

Three-power operated portable AM superheterodyne receiver

TUBES
4

POWER SUPPLY
105-120 volts AC/DC or 3 volts A & 75 volts B supply

TUNING RANGE
540-1600KC

MFR/SUPPLIER
Olympic Radio & Television

PHOTOFACT SET
264-15

PUBLISHED
1955

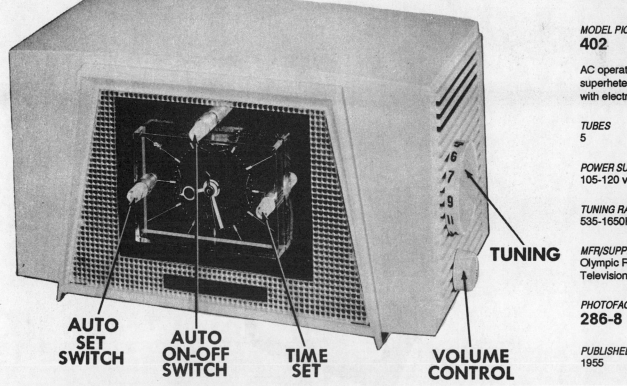

AUTO SET SWITCH

AUTO ON-OFF SWITCH

TIME SET

VOLUME CONTROL

TUNING

OLYMPIC

MODEL PICTURED
402

AC operated AM superheterodyne receiver with electric clock

TUBES
5

POWER SUPPLY
105-120 volts AC

TUNING RANGE
535-1650KC

MFR/SUPPLIER
Olympic Radio & Television

PHOTOFACT SET
286-8

PUBLISHED
1955

OLYMPIC

MODEL PICTURED
403

AC operated AM
superheterodyne receiver
with three-speed auto
record changer

TUBES
5

POWER SUPPLY
105-120 volts AC

TUNING RANGE
540-1620KC

MFR/SUPPLIER
Olympic Radio &
Television

PHOTOFACT SET
315-8

PUBLISHED
1956

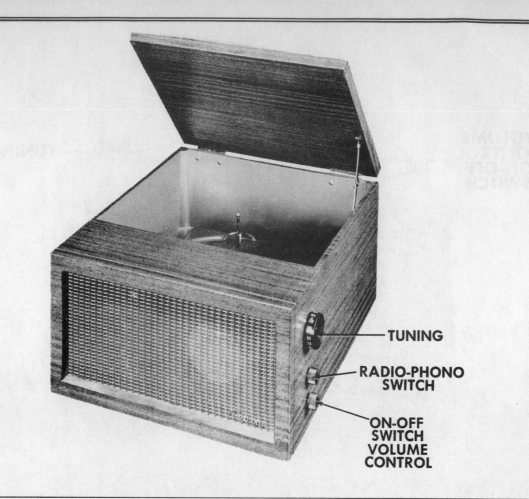

— TUNING

RADIO-PHONO
SWITCH

ON-OFF
SWITCH
VOLUME
CONTROL

OLYMPIC

MODEL PICTURED
575

AC operated AM-FM
receiver with three-speed
auto record changer

TUBES
7

POWER SUPPLY
110-120 volts AC

TUNING RANGE
535-1700KC,
88-108MC

MFR/SUPPLIER
Olympic Radio &
Television

PHOTOFACT SET
317-11

PUBLISHED
1956

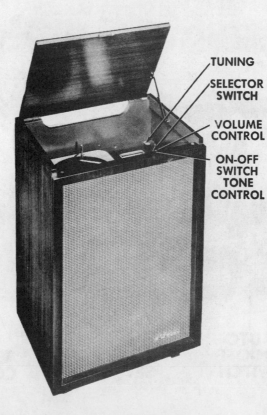

TUNING
SELECTOR
SWITCH

VOLUME
CONTROL

ON-OFF
SWITCH
TONE
CONTROL

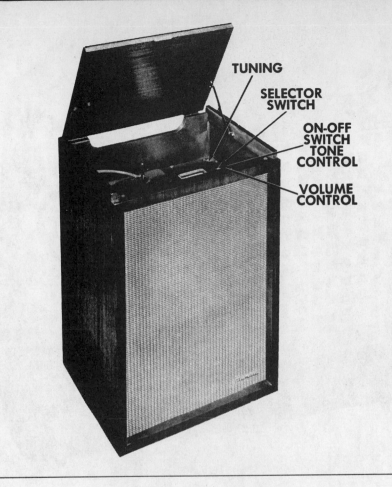

TUNING

SELECTOR
SWITCH

ON-OFF
SWITCH
TONE
CONTROL

VOLUME
CONTROL

OLYMPIC

MODEL PICTURED
574

AC operated AM
superheterodyne receiver
with three-speed auto
record changer

TUBES
5

POWER SUPPLY
105-120 volts AC

TUNING RANGE
540-1620KC

MFR/SUPPLIER
Olympic Radio &
Television

PHOTOFACT SET
319-9

PUBLISHED
1956

OLYMPIC

MODEL PICTURED
576

AC operated AM-FM
superheterodyne receiver
with three-speed auto
record changer

TUBES
12

POWER SUPPLY
105-120 volts AC

TUNING RANGE
535-1700KC,
88-108MC

MFR/SUPPLIER
Olympic Radio &
Television

PHOTOFACT SET
321-9

PUBLISHED
1956

OLYMPIC

MODEL PICTURED
447

Battery operated portable
AM transistorized
receiver

TUBES
4 transistor

POWER SUPPLY
9 volts DC

TUNING RANGE
540-1620KC

MFR/SUPPLIER
Olympic Radio &
Television

PHOTOFACT SET
349-7

PUBLISHED
1957

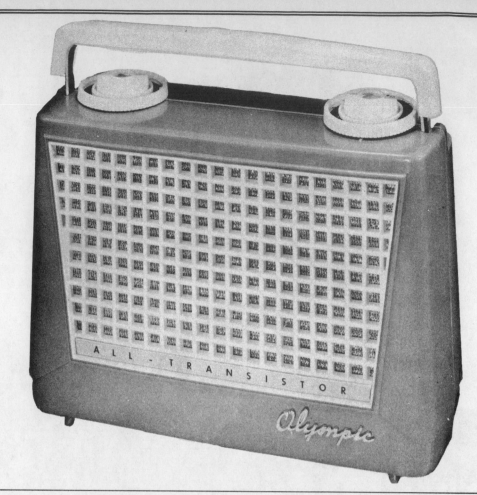

OLYMPIC

MODEL PICTURED
450-V

Battery operated portable
AM receiver

TUBES
4

POWER SUPPLY
1.5 volts A & 67.5 volts B
supply

TUNING RANGE
540-1620KC

MFR/SUPPLIER
Olympic Radio &
Television

PHOTOFACT SET
363-16

PUBLISHED
1957

OLYMPIC

MODEL PICTURED
Kobold 5720W

AC operated AM-FM
receiver

TUBES
5

POWER SUPPLY
110-120 volts AC

TUNING RANGE
535-1620KC,
88-108MC

MFR/SUPPLIER
Olympic Radio &
Television

PHOTOFACT SET
390-8

PUBLISHED
1958

OLYMPIC

MODEL PICTURED
Moderna
5783W

AC operated FM-AM-SW
receiver

TUBES
6

POWER SUPPLY
110-120 volts AC

TUNING RANGE
530-1620KC,
88-108MC,
6-16MC

MFR/SUPPLIER
Olympic Radio &
Television

PHOTOFACT SET
392-9

PUBLISHED
1958

OLYMPIC

MODEL PICTURED
544

AC/DC operated AM
receiver

TUBES
5

POWER SUPPLY
110-120 volts AC/DC

TUNING RANGE
535-1650KC

MFR/SUPPLIER
Olympic Radio &
Television

PHOTOFACT SET
428-10

PUBLISHED
1959

OLYMPIC

MODEL PICTURED
455

AC/DC operated AM
receiver

TUBES
4

POWER SUPPLY
110-120 volts AC/DC

TUNING RANGE
535-1650KC

MFR/SUPPLIER
Olympic Radio &
Television

PHOTOFACT SET
429-9

PUBLISHED
1959

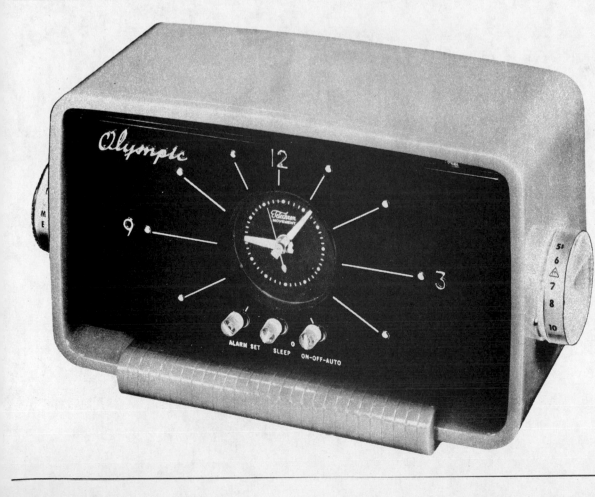

OLYMPIC

MODEL PICTURED
441

AC/DC operated AM
receiver

TUBES
5

POWER SUPPLY
105-125 volts AC/DC

TUNING RANGE
540-1620KC

MFR/SUPPLIER
Olympic Radio &
Television

PHOTOFACT SET
430-10

PUBLISHED
1959

OLYMPIC

MODEL PICTURED
509

AC operated AM receiver
with electric clock

TUBES
5

POWER SUPPLY
105-125 volts AC

TUNING RANGE
535-1650KC

MFR/SUPPLIER
Olympic Radio &
Television

PHOTOFACT SET
431-16

PUBLISHED
1959

OLYMPIC

MODEL PICTURED
666

Battery operated portable transistorized AM receiver

TUBES
6 transistor

POWER SUPPLY
9 volts DC

TUNING RANGE
530-1750KC

MFR/SUPPLIER
Olympic Radio & Television

PHOTOFACT SET
434-9

PUBLISHED
1959

OLYMPIC

MODEL PICTURED
593

AC operated AM-FM receiver with four-speed auto record changer

TUBES
12

POWER SUPPLY
105-120 volts AC

TUNING RANGE
535-1700KC,
88-108MC

MFR/SUPPLIER
Olympic Radio & Television

PHOTOFACT SET
435-7

PUBLISHED
1959

OLYMPIC

MODEL PICTURED
766

Battery operated portable transistorized AM receiver

TUBES
6 transistor

POWER SUPPLY
6 volts DC

TUNING RANGE
540-1615KC

MFR/SUPPLIER
Olympic Radio & Television

PHOTOFACT SET
436-11

PUBLISHED
1959

OLYMPIC

MODEL PICTURED
683

AC operated AM-FM receiver with four-speed auto record changer

TUBES
12

POWER SUPPLY
110-120 volts AC

TUNING RANGE
535-1700KC, 88-108MC

MFR/SUPPLIER
Olympic Radio & Television

PHOTOFACT SET
437-9

PUBLISHED
1959

OLYMPIC

MODEL PICTURED
768

Battery operated portable
transistorized AM
receiver

TUBES
6 transistor

POWER SUPPLY
9 volts DC

TUNING RANGE
540-1615KC

MFR/SUPPLIER
Olympic Radio &
Television

PHOTOFACT SET
449-16

PUBLISHED
1959

OLYMPIC

MODEL PICTURED
771

Battery operated portable
transistorized AM
receiver

TUBES
6 transistor

POWER SUPPLY
6 volts DC

TUNING RANGE
535-1650KC

MFR/SUPPLIER
Olympic Radio &
Television

PHOTOFACT SET
450-11

PUBLISHED
1959

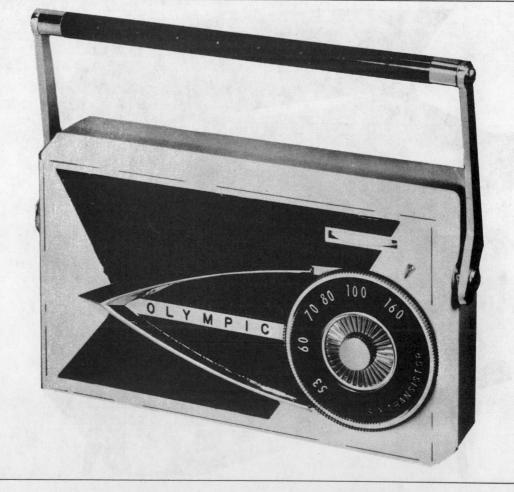

OLYMPIC

MODEL PICTURED
551

AC/DC operated AM
receiver

TUBES
5

POWER SUPPLY
110-120 volts AC/DC

TUNING RANGE
535-1650KC

MFR/SUPPLIER
Olympic Radio &
Television

PHOTOFACT SET
452-14

PUBLISHED
1959

OLYMPIC

MODEL PICTURED
688

AC operated AM receiver
with four-speed auto
record changer

TUBES
5

POWER SUPPLY
105-120 volts AC

TUNING RANGE
540-1620KC

MFR/SUPPLIER
Olympic Radio &
Television

PHOTOFACT SET
453-10

PUBLISHED
1959

OLYMPIC

MODEL PICTURED
412

AC/DC operated AM
receiver with four-speed
auto record changer

TUBES
5

POWER SUPPLY
105-120 volts AC

TUNING RANGE
540-1620KC

MFR/SUPPLIER
Olympic Radio &
Television

PHOTOFACT SET
454-15

PUBLISHED
1959

OLYMPIC

MODEL PICTURED
461

AC/DC operated portable
AM receiver

TUBES
4

POWER SUPPLY
110-120 volts AC or 1.5
volts A & 67.5 volts B
supply

TUNING RANGE
535-1650KC

MFR/SUPPLIER
Olympic Radio &
Television

PHOTOFACT SET
455-14

PUBLISHED
1959

OLYMPIC

MODEL PICTURED
689M

AC operated FM-AM
receiver with four-speed
auto record changer

TUBES
7

POWER SUPPLY
110-120 volts AC

TUNING RANGE
535-1700KC,
88-108MC

MFR/SUPPLIER
Olympic Radio &
Television

PHOTOFACT SET
456-17

PUBLISHED
1959

OLYMPIC

MODEL PICTURED
694

AC operated FM-AM
receiver with stereo
amplifier and four-speed
auto record changer

TUBES
8

POWER SUPPLY
110-120 volts AC

TUNING RANGE
535-1700KC,
87-109MC

MFR/SUPPLIER
Olympic Radio &
Television

PHOTOFACT SET
459-10

PUBLISHED
1959

OLYMPIC

MODEL PICTURED
770

Battery operated portable
transistorized AM
receiver

TUBES
6 transistor

POWER SUPPLY
9 volts DC

TUNING RANGE
530-1650KC

MFR/SUPPLIER
Olympic Radio &
Television

PHOTOFACT SET
460-15

PUBLISHED
1959

OLYMPIC

MODEL PICTURED
465

AC operated AM receiver
with electric clock

TUBES
4

POWER SUPPLY
110-120 volts AC

TUNING RANGE
535-1650KC

MFR/SUPPLIER
Olympic Radio &
Television

PHOTOFACT SET
464-14

PUBLISHED
1959

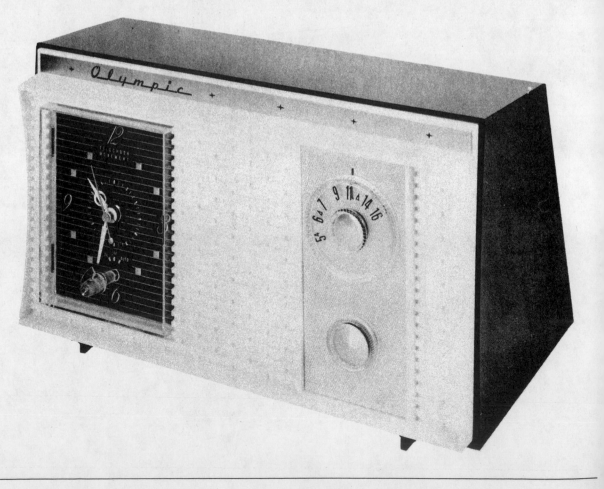

OLYMPIC

MODEL PICTURED
555

AC operated AM receiver
with electric clock

TUBES
5

POWER SUPPLY
110-120 volts AC

TUNING RANGE
535-1650KC

MFR/SUPPLIER
Olympic Radio &
Television

PHOTOFACT SET
464-15

PUBLISHED
1959

OLYMPIC

MODEL PICTURED
697

AC operated FM-AM
receiver with four-speed
auto record changer

TUBES
12

POWER SUPPLY
105-120 volts AC

TUNING RANGE
535-1700KC,
88-108MC

MFR/SUPPLIER
Olympic Radio &
Television

PHOTOFACT SET
472-8

PUBLISHED
1960

OLYMPIC

MODEL PICTURED
552

AC/DC operated AM
receiver

TUBES
5

POWER SUPPLY
110-120 volts AC/DC

TUNING RANGE
540-1650KC

MFR/SUPPLIER
Olympic Radio &
Television

PHOTOFACT SET
477-8

PUBLISHED
1960

OLYMPIC

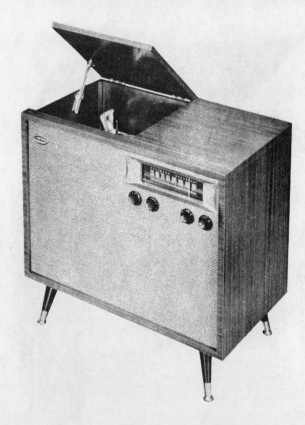

MODEL PICTURED
730

AC operated AM receiver
with stereo output and
four-speed auto record
changer

TUBES
5

POWER SUPPLY
105-120 volts AC

TUNING RANGE
540-1620KC

MFR/SUPPLIER
Olympic Radio &
Television

PHOTOFACT SET
479-13

PUBLISHED
1960

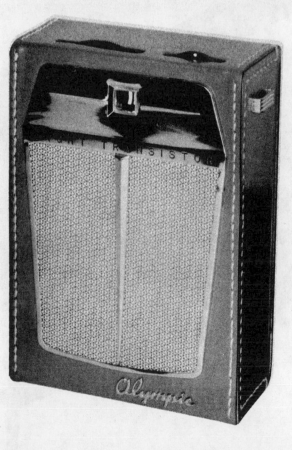

OLYMPIC

MODEL PICTURED
808

Battery operated
transistorized portable
AM receiver

TUBES
8 transistor

POWER SUPPLY
9 volts DC

TUNING RANGE
540-1650KC

MFR/SUPPLIER
Olympic Radio &
Television

PHOTOFACT SET
480-8

PUBLISHED
1960

OLYMPIC

MODEL PICTURED
859

Battery operated
transistorized portable
AM receiver

TUBES
8 transistor

POWER SUPPLY
6 volts DC

TUNING RANGE
520-1650KC

MFR/SUPPLIER
Olympic, Div. of Siegler
Corp.

PHOTOFACT SET
497-15

PUBLISHED
1960

OLYMPIC

MODEL PICTURED
777

Battery operated
transistorized portable
AM receiver

TUBES
6 transistor

POWER SUPPLY
9 volts DC

TUNING RANGE
540-1615KC

MFR/SUPPLIER
Olympic, Div. of Siegler
Corp.

PHOTOFACT SET
500-11

PUBLISHED
1960

OLYMPIC

MODEL PICTURED
7511

AC operated FM-AM
receiver with stereo
output and four-speed
auto record changer

TUBES
11

POWER SUPPLY
110-120 volts AC

TUNING RANGE
535-1620KC,
88-108MC

MFR/SUPPLIER
Olympic, Div. of Siegler
Corp.

PHOTOFACT SET
501-12

PUBLISHED
1960

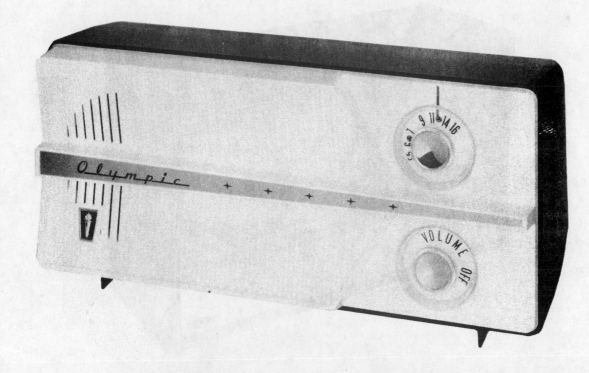

OLYMPIC

MODEL PICTURED
442W-1

AC/DC operated AM-SW
receiver

TUBES
5

POWER SUPPLY
105-125 volts AC/DC

TUNING RANGE
540-1620KC,
4.75-16.1MC

MFR/SUPPLIER
Olympic, Div. of Siegler
Corp.

PHOTOFACT SET
502-14

PUBLISHED
1960

OLYMPIC

MODEL PICTURED
557

AC/DC operated AM
receiver

TUBES
5

POWER SUPPLY
110-120 volts AC/DC

TUNING RANGE
540-1650KC

MFR/SUPPLIER
Olympic, Div. of Siegler
Corp.

PHOTOFACT SET
510-14

PUBLISHED
1960

OLYMPIC-CONTINENTAL

MODEL PICTURED
GB374

AC operated FM-AM-SW
receiver with four-speed
auto record changer

TUBES
8

POWER SUPPLY
105-120 volts AC

TUNING RANGE
550-1600KC,
88-108MC,
6-18MC

MFR/SUPPLIER
Olympic Radio &
Television

PHOTOFACT SET
451-13

PUBLISHED
1959

OLYMPIC-CONTINENTAL

MODEL PICTURED
GB376

AC operated FM-AM-SW
receiver with four-speed
auto record changer

TUBES
9

POWER SUPPLY
105-120 volts AC

TUNING RANGE
535-1620KC,
88-108MC,
6-18MC

MFR/SUPPLIER
Olympic Radio &
Television

PHOTOFACT SET
465-9

PUBLISHED
1959

OLYMPIC-CONTINENTAL

MODEL PICTURED
GB375

AC operated FM-AM-SW
receiver with four-speed
auto record changer

TUBES
9

POWER SUPPLY
110-120 volts AC

TUNING RANGE
535-1620KC,
88-108MC

MFR/SUPPLIER
Olympic Radio &
Television

PHOTOFACT SET
474-10

PUBLISHED
1960

OLYMPIC-CONTINENTAL

MODEL PICTURED
Tivoli 300

AC operated FM-AM-SW
receiver

TUBES
6

POWER SUPPLY
110-120 volts AC

TUNING RANGE
510-1640KC,
88-108MC,
5.77-18.8MC

MFR/SUPPLIER
Olympic, Div. of Siegler
Corp.

PHOTOFACT SET
494-13

PUBLISHED
1960

OLYMPIC-CONTINENTAL

MODEL PICTURED
Goldy 250

AC operated FM-AM-SW
receiver

TUBES
6

POWER SUPPLY
110-120 volts AC

TUNING RANGE
510-1640KC,
88-108MC,
5.77-18.8MC

MFR/SUPPLIER
Olympic, Div. of Siegler
Corp.

PHOTOFACT SET
495-14

PUBLISHED
1960

OLYMPIC-CONTINENTAL

MODEL PICTURED
GBS388

AC operated FM-AM-SW
receiver with four-speed
auto record changer

TUBES
13

POWER SUPPLY
110-120 volts AC

TUNING RANGE
550-1600KC,
88-108MC,
6-18MC

MFR/SUPPLIER
Olympic, Div. of Siegler
Corp.

PHOTOFACT SET
508-16

PUBLISHED
1960

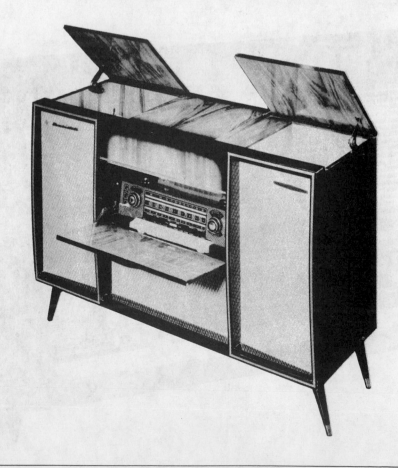

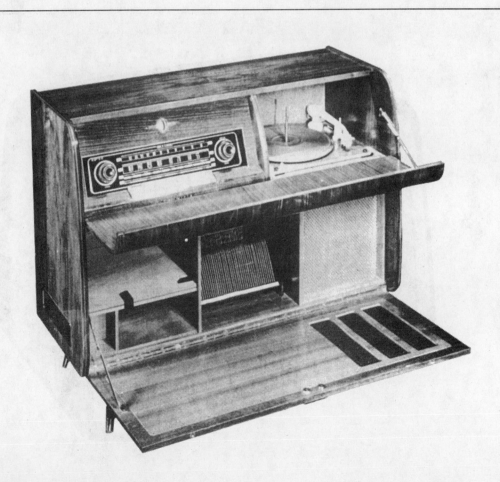

OLYMPIC-OPTA

MODEL PICTURED
**Cremona
5804T/W**

AC operated FM-AM-SW
receiver

TUBES
6

POWER SUPPLY
110-120 volts AC

TUNING RANGE
535-1620KC,
88-108MC,
6-15MC

MFR/SUPPLIER
Olympic Radio &
Television

PHOTOFACT SET
389-12

PUBLISHED
1958

OLYMPIC-OPTA

MODEL PICTURED
**Domino
5806T/W**

AC operated FM-AM-SW
receiver with three-speed
auto record changer

TUBES
6

POWER SUPPLY
110/127/150/220 volts AC

TUNING RANGE
540-1600KC, 88-108MC,
6-18MC, 140-390KC

MFR/SUPPLIER
Olympic Radio &
Television

PHOTOFACT SET
398-9

PUBLISHED
1958

OLYMPIC-OPTA

MODEL PICTURED
52804

AC operated FM-AM-SW
receiver with four-speed
automatic record changer

TUBES
7

POWER SUPPLY
110-120 volts AC

TUNING RANGE
535-1630KC,
88-108MC,
5.9-19.5MC

MFR/SUPPLIER
Olympic Radio &
Television

PHOTOFACT SET
420-6

PUBLISHED
1958

OLYMPIC-OPTA

MODEL PICTURED
5711W

AC operated FM-AM-SW
receiver

TUBES
6

POWER SUPPLY
110-120 volts AC

TUNING RANGE
535-1620KC,
88-108MC,
5.9-18.5MC

MFR/SUPPLIER
Olympic Radio &
Television

PHOTOFACT SET
424-11

PUBLISHED
1958

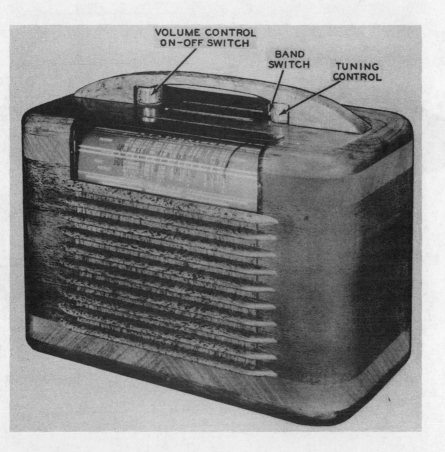

OLYMPIC-OPTA

MODEL PICTURED
5920

AC operated
FM-AM-SW-LW receiver
with four-speed auto
record changer

TUBES
8

POWER SUPPLY
110-120 volts AC

TUNING RANGE
525-1620KC, 150-275KC,
99-108MC,
6-19MC

MFR/SUPPLIER
Olympic Radio &
Television

PHOTOFACT SET
441-10

PUBLISHED
1959

PACKARD-BELL

MODEL PICTURED
651

AC operated two-band
superheterodyne receiver
with self-contained loop
antenna

TUBES
6

POWER SUPPLY
110-120 volts AC

TUNING RANGE
540-1740KC,
5.7-18.2MC

MFR/SUPPLIER
Packard-Bell Company

PHOTOFACT SET
4-42

PUBLISHED
1946

VOLUME CONTROL
ON-OFF SWITCH

BAND
SWITCH

TUNING
CONTROL

PACKARD-BELL

MODEL PICTURED
661

AC operated phono-radio combo AM superheterodyne receiver with self-contained loop antenna

TUBES
6

POWER SUPPLY
110-120 volts AC

TUNING RANGE
540-1740KC

MFR/SUPPLIER
Packard-Bell Company

PHOTOFACT SET
8-25

PUBLISHED
1946

MOTOR ON-OFF SWITCH

TUNING CONTROL

PHONO-RADIO SWITCH

TONE CONTROL

VOLUME CONTROL ON-OFF SWITCH

PACKARD-BELL

MODEL PICTURED
1052A

AC operated phono-recorder-PA combo two-band superheterodyne receiver with loop antenna

TUBES
10

POWER SUPPLY
110-120 volts AC

TUNING RANGE
540-1740KC,
6.0-18.0MC

MFR/SUPPLIER
Packard-Bell Company

PHOTOFACT SET
8-26

PUBLISHED
1946

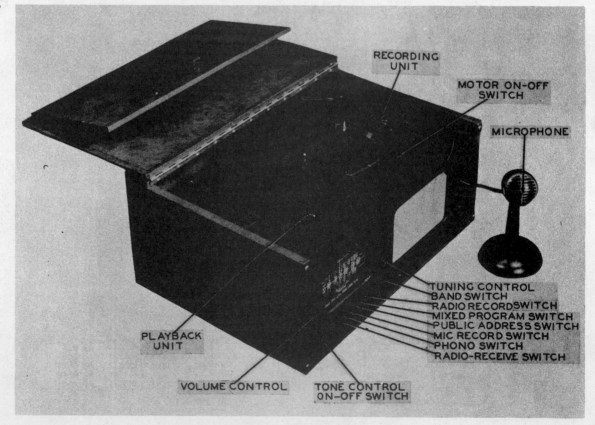

RECORDING UNIT

MOTOR ON-OFF SWITCH

MICROPHONE

TUNING CONTROL
BAND SWITCH
RADIO RECORD SWITCH
MIXED PROGRAM SWITCH
PUBLIC ADDRESS SWITCH
MIC RECORD SWITCH
PHONO SWITCH
RADIO-RECEIVE SWITCH

PLAYBACK UNIT

VOLUME CONTROL

TONE CONTROL ON-OFF SWITCH

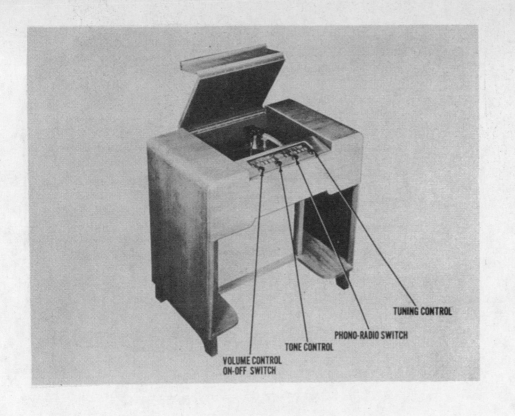

MODEL PICTURED
662

AC operated phono-radio combo AM superheterodyne receiver with self-contained loop antenna

TUBES
6

POWER SUPPLY
110-120 volts AC

TUNING RANGE
540-1740KC

MFR/SUPPLIER
Packard-Bell Company

PHOTOFACT SET
13-22

PUBLISHED
1947

TUNING CONTROL

PHONO-RADIO SWITCH

TONE CONTROL

VOLUME CONTROL
ON-OFF SWITCH

PACKARD-BELL

RADIO-RECORD
SWITCH

TUNING CONTROL

MIXED PROGRAM
SWITCH

BAND SWITCH

MIC RECORD
SWITCH

PUBLIC ADDRESS
SWITCH

PHONO SWITCH

VOLUME CONTROL

RADIO-RECEIVE
SWITCH

MOTOR ON-OFF
SWITCH

TONE CONTROL
ON-OFF SWITCH

PACKARD-BELL MODEL 1054-B

MODEL PICTURED
1054-B

AC operated phono-recorder-PA combo two-band superheterodyne receiver with self-contained loop antenna

TUBES
10

POWER SUPPLY
110-120 volts AC

TUNING RANGE
540-1740KC,
6.0-18.0MC

MFR/SUPPLIER
Packard-Bell Company

PHOTOFACT SET
13-23

PUBLISHED
1947

PACKARD-BELL

MODEL PICTURED
5DA

AC/DC operated AM
superheterodyne receiver
with self-contained loop
antenna

TUBES
5

POWER SUPPLY
110-120 volts AC/DC

TUNING RANGE
540-1750

MFR/SUPPLIER
Packard-Bell Company

PHOTOFACT SET
16-29

PUBLISHED
1947

VOLUME CONTROL
ON-OFF SWITCH

TUNING CONTROL

PACKARD-BELL

MODEL PICTURED
861

AC operated
phono-recorder-PA
combo AM
superheterodyne receiver
with loop antenna

TUBES
8

POWER SUPPLY
110-120 volts AC

TUNING RANGE
540-1620KC

MFR/SUPPLIER
Packard-Bell Company

PHOTOFACT SET
17-23

PUBLISHED
1947

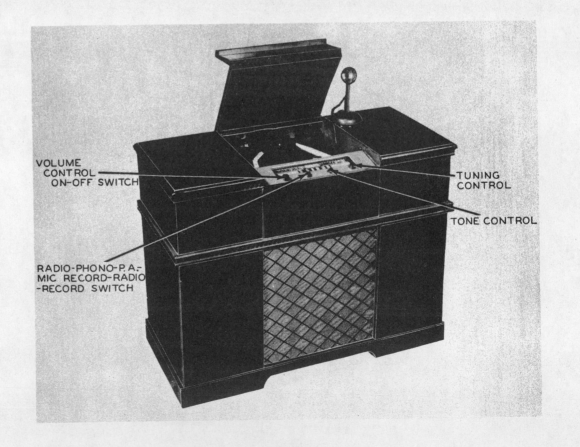

VOLUME
CONTROL
ON-OFF SWITCH

TUNING
CONTROL

TONE CONTROL

RADIO-PHONO-P.A.-
MIC RECORD-RADIO
-RECORD SWITCH

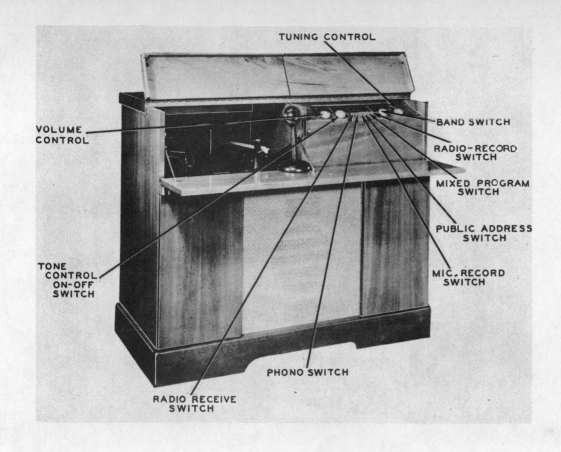

TUNING CONTROL

VOLUME
CONTROL

BAND SWITCH

RADIO—RECORD
SWITCH

MIXED PROGRAM
SWITCH

PUBLIC ADDRESS
SWITCH

MIC. RECORD
SWITCH

TONE
CONTROL
ON-OFF
SWITCH

PHONO SWITCH

RADIO RECEIVE
SWITCH

PACKARD-BELL

MODEL PICTURED
1063

AC operated
phono-radio-PA combo
two-band
superheterodyne receiver
with loop antenna

TUBES
10

POWER SUPPLY
110-120 volts AC

TUNING RANGE
540-1740KC,
6-18MC

MFR/SUPPLIER
Packard-Bell Company

PHOTOFACT SET
18-25

PUBLISHED
1947

MOTOR ON-OFF
SWITCH

TUNING
CONTROL

VOLUME CONTROL
ON-OFF SWITCH

RADIO-PHONO
SWITCH

PACKARD-BELL

MODEL PICTURED
568

AC operated portable
phono-radio combo AM
superheterodyne receiver
with loop antenna

TUBES
5

POWER SUPPLY
110-120 volts AC

TUNING RANGE
540-1740KC

MFR/SUPPLIER
Packard-Bell Company

PHOTOFACT SET
19-24

PUBLISHED
1947

PACKARD-BELL

MODEL PICTURED
572

AC/DC operated AM
superheterodyne receiver
with loop antenna

TUBES
5

POWER SUPPLY
110-120 volts AC

TUNING RANGE
540-1650KC

MFR/SUPPLIER
Packard-Bell Company

PHOTOFACT SET
22-22

PUBLISHED
1947

PACKARD-BELL

MODEL PICTURED
471

Three-power operated
portable AM
superheterodyne receiver
with loop antenna

TUBES
4

POWER SUPPLY
105-120 volts AC/DC or
7.5 volts A & 90 volts B
supply

TUNING RANGE
540-1620KC

MFR/SUPPLIER
Packard-Bell Company

PHOTOFACT SET
30-22

PUBLISHED
1947

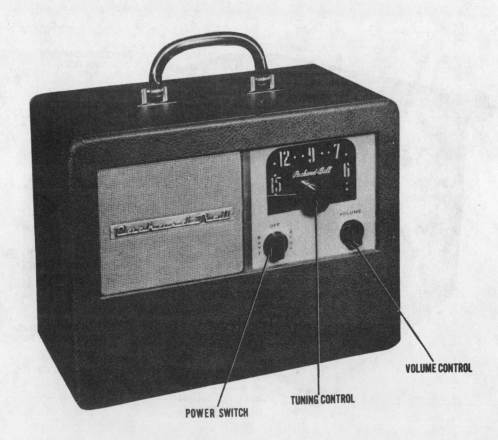

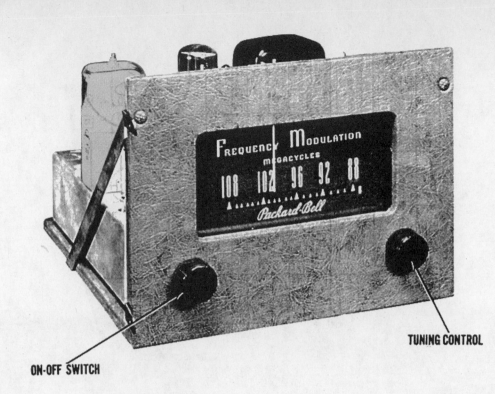

ON-OFF SWITCH

TUNING CONTROL

PACKARD-BELL

MODEL PICTURED
872

AC operated FM receiver

TUBES
8

POWER SUPPLY
110-120 volts AC

TUNING RANGE
88-108MC

MFR/SUPPLIER
Packard-Bell Company

PHOTOFACT SET
31-23

PUBLISHED
1948

VOLUME CONTROL
ON-OFF SWITCH

TONE CONTROL

PHONO-BAND
SWITCH

TUNING CONTROL

MODEL 880

MODEL 880-A

TUNING CONTROL

PHONO-
BAND
SWITCH

TONE CONTROL

VOLUME CONTROL
ON-OFF SWITCH

PACKARD-BELL

MODEL PICTURED
673A

AC operated phono-radio
combo AM
superheterodyne receiver
with loop antenna

TUBES
8

POWER SUPPLY
110-102 volts AC

TUNING RANGE
540-1620KC

MFR/SUPPLIER
Packard-Bell Company

PHOTOFACT SET
46-18

PUBLISHED
1948

PACKARD-BELL

MODEL PICTURED
1273

AC operated phono-radio
combo AM-FM
superheterodyne receiver
with loop antenna

TUBES
13

POWER SUPPLY
110-120 volts AC/DC

TUNING RANGE
540-1620KC,
87.5-108.5MC

MFR/SUPPLIER
Packard-Bell Company

PHOTOFACT SET
46-19

PUBLISHED
1948

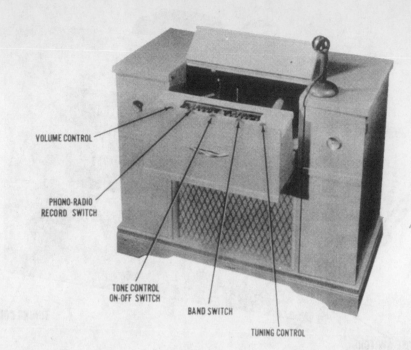

VOLUME CONTROL

PHONO-RADIO
RECORD SWITCH

TONE CONTROL
ON-OFF SWITCH

BAND SWITCH

TUNING CONTROL

PACKARD-BELL

MODEL PICTURED
881-A, 881-B

AC operated
phono-recorder-radio
combo AM
superheterodyne receiver

TUBES
8

POWER SUPPLY
110-120 volts AC

TUNING RANGE
540-1620KC

MFR/SUPPLIER
Packard-Bell Company

PHOTOFACT SET
47-17

PUBLISHED
1948

TUNING CONTROL

TONE CONTROL

VOLUME CONTROL
ON-OFF SWITCH

FUNCTION
SELECTOR
SWITCH

MODEL 881 - B

VOLUME CONTROL
ON-OFF SWITCH

FUNCTION
SLECTOR
SWITCH

TONE CONTROL

TUNING CONTROL

MODEL 881-A

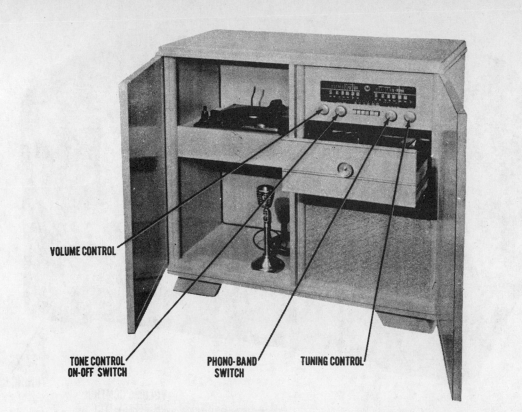

VOLUME CONTROL

TONE CONTROL
ON-OFF SWITCH

PHONO-BAND
SWITCH

TUNING CONTROL

PACKARD-BELL

MODEL PICTURED
1472

AC operated
phono-recorder-radio
combo AM-FM
superheterodyne receiver

TUBES
14

POWER SUPPLY
110-120 volts AC

TUNING RANGE
540-1700KC,
88-108MC

MFR/SUPPLIER
Packard-Bell Company

PHOTOFACT SET
48-17

PUBLISHED
1948

VOLUME CONTROL
ON-OFF SWITCH

TUNING CONTROL

PACKARD-BELL

MODEL PICTURED
100

AC/DC operated AM
superheterodyne receiver
with loop antenna

TUBES
5

POWER SUPPLY
110-120 volts AC/DC

TUNING RANGE
540-1620KC

MFR/SUPPLIER
Packard-Bell Company

PHOTOFACT SET
53-16

PUBLISHED
1949

PACKARD-BELL

MODEL PICTURED
682

AC/DC operated AM
superheterodyne receiver
with loop antenna

TUBES
6

POWER SUPPLY
110-120 volts AC/DC

TUNING RANGE
540-1620KC

MFR/SUPPLIER
Packard-Bell Company

PHOTOFACT SET
54-16

PUBLISHED
1949

TONE CONTROL

VOLUME CONTROL
ON-OFF SWITCH

TUNING CONTROL

PACKARD-BELL

MODEL PICTURED
884, 892

AC operated phono-radio
combo AM-FM
superheterodyne receiver
with loop antenna

TUBES
8

POWER SUPPLY
110-120 volts AC

TUNING RANGE
540-1620KC,
87.5-108.5MC

MFR/SUPPLIER
Packard-Bell Company

PHOTOFACT SET
74-6

PUBLISHED
1949

MODEL   892

MODEL   884

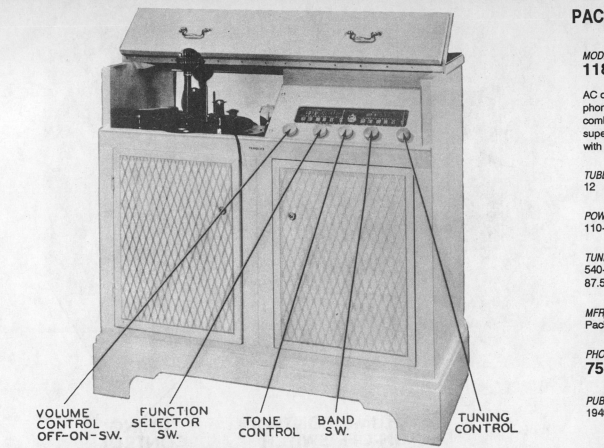

PACKARD-BELL

MODEL PICTURED
1181

AC operated
phono-recorder-radio
combo AM-FM
superheterodyne receiver
with loop antenna

TUBES
12

POWER SUPPLY
110-120 volts AC

TUNING RANGE
540-1620KC,
87.5-108.5MC

MFR/SUPPLIER
Packard-Bell Company

PHOTOFACT SET
75-12

PUBLISHED
1949

VOLUME
CONTROL
OFF—ON—SW.

FUNCTION
SELECTOR
SW.

TONE
CONTROL

BAND
SW.

TUNING
CONTROL

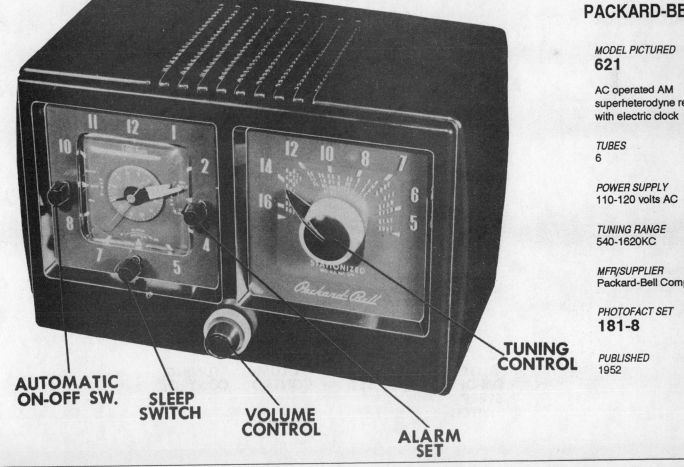

PACKARD-BELL

MODEL PICTURED
621

AC operated AM
superheterodyne receiver
with electric clock

TUBES
6

POWER SUPPLY
110-120 volts AC

TUNING RANGE
540-1620KC

MFR/SUPPLIER
Packard-Bell Company

PHOTOFACT SET
181-8

PUBLISHED
1952

**TUNING
CONTROL**

**AUTOMATIC
ON-OFF SW.**

**SLEEP
SWITCH**

**VOLUME
CONTROL**

**ALARM
SET**

PACKARD-BELL

MODEL PICTURED
531

AC/DC operated AM
superheterodyne receiver

TUBES
5

POWER SUPPLY
110-120 volts AC/DC

TUNING RANGE
540-1620KC

MFR/SUPPLIER
Packard-Bell Company

PHOTOFACT SET
231-11

PUBLISHED
1954

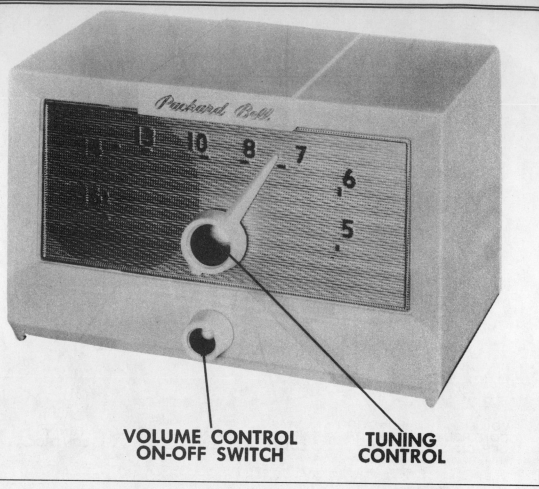

**VOLUME CONTROL
ON-OFF SWITCH**

**TUNING
CONTROL**

PACKARD-BELL

MODEL PICTURED
532

AC operated AM
superheterodyne receiver
with electric clock

TUBES
5

POWER SUPPLY
110-120 volts AC

TUNING RANGE
540-1620KC

MFR/SUPPLIER
Packard-Bell Company

PHOTOFACT SET
232-4

PUBLISHED
1954

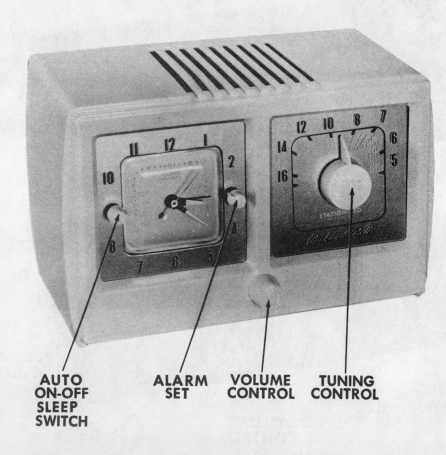

**AUTO
ON-OFF
SLEEP
SWITCH**

**ALARM
SET**

**VOLUME
CONTROL**

**TUNING
CONTROL**

PACKARD-BELL

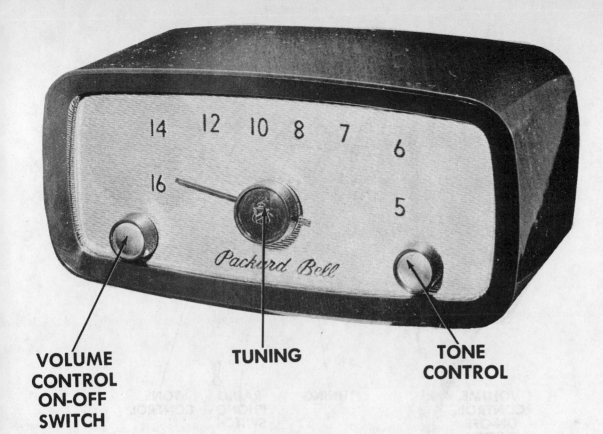

VOLUME CONTROL ON-OFF SWITCH

TUNING

TONE CONTROL

MODEL PICTURED
631

AC/DC operated AM superheterodyne receiver

TUBES
6

POWER SUPPLY
110-120 volts AC/DC

TUNING RANGE
540-1620KC

MFR/SUPPLIER
Packard-Bell Company

PHOTOFACT SET
256-12

PUBLISHED
1954

PACKARD-BELL

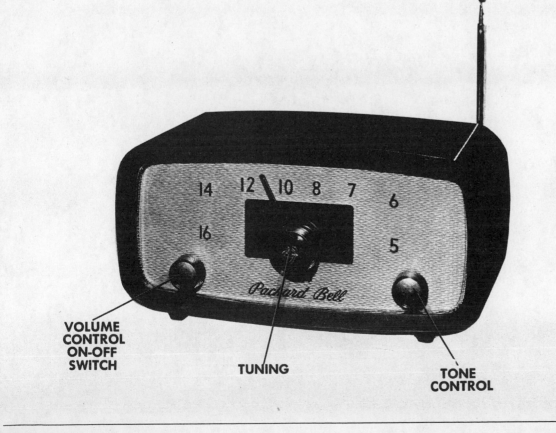

VOLUME CONTROL ON-OFF SWITCH

TUNING

TONE CONTROL

MODEL PICTURED
632

AC/DC operated two-band superheterodyne receiver

TUBES
6

POWER SUPPLY
110-120 volts AC/DC

TUNING RANGE
540-1620KC, 1.7-5.6MC

MFR/SUPPLIER
Packard-Bell Company

PHOTOFACT SET
266-11

PUBLISHED
1955

PACKARD-BELL

MODEL PICTURED
543

AC operated AM
superheterodyne receiver

TUBES
5

POWER SUPPLY
110-120 volts AC

TUNING RANGE
540-1620KC

MFR/SUPPLIER
Packard-Bell Company

PHOTOFACT SET
270-12

PUBLISHED
1955

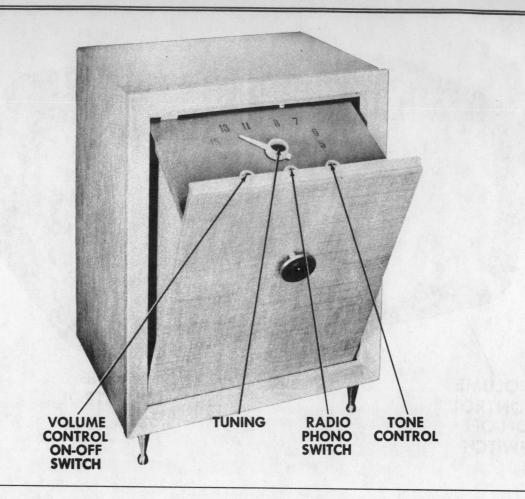

VOLUME CONTROL ON-OFF SWITCH   **TUNING**   **RADIO PHONO SWITCH**   **TONE CONTROL**

PACKARD-BELL

MODEL PICTURED
10RP1

AC operated FM-AM
superheterodyne receiver
with three-speed auto
record changer

TUBES
11

POWER SUPPLY
110-120 volts AC

TUNING RANGE
530-1620KC,
88-108MC

MFR/SUPPLIER
Packard-Bell Company

PHOTOFACT SET
327-7

PUBLISHED
1956

PACKARD-BELL

MODEL PICTURED
5R1

AC/DC operated AM
receiver

TUBES
5

POWER SUPPLY
110-120 volts AC/DC

TUNING RANGE
540-1620KC

MFR/SUPPLIER
Packard-Bell Company

PHOTOFACT SET
343-9

PUBLISHED
1957

PACKARD-BELL

MODEL PICTURED
4RB1

Battery operated portable
AM receiver

TUBES
4

POWER SUPPLY
1.5 volts A & 45 volts B
supply

TUNING RANGE
535-1620KC

MFR/SUPPLIER
Packard-Bell Company

PHOTOFACT SET
369-17

PUBLISHED
1957

PACKARD-BELL

MODEL PICTURED
5RC1

AC operated AM receiver
with electric clock

TUBES
5

POWER SUPPLY
110-120 volts AC

TUNING RANGE
540-1620KC

MFR/SUPPLIER
Packard-Bell Company

PHOTOFACT SET
372-11

PUBLISHED
1957

PACKARD-BELL

MODEL PICTURED
10RP2

AC operated AM-FM
receiver with four-speed
auto record changer

TUBES
11

POWER SUPPLY
110-120 volts AC

TUNING RANGE
530-1620KC,
88-108MC

MFR/SUPPLIER
Packard-Bell Company

PHOTOFACT SET
377-13

PUBLISHED
1957

PACKARD-BELL

MODEL PICTURED
6RC1

AC operated AM receiver
with clock

TUBES
6

POWER SUPPLY
110-120 volts AC

TUNING RANGE
540-1620KC

MFR/SUPPLIER
Packard-Bell Company

PHOTOFACT SET
417-8

PUBLISHED
1958

PACKARD-BELL

MODEL PICTURED
5R5

AC/DC operated AM
receiver

TUBES
5

POWER SUPPLY
110-120 volts AC/DC

TUNING RANGE
540-1620KC

MFR/SUPPLIER
Packard-Bell Company

PHOTOFACT SET
457-16

PUBLISHED
1959

PACKARD-BELL

MODEL PICTURED
11RP7S

AC operated FM-AM
receiver with four-speed
auto record changer

TUBES
11

POWER SUPPLY
110-120 volts AC

TUNING RANGE
530-1620KC,
88-108MC

MFR/SUPPLIER
Packard-Bell Company

PHOTOFACT SET
459-11

PUBLISHED
1959

PACKARD-BELL

MODEL PICTURED
7R3

AC/DC operated FM-AM
receiver

TUBES
7

POWER SUPPLY
110-120 volts AC/DC

TUNING RANGE
530-1620KC,
88-108MC

MFR/SUPPLIER
Packard-Bell Company

PHOTOFACT SET
463-10

PUBLISHED
1959

PACKARD-BELL

MODEL PICTURED
5R6

AC operated AM receiver

TUBES
5

POWER SUPPLY
110-120 volts AC

TUNING RANGE
540-1620KC

MFR/SUPPLIER
Packard-Bell Company

PHOTOFACT SET
473-10

PUBLISHED
1960

PACKARD-BELL

MODEL PICTURED
RPC-3

AC operated FM-AM
tuner with stereo amplifier
and four-speed auto
record changer

TUBES
9

POWER SUPPLY
110-120 volts AC

TUNING RANGE
530-1620KC,
88-108MC

MFR/SUPPLIER
Packard-Bell Company

PHOTOFACT SET
480-9

PUBLISHED
1960

PACKARD-BELL

MODEL PICTURED
4RC1

AC operated AM receiver
with electric clock

TUBES
4

POWER SUPPLY
110-120 volts AC

TUNING RANGE
540-1620KC

MFR/SUPPLIER
Packard-Bell Company

PHOTOFACT SET
496-13

PUBLISHED
1960

PHILCO

MODEL PICTURED
46-132

Battery operated AM
superheterodyne receiver

TUBES
5

POWER SUPPLY
1.5 volts A & 90 volts B
supply

TUNING RANGE
540-1720KC

MFR/SUPPLIER
Philco Corp.

PHOTOFACT SET
4-20

PUBLISHED
1946

PHILCO

MODEL PICTURED
46-1201

AC operated phono-radio combo AM superheterodyne receiver with self-contained loop antenna

TUBES
5

POWER SUPPLY
105-120 volts AC

TUNING RANGE
540-1600KC

MFR/SUPPLIER
Philco Corp.

PHOTOFACT SET
4-35

PUBLISHED
1946

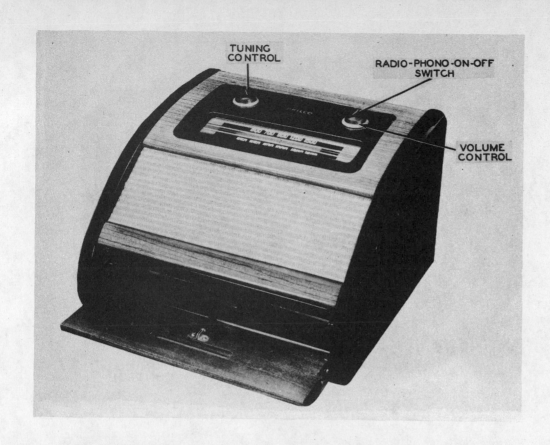

PHILCO

MODEL PICTURED
46-421

AC/DC operated superheterodyne receiver with self-contained loop antenna

TUBES
6

POWER SUPPLY
105-120 volts AC/DC

TUNING RANGE
540-1620KC

MFR/SUPPLIER
Philco Corp.

PHOTOFACT SET
5-12

PUBLISHED
1946

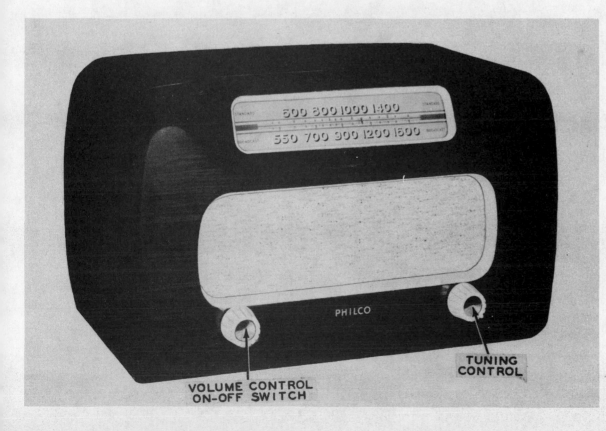

PHILCO

MODEL PICTURED
46-131

Battery operated AM
superheterodyne receiver

TUBES
4

POWER SUPPLY
1.5 volts A & 90 volts B
supply

TUNING RANGE
540-1720KC

MFR/SUPPLIER
Philco Corp.

PHOTOFACT SET
5-13

PUBLISHED
1946

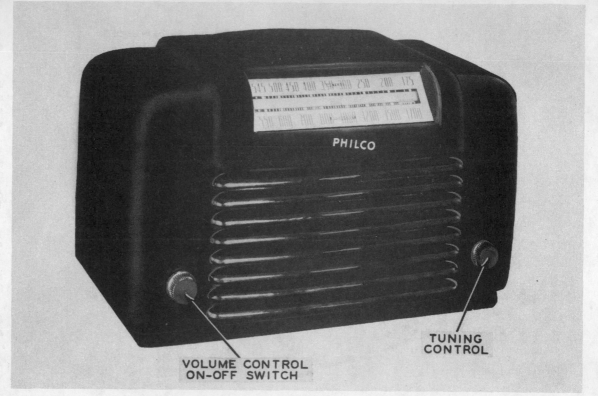

VOLUME CONTROL
ON-OFF SWITCH

TUNING
CONTROL

PHILCO

MODEL PICTURED
46-420

AC/DC operated AM
superheterodyne receiver
with self-contained loop
antenna

TUBES
6

POWER SUPPLY
105-125 volts AC/DC

TUNING RANGE
540-1620KC

MFR/SUPPLIER
Philco Corp.

PHOTOFACT SET
6-22

PUBLISHED
1946

VOLUME CONTROL
ON-OFF SWITCH

TUNING CONTROL

VOLUME CONTROL
ON-OFF SWITCH

TUNING CONTROL

PHILCO

MODEL PICTURED
46-350

Three-power operated
portable AM
superheterodyne receiver
with self-contained loop
antenna

TUBES
6

POWER SUPPLY
105-120 volts AC/DC or 9
volts A & 90 volts B
supply

TUNING RANGE
540-1600KC

MFR/SUPPLIER
Philco Corp.

PHOTOFACT SET
10-24

PUBLISHED
1946

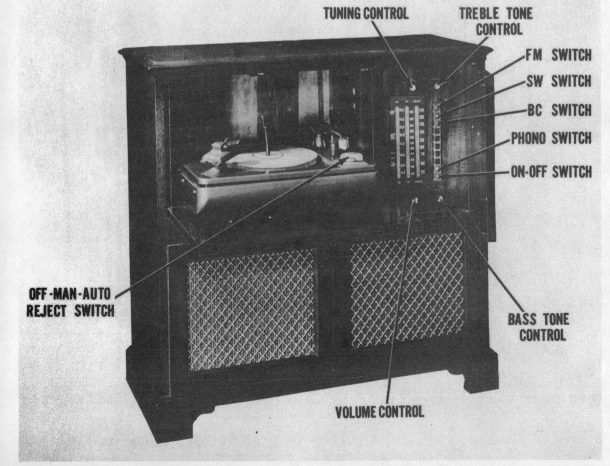

TUNING CONTROL

TREBLE TONE
CONTROL

FM SWITCH

SW SWITCH

BC SWITCH

PHONO SWITCH

ON-OFF SWITCH

OFF-MAN-AUTO
REJECT SWITCH

BASS TONE
CONTROL

VOLUME CONTROL

PHILCO

MODEL PICTURED
46-1213

AC operated three-band
superheterodyne receiver
with auto phono

TUBES
11

POWER SUPPLY
105-120 volts AC

TUNING RANGE
540-1720KC,
88-108MC,
9.3-15.5MC

MFR/SUPPLIER
Philco Corp.

PHOTOFACT SET
12-33

PUBLISHED
1947

PHILCO

MODEL PICTURED
46-1209

AC operated phono-radio
combo two-band
superheterodyne receiver
with self-contained loop
antenna

TUBES
8

POWER SUPPLY
110-120 volts AC

TUNING RANGE
540-1720KC,
9.3-15.5MC

MFR/SUPPLIER
Philco Corp.

PHOTOFACT SET
13-24

PUBLISHED
1947

PHILCO

MODEL PICTURED
46-1226

AC operated phono-radio
combo two-band
superheterodyne receiver
with loop antenna

TUBES
8

POWER SUPPLY
110-120 volts AC

TUNING RANGE
540-1720KC,
9.3-15.5MC

MFR/SUPPLIER
Philco Corp.

PHOTOFACT SET
15-24

PUBLISHED
1947

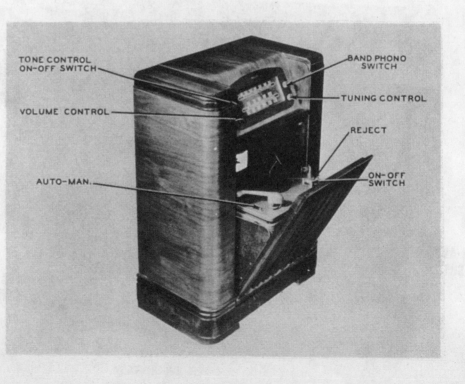

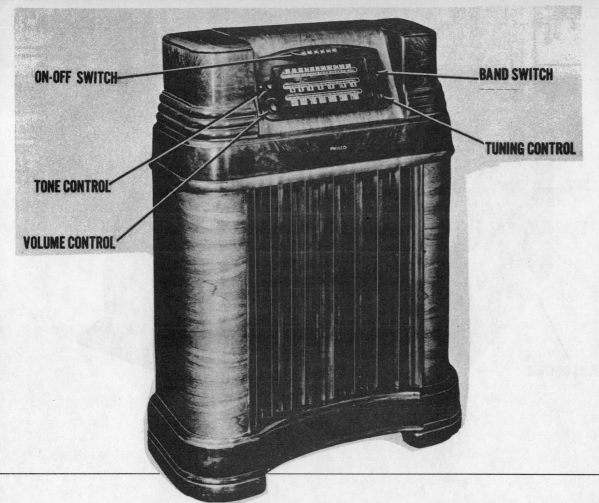

ON-OFF SWITCH

BAND SWITCH

TONE CONTROL

TUNING CONTROL

VOLUME CONTROL

PHILCO

MODEL PICTURED
46-480

AC operated FM-AM
superheterodyne receiver
with loop antenna

TUBES
7

POWER SUPPLY
110-120 volts AC

TUNING RANGE
540-1720KC,
88-108MC,
9.3-15.5MC

MFR/SUPPLIER
Philco Corp.

PHOTOFACT SET
19-25

PUBLISHED
1947

PHILCO

MODEL PICTURED
48-464

AC/DC operated
two-band
superheterodyne receiver
with loop antenna

TUBES
6

POWER SUPPLY
110-120 volts AC/DC

TUNING RANGE
540-1720KC,
9-15.5MC

MFR/SUPPLIER
Philco Corp.

PHOTOFACT SET
26-20

PUBLISHED
1947

TONE CONTROL
ON-OFF SWITCH

VOLUME CONTROL

TUNING CONTROL

BAND SWITCH

PHILCO

MODEL PICTURED
46-1201
(Revised)

AC operated phono-radio combo AM superheterodyne receiver with loop antenna

TUBES
5

POWER SUPPLY
105-120 volts AC

TUNING RANGE
540-1600KC

MFR/SUPPLIER
Philco Corp.

PHOTOFACT SET
29-21

PUBLISHED
1947

**PHONO-RADIO
VOLUME CONTROL
ON-OFF SWITCH**

TUNING CONTROL

PHILCO

MODEL PICTURED
48-482

AC operated multi-band superheterodyne receiver with loop antenna

TUBES
9

POWER SUPPLY
105-120 volts AC

TUNING RANGE
540-1720KC,
88-108MC,
9.3-115.5MC

MFR/SUPPLIER
Philco Corp.

PHOTOFACT SET
30-24

PUBLISHED
1947

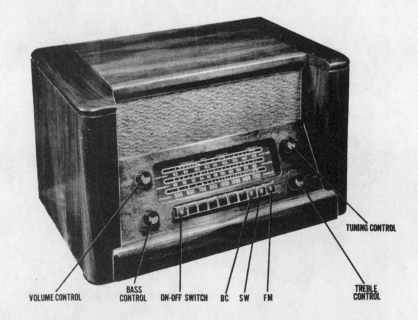

VOLUME CONTROL   **BASS CONTROL**   **ON-OFF SWITCH**   **BC**   **SW**   **FM**   **TUNING CONTROL**   **TREBLE CONTROL**

PHILCO

MODEL PICTURED
48-1260

AC operated phono-radio combo AM superheterodyne receiver with loop antenna

TUBES
5

POWER SUPPLY
105-120 volts AC

TUNING RANGE
540-1600KC

MFR/SUPPLIER
Philco Corp.

PHOTOFACT SET
31-25

PUBLISHED
1948

VOLUME CONTROL
ON-OFF SWITCH
PHONO-RADIO SWITCH

TUNING CONTROL

PHILCO

MODEL PICTURED
46-131

Battery operated AM superheterodyne receiver with loop antenna

TUBES
4

POWER SUPPLY
1.5 volts A & 90 volts B supply

TUNING RANGE
540-1620KC

MFR/SUPPLIER
Philco Corp.

PHOTOFACT SET
32-16

PUBLISHED
1948

VOLUME CONTROL
ON-OFF SWITCH

TUNING CONTROL

PHILCO

MODEL PICTURED
48-250-I

AC/DC operated AM
superheterodyne receiver
with loop antenna

TUBES
5

POWER SUPPLY
105-120 volts AC/DC

TUNING RANGE
540-1620KC

MFR/SUPPLIER
Philco Corp.

PHOTOFACT SET
32-17

PUBLISHED
1948

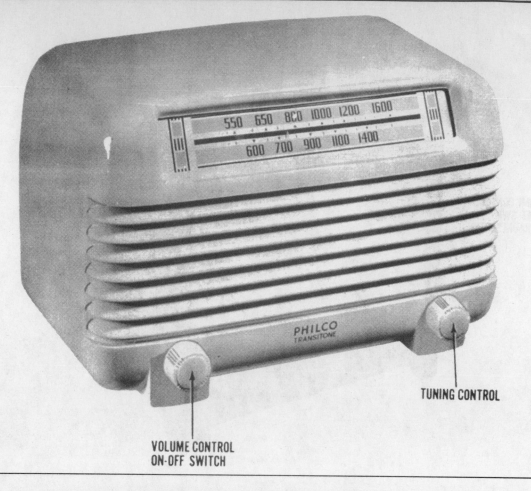

VOLUME CONTROL
ON-OFF SWITCH

TUNING CONTROL

PHILCO

MODEL PICTURED
48-1263

AC operated phono-radio
combo two-band
superheterodyne receiver
with loop antenna

TUBES
8

POWER SUPPLY
110-120 volts AC

TUNING RANGE
540-1720KC,
9.3-15.5MC

MFR/SUPPLIER
Philco Corp.

PHOTOFACT SET
32-18

PUBLISHED
1948

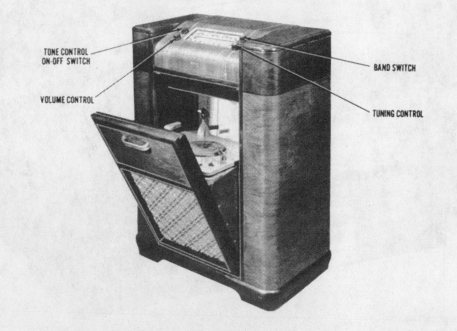

TONE CONTROL
ON-OFF SWITCH

VOLUME CONTROL

BAND SWITCH

TUNING CONTROL

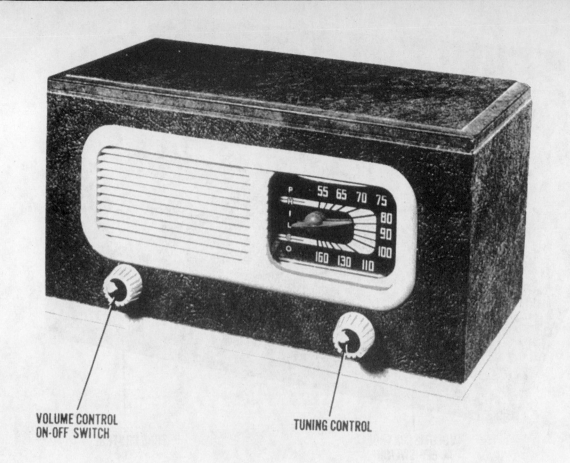

VOLUME CONTROL
ON-OFF SWITCH

TUNING CONTROL

PHILCO

MODEL PICTURED
47-204

AC/DC operated AM
superheterodyne receiver
with loop antenna

TUBES
5

POWER SUPPLY
105-120 volts AC/DC

TUNING RANGE
540-1620KC

MFR/SUPPLIER
Philco Corp.

PHOTOFACT SET
33-18

PUBLISHED
1948

PHILCO

MODEL PICTURED
48-200-I, 48-214

AC/DC operated AM
superheterodyne receiver
with loop antenna

TUBES
5

POWER SUPPLY
105-120 volts AC/DC

TUNING RANGE
540-1620KC

MFR/SUPPLIER
Philco Corp.

PHOTOFACT SET
33-19

PUBLISHED
1948

PHILCO

MODEL PICTURED
48-150

Battery operated AM
superheterodyne receiver

TUBES
5

POWER SUPPLY
1.5 volts A & 90 volts B
supply

TUNING RANGE
540-1720KC

MFR/SUPPLIER
Philco Corp.

PHOTOFACT SET
34-16

PUBLISHED
1948

VOLUME CONTROL
ON-OFF SWITCH   TONE CONTROL

PHILCO

MODEL PICTURED
48-460

AC/DC operated AM
superheterodyne receiver
with loop antenna

TUBES
6

POWER SUPPLY
105-120 volts AC/DC

TUNING RANGE
540-1620KC

MFR/SUPPLIER
Philco Corp.

PHOTOFACT SET
34-17

PUBLISHED
1948

VOLUME CONTROL
ON-OFF SWITCH   TUNING CONTROL

TUNING CONTROL

TONE SWITCH

VOLUME CONTROL PHONO-RADIO SWITCH

PHILCO

MODEL PICTURED
48-1256

AC operated phono-radio combo AM superheterodyne receiver with loop antenna

TUBES
6

POWER SUPPLY
105-120 volts AC

TUNING RANGE
540-1620KC

MFR/SUPPLIER
Philco Corp.

PHOTOFACT SET
34-18

PUBLISHED
1948

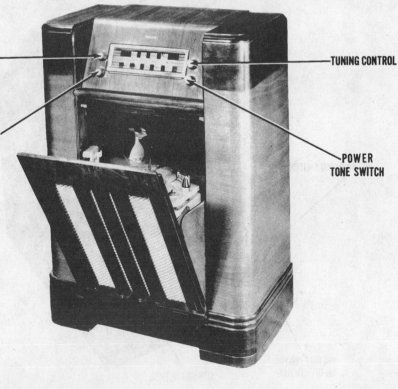

VOLUME CONTROL

TUNING CONTROL

PHONO-RADIO SWITCH

POWER TONE SWITCH

PHILCO

MODEL PICTURED
48-1262

AC operated phono-radio combo AM superheterodyne receiver with loop antenna

TUBES
6

POWER SUPPLY
105-120 volts AC

TUNING RANGE
540-1620KC

MFR/SUPPLIER
Philco Corp.

PHOTOFACT SET
35-18

PUBLISHED
1948

PHILCO

MODEL PICTURED
46-142

Battery operated AM
superheterodyne receiver

TUBES
5

POWER SUPPLY
1.5 volts A & 90 volts B
supply

TUNING RANGE
540-1600KC

MFR/SUPPLIER
Philco Corp.

PHOTOFACT SET
36-16

PUBLISHED
1948

**VOLUME CONTROL
ON-OFF SWITCH**

TUNING CONTROL

PHILCO

MODEL PICTURED
48-1253

AC operated phono-radio
combo AM
superheterodyne receiver
with loop antenna

TUBES
5

POWER SUPPLY
105-120 volts AC

TUNING RANGE
540-1600KC

MFR/SUPPLIER
Philco Corp.

PHOTOFACT SET
36-17

PUBLISHED
1948

PHONO-RADIO SWITCH

**VOLUME CONTROL
ON-OFF SWITCH**

TUNING CONTROL

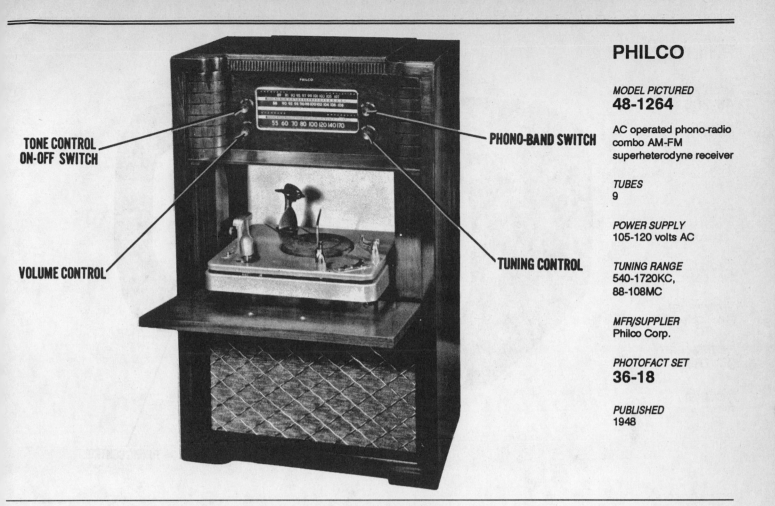

TONE CONTROL
ON-OFF SWITCH

PHONO-BAND SWITCH

VOLUME CONTROL

TUNING CONTROL

PHILCO

MODEL PICTURED
48-1264

AC operated phono-radio
combo AM-FM
superheterodyne receiver

TUBES
9

POWER SUPPLY
105-120 volts AC

TUNING RANGE
540-1720KC,
88-108MC

MFR/SUPPLIER
Philco Corp.

PHOTOFACT SET
36-18

PUBLISHED
1948

VOLUME CONTROL
ON-OFF SWITCH

TUNING CONTROL

MODEL 48-225

VOLUME CONTROL
ON-OFF SWITCH

TUNING CONTROL

MODEL 48-230

PHILCO

MODEL PICTURED
48-225, 48-230

AC/DC operated AM
superheterodyne receiver
with loop antenna

TUBES
5

POWER SUPPLY
110-120 volts AC/DC

TUNING RANGE
540-1700KC

MFR/SUPPLIER
Philco Corp.

PHOTOFACT SET
37-15

PUBLISHED
1948

167

PHILCO

MODEL PICTURED
48-206

AC/DC operated AM
superheterodyne receiver
with loop antenna

TUBES
5

POWER SUPPLY
105-120 volts AC/DC

TUNING RANGE
540-1620KC

MFR/SUPPLIER
Philco Corp.

PHOTOFACT SET
37-16

PUBLISHED
1948

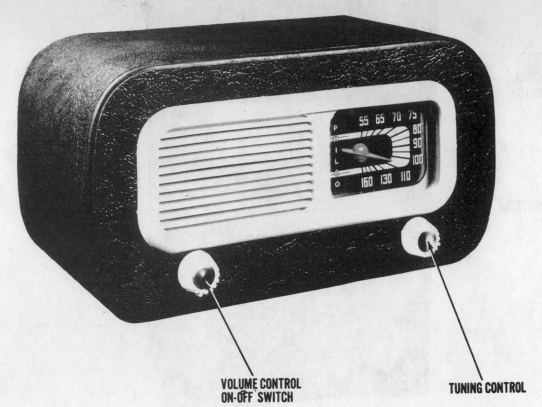

**VOLUME CONTROL
ON-OFF SWITCH**

TUNING CONTROL

PHILCO

MODEL PICTURED
48-300

Three-power operated
portable AM
superheterodyne receiver
with loop antenna

TUBES
5

POWER SUPPLY
105-120 volts AC/DC or
7.5 volts A & 90 volts B
supply

TUNING RANGE
540-1620KC

MFR/SUPPLIER
Philco Corp.

PHOTOFACT SET
37-17

PUBLISHED
1948

**VOLUME CONTROL
ON-OFF SWITCH**

TUNING CONTROL

VOLUME CONTROL
ON-OFF SWITCH

TUNING CONTROL

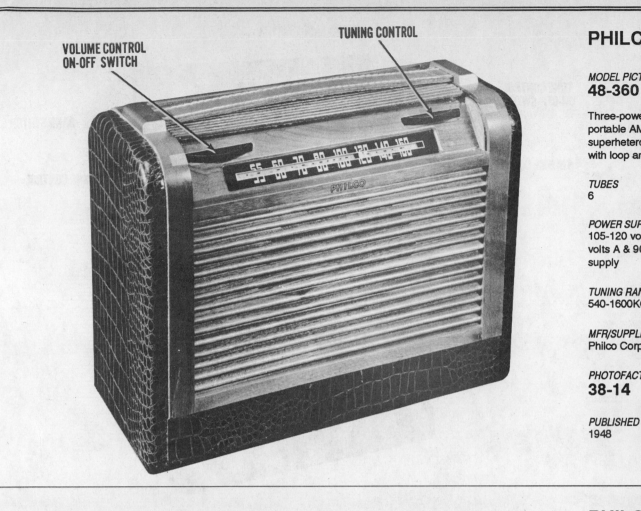

PHILCO

MODEL PICTURED
48-360

Three-power operated
portable AM
superheterodyne receiver
with loop antenna

TUBES
6

POWER SUPPLY
105-120 volts AC/DC or 9
volts A & 90 volts B
supply

TUNING RANGE
540-1600KC

MFR/SUPPLIER
Philco Corp.

PHOTOFACT SET
38-14

PUBLISHED
1948

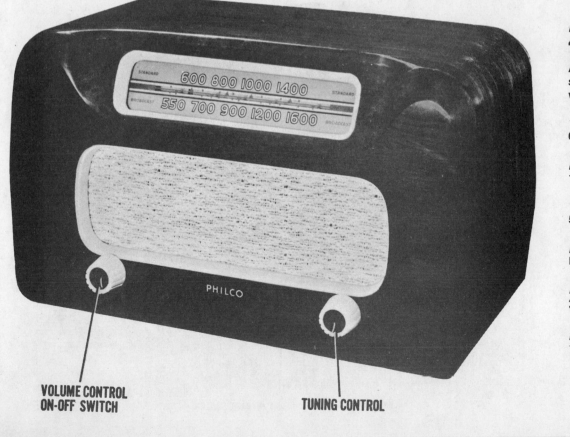

VOLUME CONTROL
ON-OFF SWITCH

TUNING CONTROL

PHILCO

MODEL PICTURED
48-461

AC/DC operated AM
superheterodyne receiver
with loop antenna

TUBES
6

POWER SUPPLY
105-120 volts AC/DC

TUNING RANGE
540-1620KC

MFR/SUPPLIER
Philco Corp.

PHOTOFACT SET
38-15

PUBLISHED
1948

PHILCO

MODEL PICTURED
48-1266

AC operated phono-radio
combo AM-FM-SW
superheterodyne receiver

TUBES
9

POWER SUPPLY
105-125 volts AC

TUNING RANGE
540-1720KC,
88-108MC,
9.3-15.5MC

MFR/SUPPLIER
Philco Corp.

PHOTOFACT SET
39-15

PUBLISHED
1948

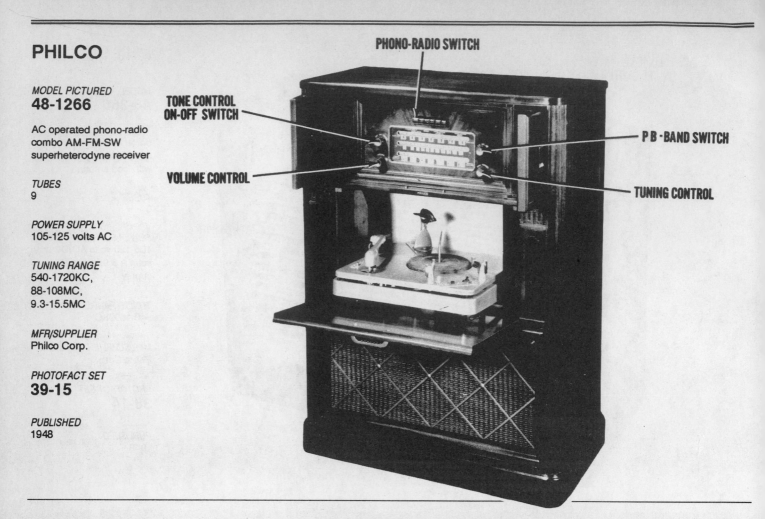

PHONO-RADIO SWITCH

TONE CONTROL
ON-OFF SWITCH

P B - BAND SWITCH

VOLUME CONTROL

TUNING CONTROL

PHILCO

MODEL PICTURED
48-475

AC operated AM-FM
superheterodyne receiver
with loop antenna

TUBES
8

POWER SUPPLY
105-120 volts AC

TUNING RANGE
540-1600KC,
88-108MC

MFR/SUPPLIER
Philco Corp.

PHOTOFACT SET
40-14

PUBLISHED
1948

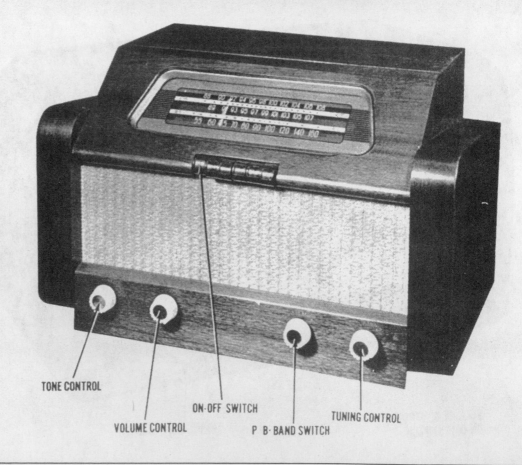

TONE CONTROL

VOLUME CONTROL

ON-OFF SWITCH

P B- BAND SWITCH

TUNING CONTROL

170

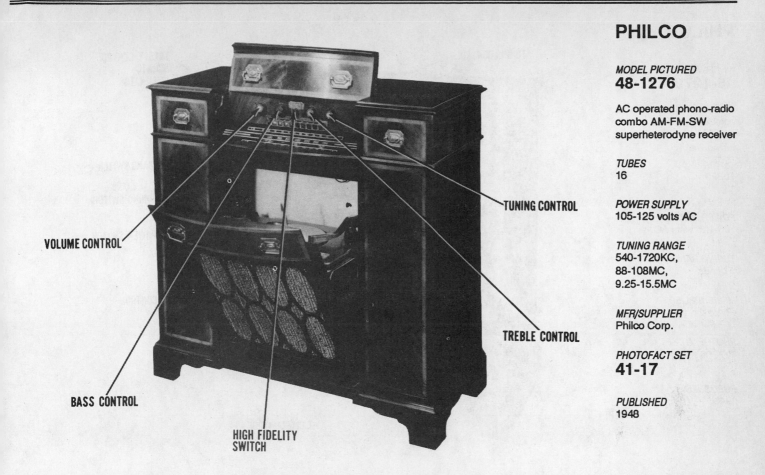

VOLUME CONTROL

TUNING CONTROL

TREBLE CONTROL

BASS CONTROL

HIGH FIDELITY
SWITCH

PHILCO

MODEL PICTURED
48-1276

AC operated phono-radio
combo AM-FM-SW
superheterodyne receiver

TUBES
16

POWER SUPPLY
105-125 volts AC

TUNING RANGE
540-1720KC,
88-108MC,
9.25-15.5MC

MFR/SUPPLIER
Philco Corp.

PHOTOFACT SET
41-17

PUBLISHED
1948

PHILCO

MODEL PICTURED
49-602

Three-power operated
portable AM
superheterodyne receiver
with loop antenna

TUBES
4

POWER SUPPLY
105-120 volts AC/DC or
7.5 volts A & 90 volts B
supply

TUNING RANGE
540-1600KC

MFR/SUPPLIER
Philco Corp.

PHOTOFACT SET
41-18

PUBLISHED
1948

VOLUME
CONTROL
OFF -ON SW.

TUNING
CONTROL

PHILCO

MODEL PICTURED
48-1270

AC operated phono-radio combo multi-band receiver with pushbutton tuning

TUBES
13

POWER SUPPLY
105-125 volts AC

TUNING RANGE
3 bands

MFR/SUPPLIER
Philco Corp.

PHOTOFACT SET
42-20

PUBLISHED
1948

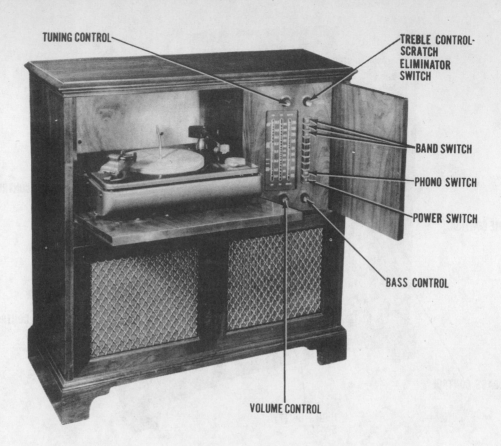

TUNING CONTROL

TREBLE CONTROL-SCRATCH ELIMINATOR SWITCH

BAND SWITCH

PHONO SWITCH

POWER SWITCH

BASS CONTROL

VOLUME CONTROL

PHILCO

MODEL PICTURED
49-601

Battery operated portable AM superheterodyne receiver with loop antenna

TUBES
4

POWER SUPPLY
7.5 volts A & 90 volts B supply

TUNING RANGE
540-1600KC

MFR/SUPPLIER
Philco Corp.

PHOTOFACT SET
42-21

PUBLISHED
1948

VOLUME CONTROL ON-OFF SWITCH

TUNING CONTROL

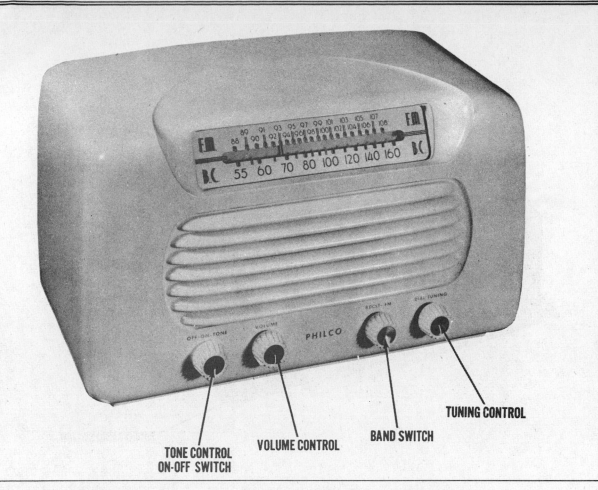

PHILCO

MODEL PICTURED
48-472-I

AC/DC operated AM-FM
superheterodyne receiver
with loop antenna

TUBES
7

POWER SUPPLY
105-125 volts AC/DC

TUNING RANGE
540-1720KC,
88-108MC

MFR/SUPPLIER
Philco Corp.

PHOTOFACT SET
43-15

PUBLISHED
1948

TONE CONTROL
ON-OFF SWITCH

VOLUME CONTROL

BAND SWITCH

TUNING CONTROL

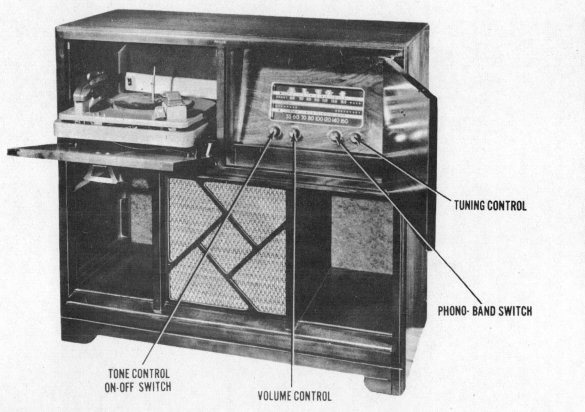

PHILCO

MODEL PICTURED
48-1284

AC operated phono-radio
combo two-band
superheterodyne receiver
with loop antenna

TUBES
7

POWER SUPPLY
105-120 volts AC

TUNING RANGE
540-1650KC,
9.3-15.7MC

MFR/SUPPLIER
Philco Corp.

PHOTOFACT SET
45-20

PUBLISHED
1948

TUNING CONTROL

PHONO- BAND SWITCH

TONE CONTROL
ON-OFF SWITCH

VOLUME CONTROL

PHILCO

MODEL PICTURED
49-1401

AC operated phono-radio combo AM superheterodyne receiver with loop antenna

TUBES
5

POWER SUPPLY
105-120 volts AC

TUNING RANGE
540-1600KC

MFR/SUPPLIER
Philco Corp.

PHOTOFACT SET
45-21

PUBLISHED
1948

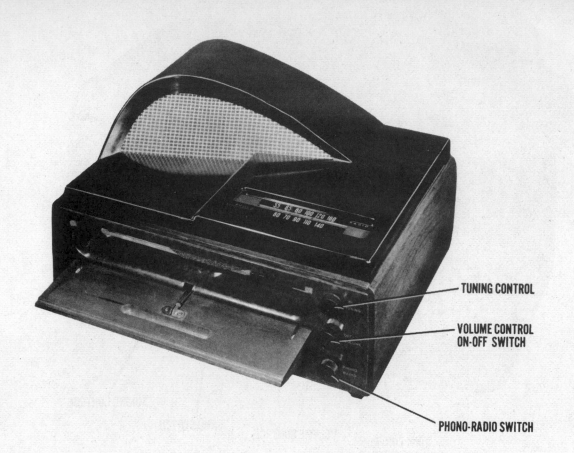

TUNING CONTROL

VOLUME CONTROL
ON-OFF SWITCH

PHONO-RADIO SWITCH

PHILCO

MODEL PICTURED
48-1290

AC operated phono-radio combo two-band AM-FM superheterodyne receiver

TUBES
13

POWER SUPPLY
105-120 volts AC

TUNING RANGE
540-1720KC,
88-108MC,
9.3-15.5MC

MFR/SUPPLIER
Philco Corp.

PHOTOFACT SET
47-18

PUBLISHED
1948

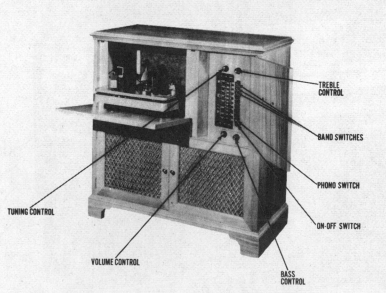

TREBLE CONTROL

BAND SWITCHES

PHONO SWITCH

ON-OFF SWITCH

TUNING CONTROL

VOLUME CONTROL

BASS CONTROL

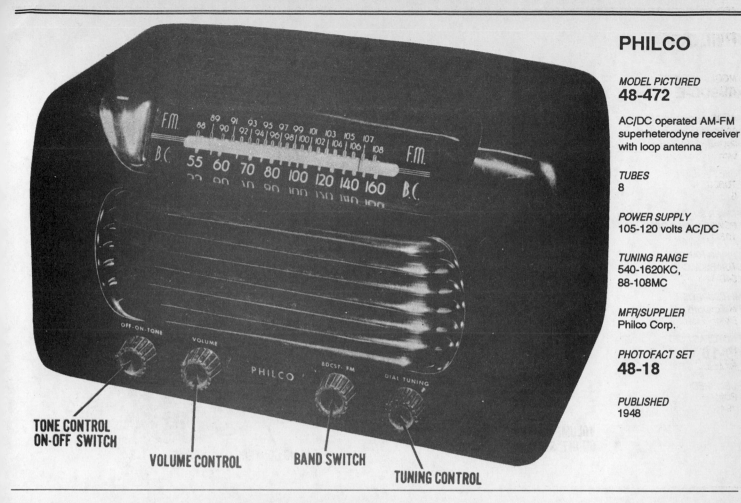

PHILCO

MODEL PICTURED
48-472

AC/DC operated AM-FM
superheterodyne receiver
with loop antenna

TUBES
8

POWER SUPPLY
105-120 volts AC/DC

TUNING RANGE
540-1620KC,
88-108MC

MFR/SUPPLIER
Philco Corp.

PHOTOFACT SET
48-18

PUBLISHED
1948

**TONE CONTROL
ON-OFF SWITCH**

VOLUME CONTROL

BAND SWITCH

TUNING CONTROL

PHILCO

MODEL PICTURED
49-500

AC/DC operated AM
superheterodyne receiver
with loop antenna

TUBES
5

POWER SUPPLY
105-120 volts AC/DC

TUNING RANGE
540-1620KC

MFR/SUPPLIER
Philco Corp.

PHOTOFACT SET
48-19

PUBLISHED
1948

**VOLUME CONTROL
ON-OFF SWITCH**

TUNING CONTROL

PHILCO

49-900-E

AC/DC operated AM
superheterodyne receiver
with loop antenna

TUBES
6

POWER SUPPLY
105-125 volts AC/DC

TUNING RANGE
540-1620KC

MFR/SUPPLIER
Philco Corp.

PHOTOFACT SET
49-16

PUBLISHED
1948

**VOLUME CONTROL
ON-OFF SWITCH**

TUNING CONTROL

PHILCO

MODEL PICTURED
49-1100

AC operated phono-radio
combo AM
superheterodyne receiver
with loop antenna

TUBES
6

POWER SUPPLY
105-120 volts AC

TUNING RANGE
540-1620KC

MFR/SUPPLIER
Philco Corp.

PHOTOFACT SET
49-19

PUBLISHED
1948

VOLUME CONTROL

TUNING CONTROL

TONE CONTROL

ON-OFF SWITCH

176

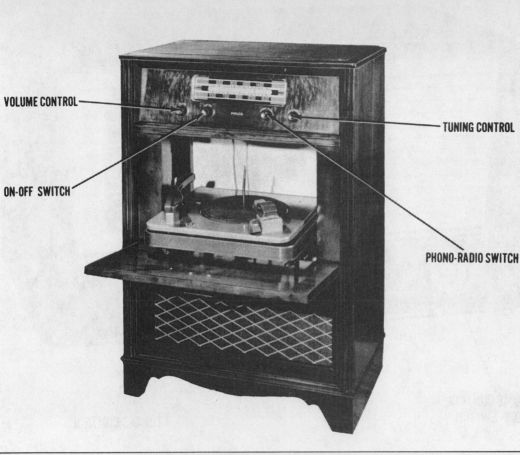

VOLUME CONTROL

ON-OFF SWITCH

TUNING CONTROL

PHONO-RADIO SWITCH

PHILCO

MODEL PICTURED
49-1600

AC operated phono-radio combo AM superheterodyne receiver with loop antenna

TUBES
5

POWER SUPPLY
110-120 volts AC

TUNING RANGE
540-1620KC

MFR/SUPPLIER
Philco Corp.

PHOTOFACT SET
50-13

PUBLISHED
1948

TONE CONTROL
ON-OFF SWITCH

VOLUME CONTROL

PHONO BAND SWITCH

TUNING CONTROL

SCRATCH ELIMINATOR SWITCH

PHILCO

MODEL PICTURED
48-1286

AC operated phono-radio combo AM-FM superheterodyne receiver with loop antenna

TUBES
11

POWER SUPPLY
105-120 volts AC

TUNING RANGE
540-1720KC,
88-108MC

MFR/SUPPLIER
Philco Corp.

PHOTOFACT SET
51-15

PUBLISHED
1948

PHILCO

MODEL PICTURED
49-902

AC/DC operated AM
superheterodyne receiver
with loop antenna

TUBES
6

POWER SUPPLY
105-125 volts AC/DC

TUNING RANGE
540-1620KC

MFR/SUPPLIER
Philco Corp.

PHOTOFACT SET
51-16

PUBLISHED
1948

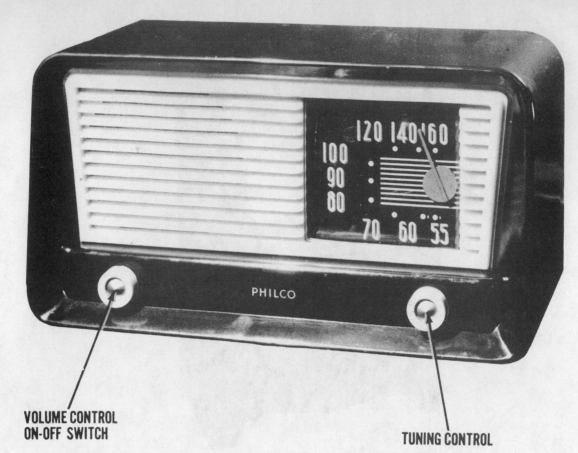

**VOLUME CONTROL
ON-OFF SWITCH**

TUNING CONTROL

PHILCO

MODEL PICTURED
49-503

AC/DC operated AM
superheterodyne receiver
with loop antenna

TUBES
5

POWER SUPPLY
105-125 volts AC/DC

TUNING RANGE
540-1620KC

MFR/SUPPLIER
Philco Corp.

PHOTOFACT SET
52-15

PUBLISHED
1948

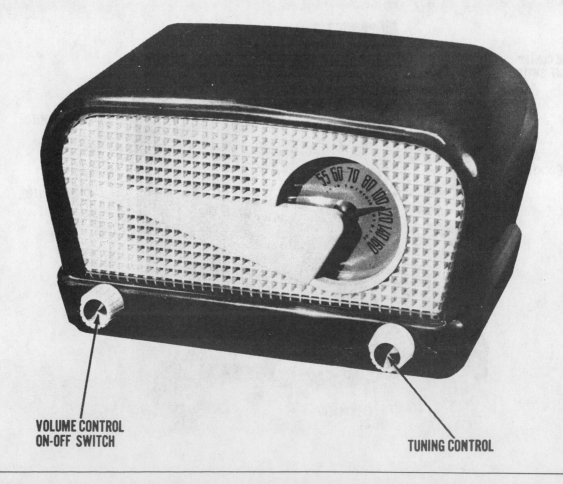

**VOLUME CONTROL
ON-OFF SWITCH**

TUNING CONTROL

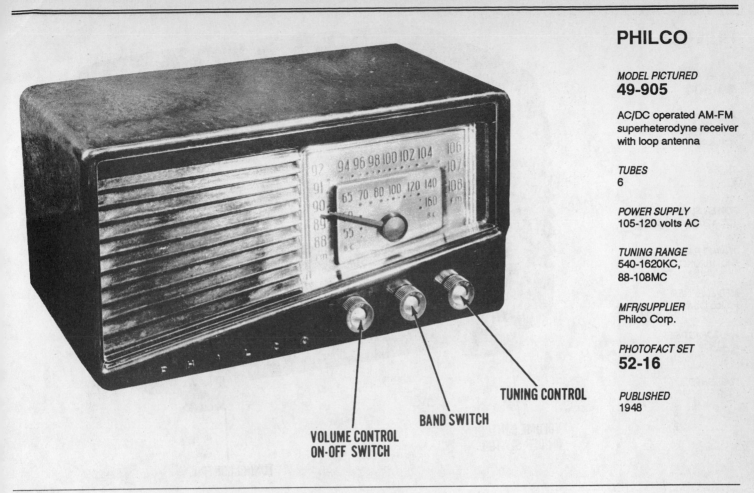

VOLUME CONTROL
ON-OFF SWITCH

BAND SWITCH

TUNING CONTROL

PHILCO

MODEL PICTURED
49-905

AC/DC operated AM-FM
superheterodyne receiver
with loop antenna

TUBES
6

POWER SUPPLY
105-120 volts AC

TUNING RANGE
540-1620KC,
88-108MC

MFR/SUPPLIER
Philco Corp.

PHOTOFACT SET
52-16

PUBLISHED
1948

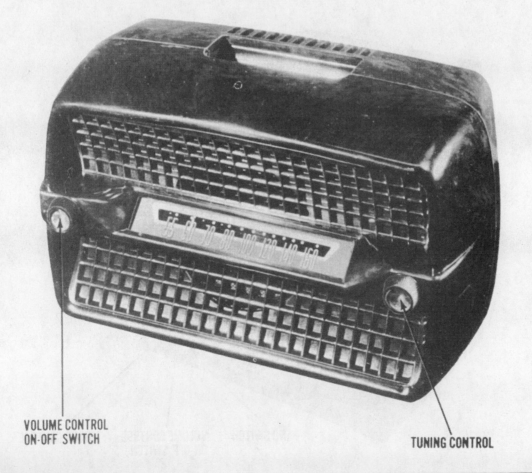

VOLUME CONTROL
ON-OFF SWITCH

TUNING CONTROL

PHILCO

MODEL PICTURED
49-505

AC/DC operated AM
superheterodyne receiver
with loop antenna

TUBES
5

POWER SUPPLY
105-120 volts AC/DC

TUNING RANGE
540-1620KC

MFR/SUPPLIER
Philco Corp.

PHOTOFACT SET
53-18

PUBLISHED
1949

PHILCO

MODEL PICTURED
49-504

AC/DC operated AM
superheterodyne receiver
with loop antenna

TUBES
5

POWER SUPPLY
105-120 volts AC/DC

TUNING RANGE
540-1620KC

MFR/SUPPLIER
Philco Corp.

PHOTOFACT SET
54-17

PUBLISHED
1949

**VOLUME CONTROL
ON-OFF SWITCH**

TUNING CONTROL

PHILCO

MODEL PICTURED
49-1405

AC operated phono-radio
combo AM
superheterodyne receiver
with loop antenna

TUBES
5

POWER SUPPLY
105-120 volts AC

TUNING RANGE
540-1600KC

MFR/SUPPLIER
Philco Corp.

PHOTOFACT SET
54-24

PUBLISHED
1949

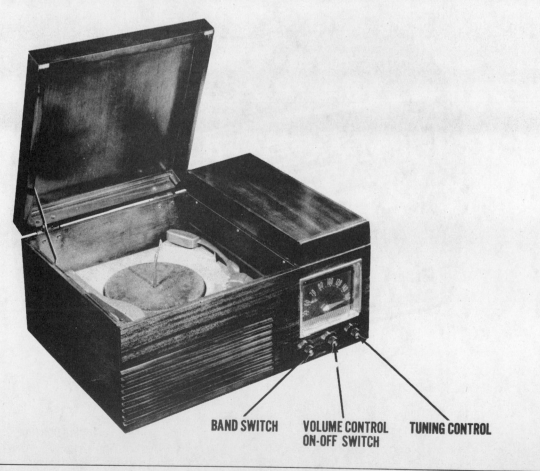

BAND SWITCH   **VOLUME CONTROL
ON-OFF SWITCH**   **TUNING CONTROL**

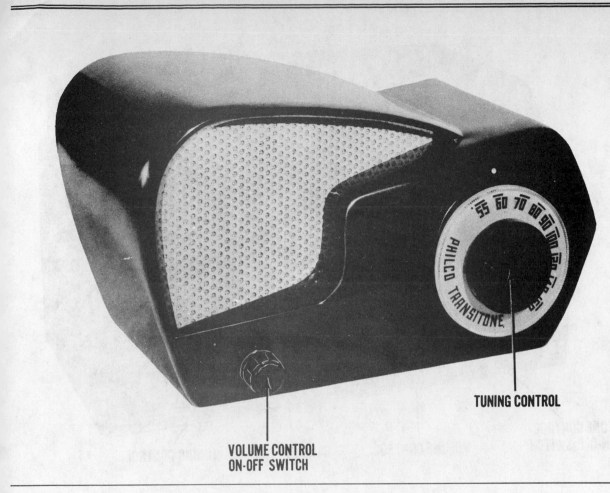

PHILCO

MODEL PICTURED
49-501

AC/DC operated AM
superheterodyne receiver
with loop antenna

TUBES
5

POWER SUPPLY
110-120 volts AC

TUNING RANGE
540-1620KC

MFR/SUPPLIER
Philco Corp.

PHOTOFACT SET
56-18

PUBLISHED
1949

TUNING CONTROL

VOLUME CONTROL
ON-OFF SWITCH

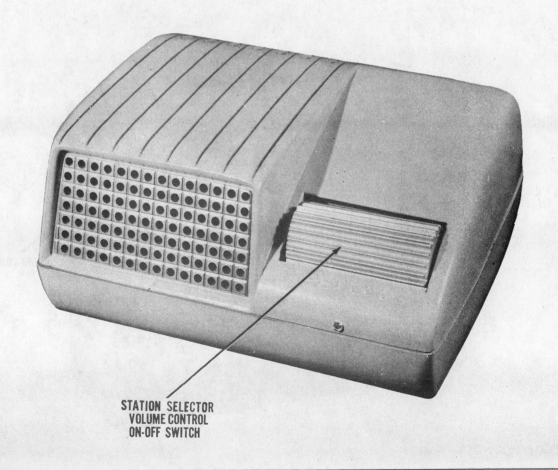

PHILCO

MODEL PICTURED
49-901

AC/DC operated AM
superheterodyne receiver
with loop antenna

TUBES
5

POWER SUPPLY
105-120 volts AC/DC

TUNING RANGE
540-1620KC

MFR/SUPPLIER
Philco Corp.

PHOTOFACT SET
56-19

PUBLISHED
1949

STATION SELECTOR
VOLUME CONTROL
ON-OFF SWITCH

PHILCO

MODEL PICTURED
46-906

AC/DC operated AM-FM
superheterodyne receiver
with loop antenna

TUBES
8

POWER SUPPLY
105-120 volts AC/DC

TUNING RANGE
540-1620KC,
88-108MC

MFR/SUPPLIER
Philco Corp.

PHOTOFACT SET
57-16

PUBLISHED
1949

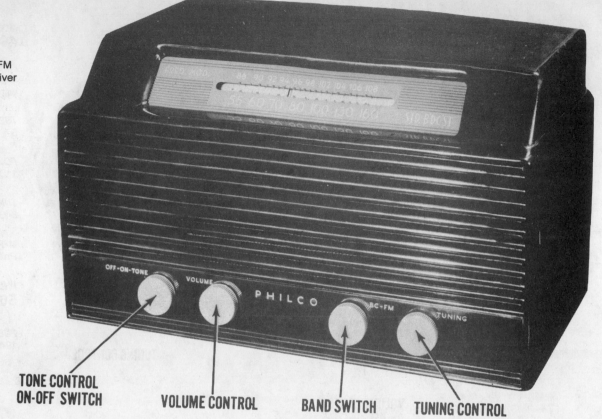

**TONE CONTROL
ON-OFF SWITCH**   **VOLUME CONTROL**   **BAND SWITCH**   **TUNING CONTROL**

PHILCO

MODEL PICTURED
49-607

Three-power operated
portable AM
superheterodyne receiver
with loop antenna

TUBES
6

POWER SUPPLY
105-120 volts AC/DC or 9
volts A & 90 volts B
supply

TUNING RANGE
540-1620KC

MFR/SUPPLIER
Philco Corp.

PHOTOFACT SET
58-15

PUBLISHED
1949

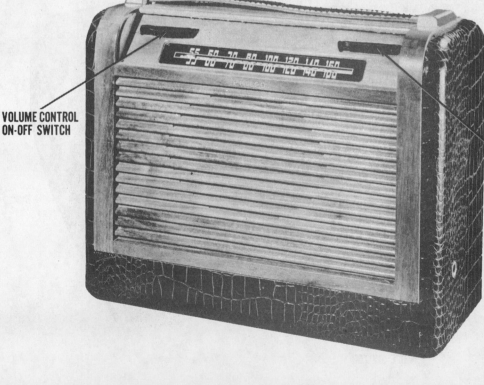

**VOLUME CONTROL
ON-OFF SWITCH**

TUNING CONTROL

PHILCO

MODEL PICTURED
49-904

AC/DC operated
two-band
superheterodyne receiver
with loop antenna

TUBES
6

POWER SUPPLY
105-1250 volts AC

TUNING RANGE
540-1620KC,
5.8-15.5MC

MFR/SUPPLIER
Philco Corp.

PHOTOFACT SET
58-16

PUBLISHED
1949

VOLUME CONTROL

TONE CONTROL
ON-OFF SWITCH

BAND SWITCH

TUNING CONTROL

PHILCO

MODEL PICTURED
49-603

AC/DC operated AM
superheterodyne receiver
with loop antenna

TUBES
5

POWER SUPPLY
105-120 volts AC/DC

TUNING RANGE
540-1620KC

MFR/SUPPLIER
Philco Corp.

PHOTOFACT SET
59-15

PUBLISHED
1949

VOLUME CONTROL
ON-OFF SWITCH

TUNING CONTROL

PHILCO

MODEL PICTURED
49-1615

AC operated phono-radio combo AM-FM superheterodyne receiver with loop antenna

TUBES
11

POWER SUPPLY
105-125 volts AC

TUNING RANGE
540-1620KC,
88-108MC

MFR/SUPPLIER
Philco Corp.

PHOTOFACT SET
64-9

PUBLISHED
1949

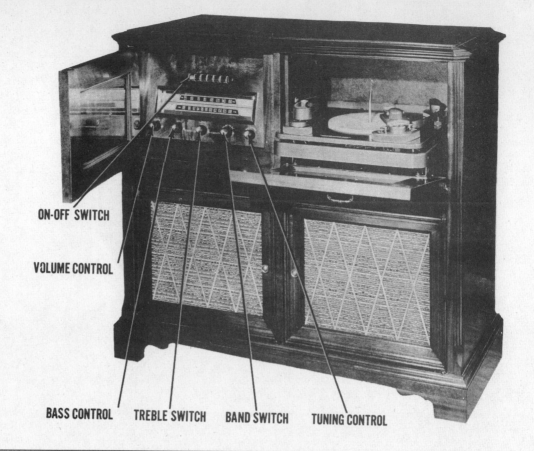

ON-OFF SWITCH

VOLUME CONTROL

BASS CONTROL   TREBLE SWITCH   BAND SWITCH   TUNING CONTROL

PHILCO

MODEL PICTURED
50-520

AC/DC operated AM superheterodyne receiver with loop antenna

TUBES
5

POWER SUPPLY
105-125 volts AC/DC

TUNING RANGE
540-1620KC

MFR/SUPPLIER
Philco Corp.

PHOTOFACT SET
73-9

PUBLISHED
1949

VOLUME CONTROL
ON-OFF SWITCH

TUNING CONTROL

PHILCO

MODEL PICTURED
50-522

AC/DC operated AM
superheterodyne receiver
with loop antenna

TUBES
5

POWER SUPPLY
105-125 volts AC/DC

TUNING RANGE
540-1620KC

MFR/SUPPLIER
Philco Corp.

PHOTOFACT SET
78-11

PUBLISHED
1949

**VOLUME CONTROL
ON-OFF SWITCH**

TUNING CONTROL

PHILCO

MODEL PICTURED
50-527

AC operated AM
superheterodyne receiver
with electric clock

TUBES
5

POWER SUPPLY
110-120 volts AC

TUNING RANGE
540-1600KC

MFR/SUPPLIER
Philco Corp.

PHOTOFACT SET
80-11

PUBLISHED
1949

**AUTO.
OFF - ON
SWITCH**   **DELAYED
OFF
SWITCH**   **VOLUME
CONTROL**   **AUTO
SET**   **TUNING
CONTROL**

PHILCO

MODEL PICTURED
50-620

Three-power operated
portable AM
superheterodyne receiver

TUBES
4

POWER SUPPLY
105-120 volts AC/DC or
7.5 volts A & 90 volts B
supply

TUNING RANGE
540-1600KC

MFR/SUPPLIER
Philco Corp.

PHOTOFACT SET
85-11

PUBLISHED
1950

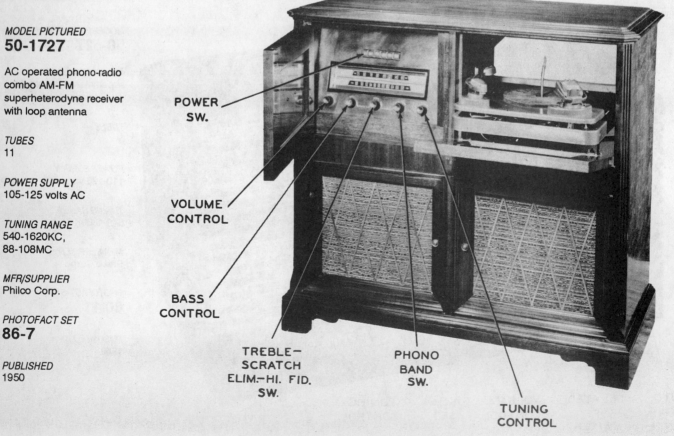

VOLUME
CONTROL
ON-OFF SW.

TUNING
CONTROL

PHILCO

MODEL PICTURED
50-1727

AC operated phono-radio
combo AM-FM
superheterodyne receiver
with loop antenna

TUBES
11

POWER SUPPLY
105-125 volts AC

TUNING RANGE
540-1620KC,
88-108MC

MFR/SUPPLIER
Philco Corp.

PHOTOFACT SET
86-7

PUBLISHED
1950

POWER
SW.

VOLUME
CONTROL

BASS
CONTROL

TREBLE-
SCRATCH
ELIM.—HI. FID.
SW.

PHONO
BAND
SW.

TUNING
CONTROL

186

**VOLUME CONTROL
ON-OFF SWITCH**

TUNING CONTROL

PHILCO

MODEL PICTURED
49-101

Three-power operated portable AM superheterodyne receiver

TUBES
4

POWER SUPPLY
105-120 volts AC/DC or 7.5 volts A & 90 volts B supply

TUNING RANGE
540-1600KC

MFR/SUPPLIER
Philco Corp.

PHOTOFACT SET
87-8

PUBLISHED
1950

**VOLUME CONTROL
ON-OFF SWITCH**

TUNING CONTROL

PHILCO

MODEL PICTURED
50-921

AC/DC operated AM superheterodyne receiver with loop antenna

TUBES
6

POWER SUPPLY
105-120 volts AC/DC

TUNING RANGE
540-1620KC

MFR/SUPPLIER
Philco Corp.

PHOTOFACT SET
88-8

PUBLISHED
1950

PHILCO

MODEL PICTURED
50-621

Three-power operated portable AM superheterodyne receiver with loop antenna

TUBES
5

POWER SUPPLY
110-120 volts AC/DC or 9 volts A & 90 volts B supply

TUNING RANGE
540-1620KC

MFR/SUPPLIER
Philco Corp.

PHOTOFACT SET
89-11

PUBLISHED
1950

VOLUME CONTROL
ON-OFF SWITCH

TUNING CONTROL

PHILCO

MODEL PICTURED
49-1613

AC operated phono-radio combo AM-FM superheterodyne receiver with loop antenna

TUBES
11

POWER SUPPLY
105-120 volts AC

TUNING RANGE
540-1720KC,
88-108MC

MFR/SUPPLIER
Philco Corp.

PHOTOFACT SET
91-9

PUBLISHED
1950

TONE CONTROL
ON-OFF SWITCH

VOLUME
CONTROL

SCRATCH
ELIMINATOR
ON-OFF SWITCH

BAND-
PHONO
SWITCH

TUNING
CONTROL

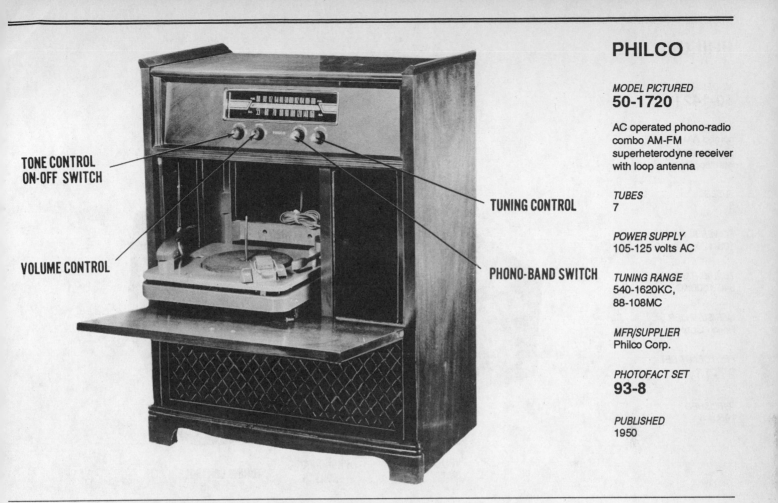

PHILCO

MODEL PICTURED
50-1720

AC operated phono-radio
combo AM-FM
superheterodyne receiver
with loop antenna

TUBES
7

POWER SUPPLY
105-125 volts AC

TUNING RANGE
540-1620KC,
88-108MC

MFR/SUPPLIER
Philco Corp.

PHOTOFACT SET
93-8

PUBLISHED
1950

TONE CONTROL
ON-OFF SWITCH

VOLUME CONTROL

TUNING CONTROL

PHONO-BAND SWITCH

PHILCO

MODEL PICTURED
50-526

AC/DC operated AM
superheterodyne receiver
with loop antenna

TUBES
5

POWER SUPPLY
105-125 volts AC/DC

TUNING RANGE
540-1620KC

MFR/SUPPLIER
Philco Corp.

PHOTOFACT SET
96-8

PUBLISHED
1950

VOLUME CONTROL
ON-OFF SWITCH

TUNING CONTROL

PHILCO

MODEL PICTURED
50-1421

AC operated phono-radio
combo AM
superheterodyne receiver
with loop antenna

TUBES
5

POWER SUPPLY
105-120 volts AC

TUNING RANGE
540-1600KC

MFR/SUPPLIER
Philco Corp.

PHOTOFACT SET
97-11

PUBLISHED
1950

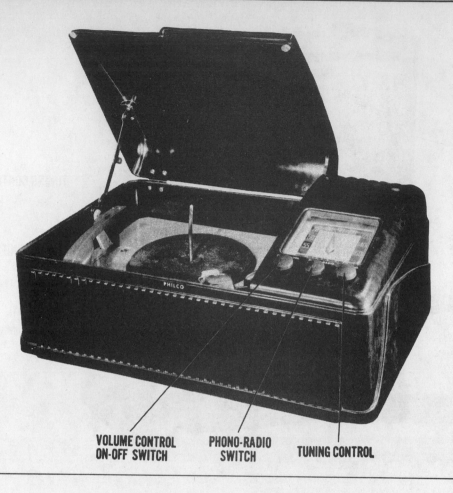

VOLUME CONTROL
ON-OFF SWITCH   PHONO-RADIO
SWITCH   TUNING CONTROL

PHILCO

MODEL PICTURED
50-1724

AC operated phono-radio
combo AM-FM
superheterodyne receiver
with loop antenna

TUBES
8

POWER SUPPLY
110-120 volts AC

TUNING RANGE
530-1630KC,
88-108MC

MFR/SUPPLIER
Philco Corp.

PHOTOFACT SET
98-9

PUBLISHED
1950

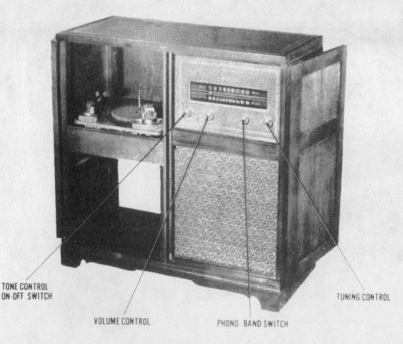

TONE CONTROL
ON-OFF SWITCH   TUNING CONTROL

VOLUME CONTROL   PHONO-BAND SWITCH

PHILCO

MODEL PICTURED
50-925

AC/DC operated AM-FM
superheterodyne receiver
with loop antenna

TUBES
6

POWER SUPPLY
105-120 volts AC/DC

TUNING RANGE
540-1620KC,
88-108MC

MFR/SUPPLIER
Philco Corp.

PHOTOFACT SET
99-12

PUBLISHED
1950

VOLUME CONTROL
ON-OFF SWITCH

AM - FM SWITCH

TUNING CONTROL

PHILCO

MODEL PICTURED
51-934

AC/DC operated AM-FM
superheterodyne receiver
with loop antenna

TUBES
6

POWER SUPPLY
105-125 volts AC/DC

TUNING RANGE
540-1630KC,
88-108MC

MFR/SUPPLIER
Philco Corp.

PHOTOFACT SET
102-10

PUBLISHED
1950

VOLUME CONTROL
ON-OFF SWITCH

TUNING CONTROL

AM-FM
SWITCH

PHILCO

TUNING CONTROL

VOLUME CONTROL
ON-OFF SWITCH

MODEL PICTURED
51-631

Three-power operated
portable AM
superheterodyne receiver
with loop antenna

TUBES
4

POWER SUPPLY
110-120 volts AC or 1.5
volts A & 67.5 volts B
supply

TUNING RANGE
540-1620KC

MFR/SUPPLIER
Philco Corp.

PHOTOFACT SET
106-12

PUBLISHED
1950

PHILCO

MODEL PICTURED
51-530

AC/DC operated AM
superheterodyne receiver
with loop antenna

TUBES
5

POWER SUPPLY
105-125 volts AC/DC

TUNING RANGE
540-1630KC

MFR/SUPPLIER
Philco Corp.

PHOTOFACT SET
122-7

PUBLISHED
1951

VOLUME CONTROL
ON-OFF SWITCH

TUNING CONTROL

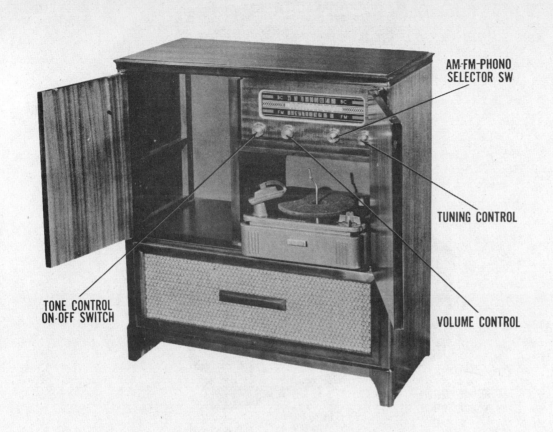

PHILCO

MODEL PICTURED
51-1732

AC operated phono-radio
combo AM-FM
superheterodyne receiver
with loop antenna

TUBES
7

POWER SUPPLY
110-120 volts AC

TUNING RANGE
540-1630KC,
88-108MC

MFR/SUPPLIER
Philco Corp.

PHOTOFACT SET
124-7

PUBLISHED
1951

AM-FM-PHONO
SELECTOR SW

TUNING CONTROL

TONE CONTROL
ON-OFF SWITCH

VOLUME CONTROL

PHILCO

MODEL PICTURED
51-537

AC operated AM
superheterodyne receiver
with loop antenna

TUBES
5

POWER SUPPLY
110-120 volts AC

TUNING RANGE
540-1600KC

MFR/SUPPLIER
Philco Corp.

PHOTOFACT SET
126-10

PUBLISHED
1951

TUNING
CONTROL

PHILCO
TRANSITONE

AUTO
ON-OFF SW

TIME
SET

VOLUME
CONTROL

RADIO OFF
TIME DELAY
SWITCH

ALARM
SET

PHILCO

MODEL PICTURED
51-1330

AC operated phono-radio combo AM superheterodyne receiver with loop antenna

TUBES
5

POWER SUPPLY
105-120 volts AC

TUNING RANGE
540-1620KC

MFR/SUPPLIER
Philco Corp.

PHOTOFACT SET
130-11

PUBLISHED
1951

VOLUME CONTROL
ON-OFF SWITCH

RADIO-PHONO
SWITCH

TUNING CONTROL

PHILCO

MODEL PICTURED
51-629

Three-power operated AM superheterodyne receiver with loop antenna

TUBES
4

POWER SUPPLY
110-120 volts AC/DC or 1.5 volts A & 90 volts B supply

TUNING RANGE
540-1620KC

MFR/SUPPLIER
Philco Corp.

PHOTOFACT SET
136-13

PUBLISHED
1951

TUNING CONTROL

VOLUME CONTROL

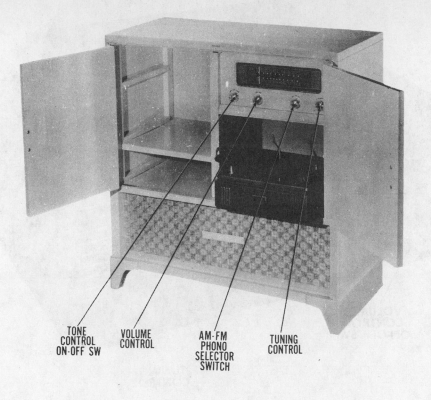

TONE
CONTROL
ON-OFF SW

VOLUME
CONTROL

AM-FM
PHONO
SELECTOR
SWITCH

TUNING
CONTROL

PHILCO

MODEL PICTURED
51-1733

AC operated phono-radio
combo AM-FM
superheterodyne receiver
with loop antenna

TUBES
8

POWER SUPPLY
110-120 volts AC

TUNING RANGE
540-1630KC,
88-108MC

MFR/SUPPLIER
Philco Corp.

PHOTOFACT SET
137-9

PUBLISHED
1951

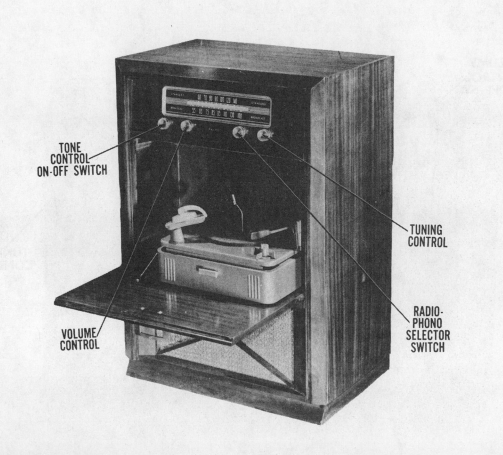

TONE
CONTROL
ON-OFF SWITCH

TUNING
CONTROL

VOLUME
CONTROL

RADIO-
PHONO
SELECTOR
SWITCH

PHILCO

MODEL PICTURED
51-1730

AC operated phono-radio
combo superheterodyne
receiver with loop
antenna

TUBES
6

POWER SUPPLY
105-120 volts AC

TUNING RANGE
540-1620KC

MFR/SUPPLIER
Philco Corp.

PHOTOFACT SET
140-8

PUBLISHED
1951

PHILCO

MODEL PICTURED
51-930

AC/DC operated AM
superheterodyne receiver
with loop antenna

TUBES
6

POWER SUPPLY
105-120 volts AC/DC

TUNING RANGE
540-1620KC

MFR/SUPPLIER
Philco Corp.

PHOTOFACT SET
153-11

PUBLISHED
1951

VOLUME
CONTROL
ON-OFF SW.

TUNING
CONTROL

PHILCO

MODEL PICTURED
52-640

Three-power operated
portable AM
superheterodyne receiver
with loop antenna

TUBES
4

POWER SUPPLY
105-125 volts AC/DC or
1.5 volts A & 67.5 volts B
supply

TUNING RANGE
540-1620KC

MFR/SUPPLIER
Philco Corp.

PHOTOFACT SET
153-12

PUBLISHED
1951

VOLUME
CONTROL
ON-OFF SW

TUNING
CONTROL

196

PHILCO

MODEL PICTURED
52-940

AC/DC operated AM
superheterodyne receiver
with loop antenna

TUBES
6

POWER SUPPLY
110-120 volts AC/DC

TUNING RANGE
540-1620KC

MFR/SUPPLIER
Philco Corp.

PHOTOFACT SET
156-9

PUBLISHED
1952

VOLUME
CONTROL
ON-OFF SW

TUNING
CONTROL

PHILCO

MODEL PICTURED
52-1340

AC operated phono-radio
combo AM receiver with
loop antenna

TUBES
5

POWER SUPPLY
105-120 volts AC

TUNING RANGE
540-1620KC

MFR/SUPPLIER
Philco Corp.

PHOTOFACT SET
160-8

PUBLISHED
1952

VOLUME
CONTROL
ON-OFF SW.

RADIO-
PHONO
SELECTOR
SWITCH

TUNING
CONTROL

PHILCO

MODEL PICTURED
52-643

Three-power operated
portable AM
superheterodyne receiver
with loop antenna

TUBES
5

POWER SUPPLY
110-120 volts AC/DC or 9
volts A & 90 volts B
supply

TUNING RANGE
540-1620KC

MFR/SUPPLIER
Philco Corp.

PHOTOFACT SET
161-7

PUBLISHED
1952

VOLUME
CONTROL
ON-OFF SW.

TUNING
CONTROL

PHILCO

MODEL PICTURED
52-544-I

AC operated AM
superheterodyne receiver
with electric clock

TUBES
5

POWER SUPPLY
110-120 volts AC

TUNING RANGE
540-1600KC

MFR/SUPPLIER
Philco Corp.

PHOTOFACT SET
163-9

PUBLISHED
1952

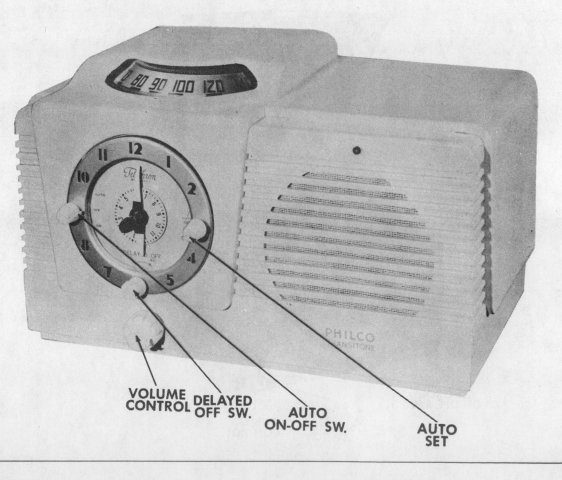

VOLUME
CONTROL

DELAYED
OFF SW.

AUTO
ON-OFF SW.

AUTO
SET

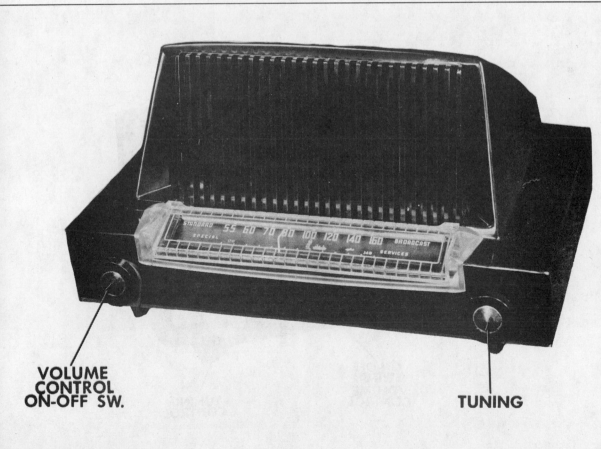

PHILCO

MODEL PICTURED
52-944

AC/DC operated AM-FM
superheterodyne receiver

TUBES
6

POWER SUPPLY
105-125 volts AC/DC

TUNING RANGE
540-1620KC,
88-108MC

MFR/SUPPLIER
Philco Corp.

PHOTOFACT SET
169-12

PUBLISHED
1952

VOLUME
CONTROL
ON-OFF SW.

TUNING
CONTROL

AM-FM
SELECTOR
SWITCH

PHILCO

MODEL PICTURED
53-566

AC/DC operated
two-band
superheterodyne receiver

TUBES
5

POWER SUPPLY
110-120 volts AC/DC

TUNING RANGE
540-1620KC,
1700-3400KC

MFR/SUPPLIER
Philco Corp.

PHOTOFACT SET
185-11

PUBLISHED
1952

VOLUME
CONTROL
ON-OFF SW.

TUNING

PHILCO

MODEL PICTURED
53-656

Three-power operated
portable two-band
superheterodyne receiver

TUBES
5

POWER SUPPLY
110-120 volts AC/DC or 9
volts A & 90 volts B
supply

TUNING RANGE
550-1600KC,
1.7-3.4MC

MFR/SUPPLIER
Philco Corp.

PHOTOFACT SET
187-10

PUBLISHED
1952

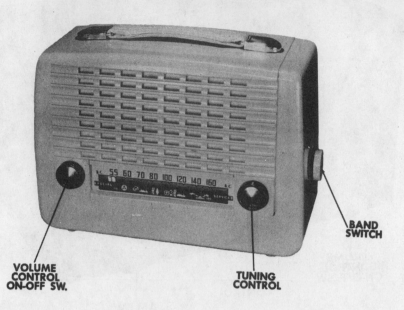

VOLUME
CONTROL
ON-OFF SW.

TUNING
CONTROL

BAND
SWITCH

PHILCO

MODEL PICTURED
53-562

AC/DC operated
two-band AM
superheterodyne receiver

TUBES
5

POWER SUPPLY
105-125 volts AC/DC

TUNING RANGE
540-1620KC,
1700-3400KC

MFR/SUPPLIER
Philco Corp.

PHOTOFACT SET
188-12

PUBLISHED
1952

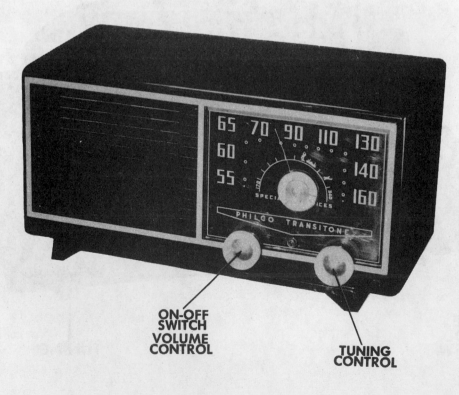

ON-OFF
SWITCH
VOLUME
CONTROL

TUNING
CONTROL

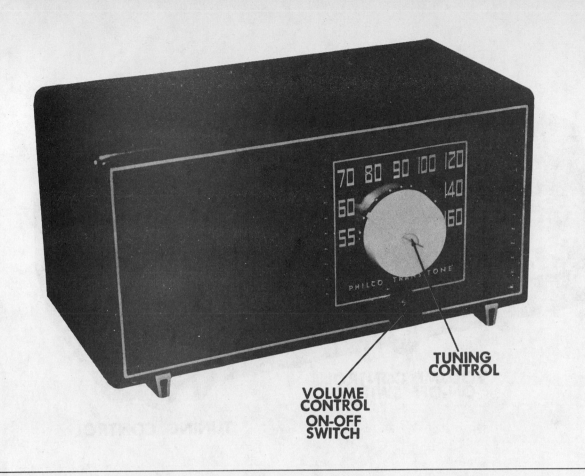

PHILCO

MODEL PICTURED
53-560

AC/DC operated AM
superheterodyne receiver

TUBES
5

POWER SUPPLY
110-120 volts AC

TUNING RANGE
540-1620KC

MFR/SUPPLIER
Philco Corp.

PHOTOFACT SET
189-13

PUBLISHED
1952

TUNING CONTROL

VOLUME CONTROL ON-OFF SWITCH

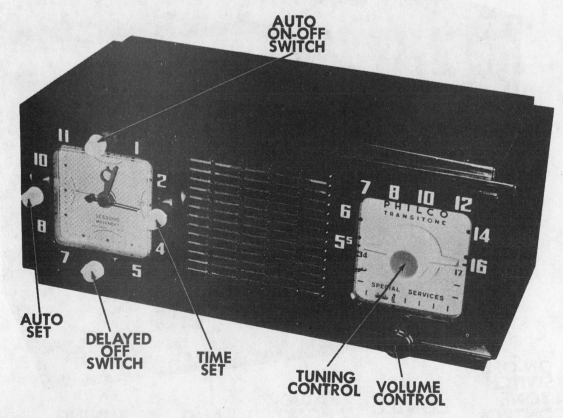

PHILCO

MODEL PICTURED
53-701

AC operated two-band
superheterodyne receiver
with electric clock

TUBES
5

POWER SUPPLY
110-120 volts AC

TUNING RANGE
540-1620KC,
1700-3400KC

MFR/SUPPLIER
Philco Corp.

PHOTOFACT SET
193-6

PUBLISHED
1953

AUTO ON-OFF SWITCH

AUTO SET

DELAYED OFF SWITCH

TIME SET

TUNING CONTROL

VOLUME CONTROL

PHILCO

MODEL PICTURED
53-563

AC/DC operated AM
superheterodyne receiver

TUBES
5

POWER SUPPLY
105-120 volts AC/DC

TUNING RANGE
540-1620KC

MFR/SUPPLIER
Philco Corp.

PHOTOFACT SET
196-12

PUBLISHED
1953

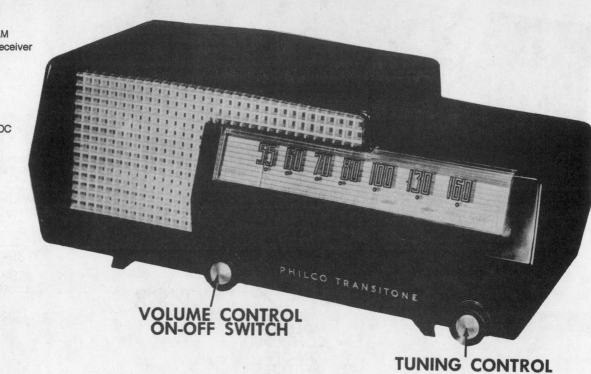

**VOLUME CONTROL
ON-OFF SWITCH**

TUNING CONTROL

PHILCO

MODEL PICTURED
53-960

AC operated multi-band
superheterodyne receiver

TUBES
8

POWER SUPPLY
110-120 volts AC

TUNING RANGE
540-1700KC,
1.7-5.3MC,
7.5-22.0MC

MFR/SUPPLIER
Philco Corp.

PHOTOFACT SET
199-7

PUBLISHED
1953

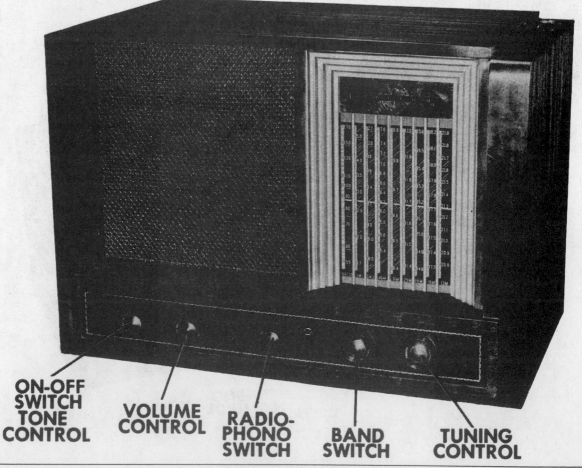

**ON-OFF
SWITCH
TONE
CONTROL**

**VOLUME
CONTROL**

**RADIO-
PHONO
SWITCH**

**BAND
SWITCH**

**TUNING
CONTROL**

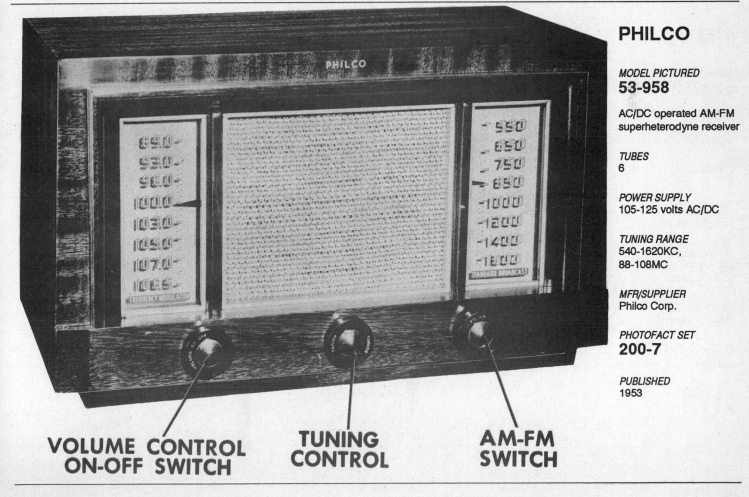

PHILCO

MODEL PICTURED
53-950

AC/DC operated
two-band
superheterodyne receiver

TUBES
6

POWER SUPPLY
105-120 volts AC/DC

TUNING RANGE
540-1620KC,
1700-3400KC

MFR/SUPPLIER
Philco Corp.

PHOTOFACT SET
200-6

PUBLISHED
1953

**VOLUME CONTROL
ON-OFF SWITCH**

**TUNING
CONTROL**

PHILCO

MODEL PICTURED
53-958

AC/DC operated AM-FM
superheterodyne receiver

TUBES
6

POWER SUPPLY
105-125 volts AC/DC

TUNING RANGE
540-1620KC,
88-108MC

MFR/SUPPLIER
Philco Corp.

PHOTOFACT SET
200-7

PUBLISHED
1953

**VOLUME CONTROL
ON-OFF SWITCH**

**TUNING
CONTROL**

**AM-FM
SWITCH**

PHILCO

MODEL PICTURED
53-702

AC operated two-band
superheterodyne receiver
with electric clock

TUBES
5

POWER SUPPLY
110-120 volts AC

TUNING RANGE
540-1620KC,
1700-3400KC

MFR/SUPPLIER
Philco Corp.

PHOTOFACT SET
202-5

PUBLISHED
1953

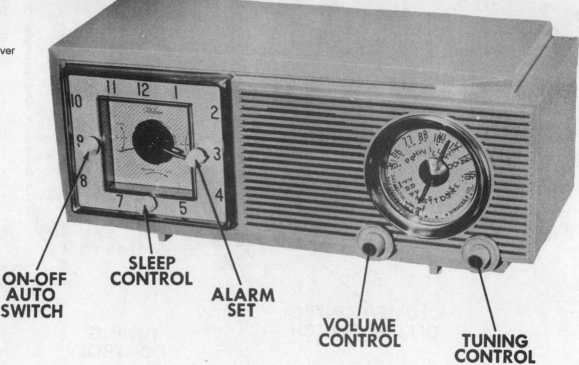

ON-OFF
AUTO
SWITCH

SLEEP
CONTROL

ALARM
SET

VOLUME
CONTROL

TUNING
CONTROL

PHILCO

MODEL PICTURED
53-1750

AC operated phono-radio
combo two-band
superheterodyne receiver

TUBES
5

POWER SUPPLY
105-120 volts AC

TUNING RANGE
540-1620KC,
1700-3400KC

MFR/SUPPLIER
Philco Corp.

PHOTOFACT SET
203-7

PUBLISHED
1953

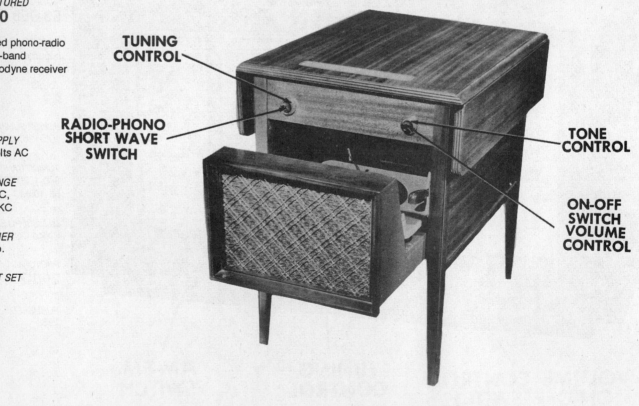

TUNING
CONTROL

RADIO-PHONO
SHORT WAVE
SWITCH

TONE
CONTROL

ON-OFF
SWITCH
VOLUME
CONTROL

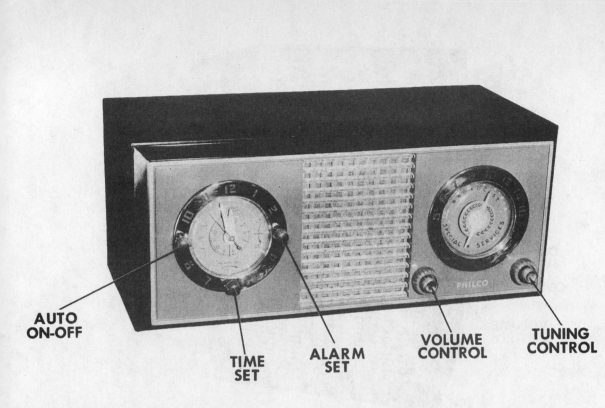

PHILCO

MODEL PICTURED
53-804

AC operated two-band
superheterodyne receiver
with electric clock

TUBES
6

POWER SUPPLY
110-120 volts AC

TUNING RANGE
540-1620KC,
1700-3400KC

MFR/SUPPLIER
Philco Corp.

PHOTOFACT SET
210-4

PUBLISHED
1953

**AUTO
ON-OFF**

**TIME
SET**

**ALARM
SET**

**VOLUME
CONTROL**

**TUNING
CONTROL**

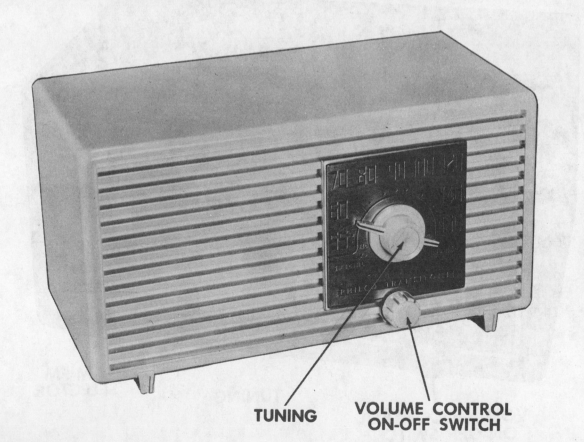

PHILCO

MODEL PICTURED
53-559

AC/DC operated
two-band
superheterodyne receiver

TUBES
5

POWER SUPPLY
110-120 volts AC/DC

TUNING RANGE
540-1620KC,
1700-3400KC

MFR/SUPPLIER
Philco Corp.

PHOTOFACT SET
213-6

PUBLISHED
1953

TUNING

**VOLUME CONTROL
ON-OFF SWITCH**

PHILCO

MODEL PICTURED
53-1754

AC operated two-band
superheterodyne receiver

TUBES
6

POWER SUPPLY
105-120 volts AC

TUNING RANGE
540-1620KC,
1700-3400KC

MFR/SUPPLIER
Philco Corp.

PHOTOFACT SET
214-8

PUBLISHED
1953

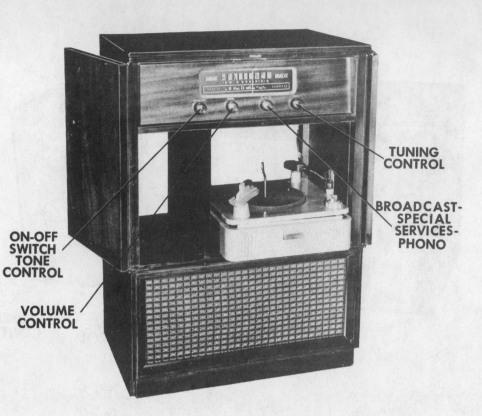

TUNING
CONTROL

BROADCAST-
SPECIAL
SERVICES-
PHONO

ON-OFF
SWITCH
TONE
CONTROL

VOLUME
CONTROL

PHILCO

MODEL PICTURED
B-956

AC/DC operated AM-FM
superheterodyne receiver

TUBES
6

POWER SUPPLY
110-120 volts AC/DC

TUNING RANGE
540-1620KC,
88-108MC

MFR/SUPPLIER
Philco Corp.

PHOTOFACT SET
218-8

PUBLISHED
1953

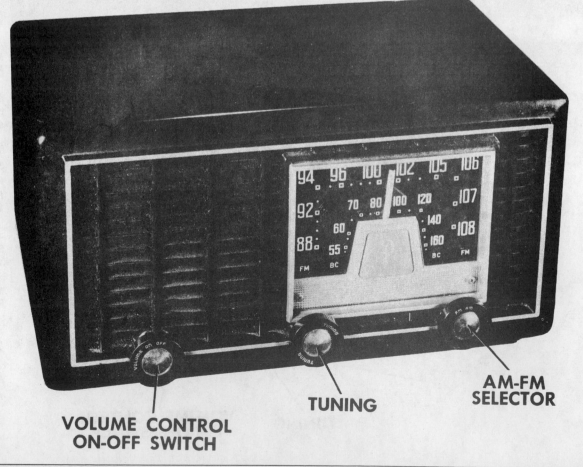

VOLUME CONTROL
ON-OFF SWITCH

TUNING

AM-FM
SELECTOR

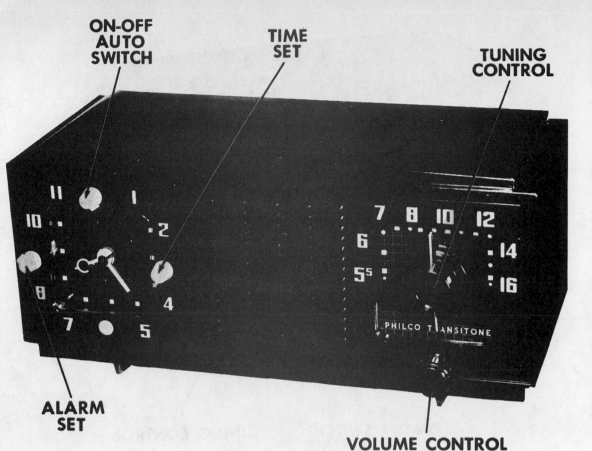

ON-OFF AUTO SWITCH

TIME SET

TUNING CONTROL

ALARM SET

VOLUME CONTROL

PHILCO

MODEL PICTURED
B710

AC operated AM superheterodyne receiver with electric clock

TUBES
5

POWER SUPPLY
110-120 volts

TUNING RANGE
540-1620KC

MFR/SUPPLIER
Philco Corp.

PHOTOFACT SET
223-8

PUBLISHED
1953

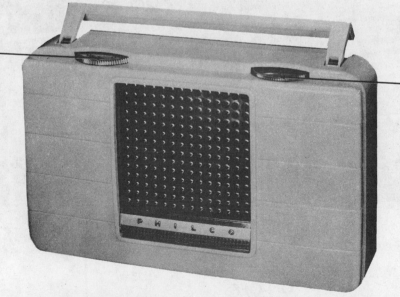

ON-OFF VOLUME CONTROL

TUNING CONTROL

PHILCO

MODEL PICTURED
B650

Battery operated portable AM superheterodyne receiver

TUBES
4

POWER SUPPLY
1.5 volts A & 75 volts B supply

TUNING RANGE
535-1620KC

MFR/SUPPLIER
Philco Corp.

PHOTOFACT SET
226-5

PUBLISHED
1954

PHILCO

MODEL PICTURED
B570

AC/DC operated AM
superheterodyne receiver

TUBES
5

POWER SUPPLY
105-120 volts AC/DC

TUNING RANGE
540-1620KC

MFR/SUPPLIER
Philco Corp.

PHOTOFACT SET
228-13

PUBLISHED
1954

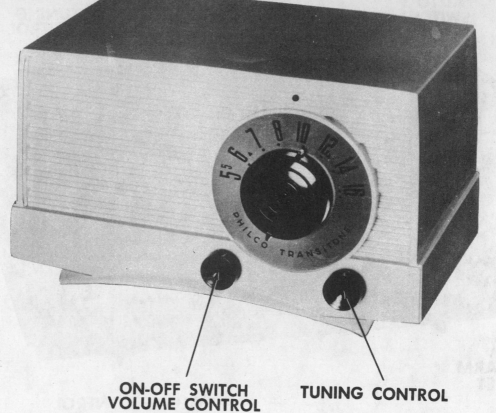

**ON-OFF SWITCH
VOLUME CONTROL**   **TUNING CONTROL**

PHILCO

MODEL PICTURED
B574

AC/DC operated
two-band
superheterodyne receiver

TUBES
5

POWER SUPPLY
105-120 volts AC/DC

TUNING RANGE
540-1620KC,
1700-3400KC

MFR/SUPPLIER
Philco Corp.

PHOTOFACT SET
229-9

PUBLISHED
1954

**ON-OFF
VOLUME
CONTROL**   **TUNING-
BAND
SWITCH**

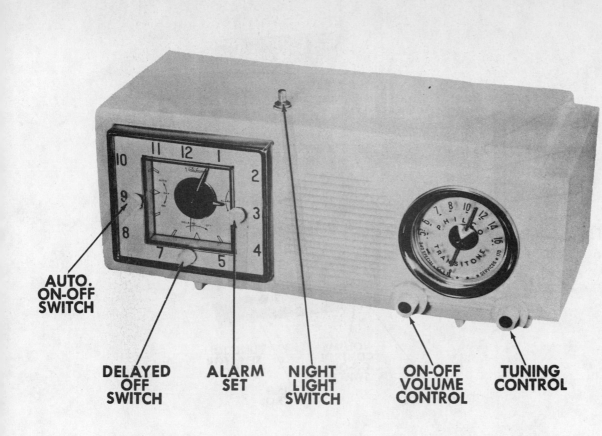

AUTO. ON-OFF SWITCH

DELAYED OFF SWITCH

ALARM SET

NIGHT LIGHT SWITCH

ON-OFF VOLUME CONTROL

TUNING CONTROL

PHILCO

MODEL PICTURED
B714

AC operated two-band superheterodyne receiver with electric clock

TUBES
5

POWER SUPPLY
110-120 volts AC

TUNING RANGE
540-1620KC,
1700-3400KC

MFR/SUPPLIER
Philco Corp.

PHOTOFACT SET
229-10

PUBLISHED
1954

VOLUME CONTROL ON-OFF SWITCH

TUNING CONTROL

PHILCO

MODEL PICTURED
B652

Three-power operated portable AM superheterodyne receiver

TUBES
4

POWER SUPPLY
110-120 volts AC/DC or 1.5 volts A & 67.5 volts B supply

TUNING RANGE
535-1620KC

MFR/SUPPLIER
Philco Corp.

PHOTOFACT SET
234-10

PUBLISHED
1954

PHILCO

MODEL PICTURED
B-1352

AC operated phono-radio
combo two-band
superheterodyne receiver
with three-speed auto
record changer

TUBES
5

POWER SUPPLY
110-120 volts AC

TUNING RANGE
540-1620KC,
1700-3400KC

MFR/SUPPLIER
Philco Corp.

PHOTOFACT SET
235-9

PUBLISHED
1954

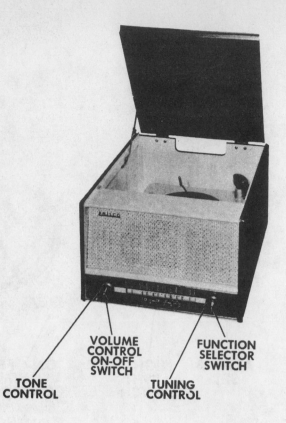

VOLUME
CONTROL
ON-OFF
SWITCH

FUNCTION
SELECTOR
SWITCH

TONE
CONTROL

TUNING
CONTROL

PHILCO

MODEL PICTURED
B-1752

AC operated phono-radio
combo two-band
superheterodyne receiver
with three-speed auto
record changer

TUBES
5

POWER SUPPLY
105-120 volts AC

TUNING RANGE
540-1620KC,
1700-3400KC

MFR/SUPPLIER
Philco Corp.

PHOTOFACT SET
240-6

PUBLISHED
1954

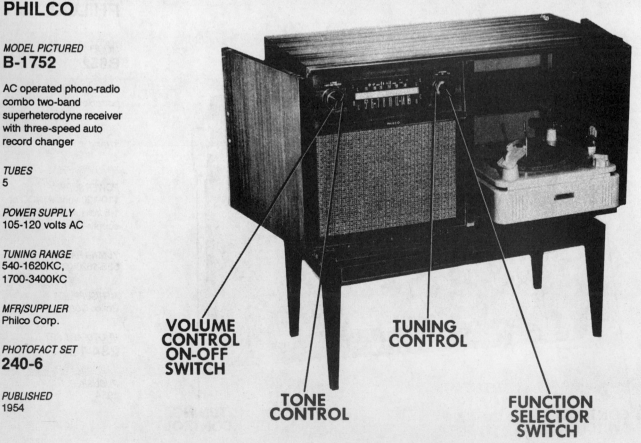

VOLUME
CONTROL
ON-OFF
SWITCH

TUNING
CONTROL

TONE
CONTROL

FUNCTION
SELECTOR
SWITCH

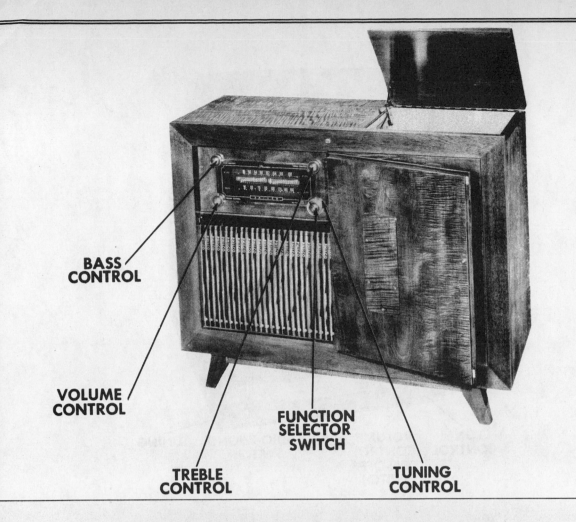

BASS
CONTROL

VOLUME
CONTROL

TREBLE
CONTROL

FUNCTION
SELECTOR
SWITCH

TUNING
CONTROL

PHILCO

MODEL PICTURED
B1756

AC operated phono-radio
combo AM-FM
superheterodyne receiver
with three-speed auto
record changer

TUBES
14

POWER SUPPLY
105-120 volts AC

TUNING RANGE
540-1620KC,
88-108MC

MFR/SUPPLIER
Philco Corp.

PHOTOFACT SET
241-10

PUBLISHED
1954

VOLUME
CONTROL
ON-OFF
SWITCH

TUNING
CONTROL

PHILCO

MODEL PICTURED
B570

AC/DC operated AM
superheterodyne receiver

TUBES
5

POWER SUPPLY
105-120 volts AC/DC

TUNING RANGE
540-1620KC

MFR/SUPPLIER
Philco Corp.

PHOTOFACT SET
257-12

PUBLISHED
1954

PHILCO

MODEL PICTURED
B1349

AC operated phono-radio
combo AM
superheterodyne receiver
with three-speed auto
record changer

TUBES
5

POWER SUPPLY
105-120 volts AC

TUNING RANGE
540-1620KC

MFR/SUPPLIER
Philco Corp.

PHOTOFACT SET
259-11

PUBLISHED
1954

TONE
CONTROL   VOLUME
CONTROL
ON-OFF
SWITCH   RADIO-PHONO   TUNING
SWITCH

PHILCO

MODEL PICTURED
B569

AC/DC operated AM
superheterodyne receiver

TUBES
5

POWER SUPPLY
105-120 volts AC/DC

TUNING RANGE
540-1620KC

MFR/SUPPLIER
Philco Corp.

PHOTOFACT SET
261-11

PUBLISHED
1954

VOLUME
CONTROL
ON-OFF
SWITCH   TUNING

PHILCO

MODEL PICTURED
C-660

Battery operated portable AM superheterodyne receiver

TUBES
4

POWER SUPPLY
1.5 volts A & 75 volts B supply

TUNING RANGE
540-1630KC

MFR/SUPPLIER
Philco Corp.

PHOTOFACT SET
271-8

PUBLISHED
1955

VOLUME CONTROL ON-OFF SWITCH　　　**TUNING**

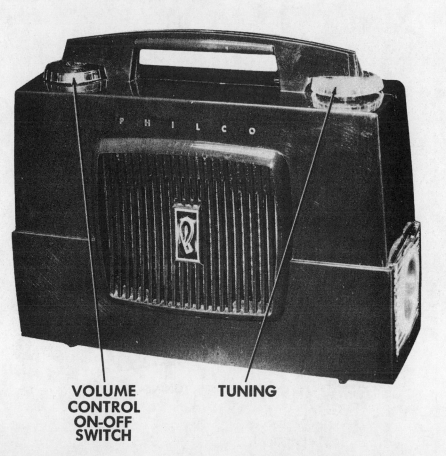

PHILCO

MODEL PICTURED
C-663

Three-power operated portable AM superheterodyne receiver

TUBES
4

POWER SUPPLY
110-120 volts AC/DC or 3 volts A & 75 volts B supply

TUNING RANGE
535-1620KC

MFR/SUPPLIER
Philco Corp.

PHOTOFACT SET
271-9

PUBLISHED
1955

VOLUME CONTROL ON-OFF SWITCH　　　**TUNING**

PHILCO

MODEL PICTURED
C-570

AC or AC/DC operated
AM superheterodyne
receiver

TUBES
5

POWER SUPPLY
110-120 volts AC/DC
(clock models AC only)

TUNING RANGE
540-1620KC

MFR/SUPPLIER
Philco Corp.

PHOTOFACT SET
272-9

PUBLISHED
1955

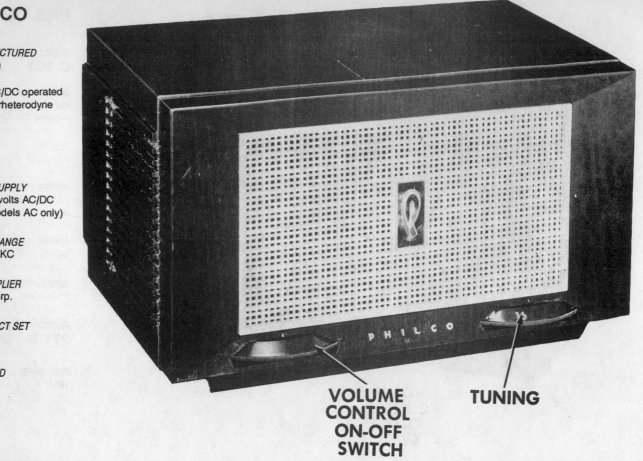

**VOLUME
CONTROL
ON-OFF
SWITCH**

TUNING

PHILCO

MODEL PICTURED
C-667

Three-power operated
portable two-band
superheterodyne receiver
with flashlight

TUBES
5

POWER SUPPLY
110-120 volts AC/DC or 9
volts A & 90 volts B
supply

TUNING RANGE
540-1620KC,
1700-3400KC

MFR/SUPPLIER
Philco Corp.

PHOTOFACT SET
279-9

PUBLISHED
1955

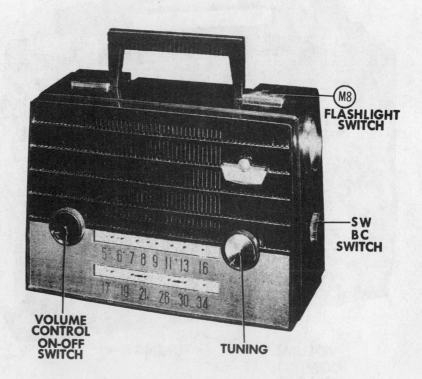

(M8)
**FLASHLIGHT
SWITCH**

**S W
B C
SWITCH**

**VOLUME
CONTROL
ON-OFF
SWITCH**

TUNING

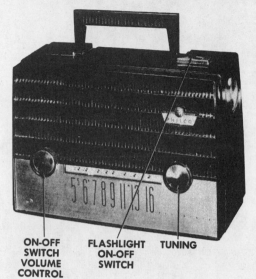

ON-OFF
SWITCH
VOLUME
CONTROL

FLASHLIGHT
ON-OFF
SWITCH

TUNING

PHILCO

MODEL PICTURED
C-666

Three-power operated
portable AM
superheterodyne receiver
with flashlight

TUBES
5

POWER SUPPLY
110-120 volts AC/DC or 9
volts A & 90 volts B
supply

TUNING RANGE
540-1620KC

MFR/SUPPLIER
Philco Corp.

PHOTOFACT SET
294-9

PUBLISHED
1955

PHILCO

MODEL PICTURED
D-592

AC/DC operated AM
superheterodyne receiver

TUBES
5

POWER SUPPLY
105-120 volts AC/DC (AC
only for clock models)

TUNING RANGE
540-1620KC

MFR/SUPPLIER
Philco Corp.

PHOTOFACT SET
321-10

PUBLISHED
1956

PHILCO

MODEL PICTURED
D-664

Three-power operated
portable AM
superheterodyne receiver

TUBES
4

POWER SUPPLY
110-120 volts AC/DC or 3
volts A & 90 volts B
supply

TUNING RANGE
540-1620KC

MFR/SUPPLIER
Philco Corp.

PHOTOFACT SET
324-11

PUBLISHED
1956

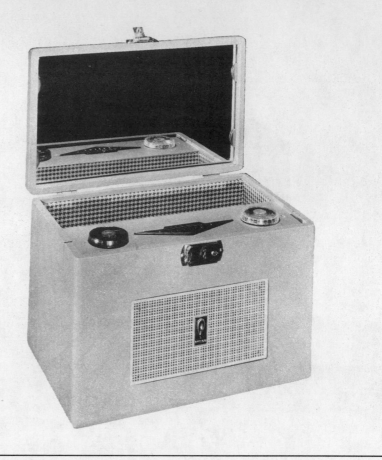

PHILCO

MODEL PICTURED
D-579

AC operated AM
superheterodyne receiver
with electric clock

TUBES
4

POWER SUPPLY
105-120 volts AC

TUNING RANGE
540-1620KC

MFR/SUPPLIER
Philco Corp.

PHOTOFACT SET
328-8

PUBLISHED
1956

PHILCO

MODEL PICTURED
E-670

Three-power operated
portable AM receiver

TUBES
4

POWER SUPPLY
110-120 volts AC/DC or
7.5 volts A & 90 volts B
supply

TUNING RANGE
540-1620KC

MFR/SUPPLIER
Philco Corp.

PHOTOFACT SET
346-8

PUBLISHED
1957

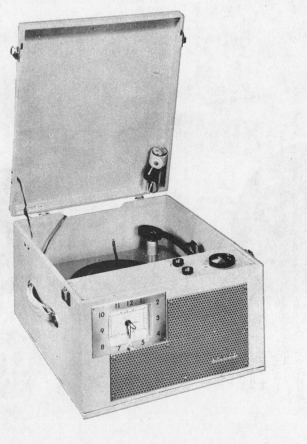

PHILCO

MODEL PICTURED
D-1345

AC operated AM receiver
with three-speed auto
record changer and
electric clock

TUBES
5

POWER SUPPLY
105-125 volts AC

TUNING RANGE
540-1620KC

MFR/SUPPLIER
Philco Corp.

PHOTOFACT SET
347-10

PUBLISHED
1957

PHILCO

MODEL PICTURED
T-7

Battery operated portable
AM transistorized
receiver

TUBES
7 transistor

POWER SUPPLY
3 volts DC

TUNING RANGE
540-1620KC

MFR/SUPPLIER
Philco Corp.

PHOTOFACT SET
347-11

PUBLISHED
1957

PHILCO

MODEL PICTURED
E-740

AC operated AM receiver
with electric clock

TUBES
5

POWER SUPPLY
105-120 volts AC

TUNING RANGE
540-1620KC

MFR/SUPPLIER
Philco Corp.

PHOTOFACT SET
351-14

PUBLISHED
1957

PHILCO

MODEL PICTURED
E-1370

AC operated AM receiver
with three-speed auto
record changer

TUBES
5

POWER SUPPLY
105-125 volts AC

TUNING RANGE
540-1620KC

MFR/SUPPLIER
Philco Corp.

PHOTOFACT SET
359-11

PUBLISHED
1957

PHILCO

MODEL PICTURED
E-976

AC/DC operated AM-FM
receiver

TUBES
7

POWER SUPPLY
105-125 volts AC/DC

TUNING RANGE
540-1620KC,
88-108MC

MFR/SUPPLIER
Philco Corp.

PHOTOFACT SET
378-11

PUBLISHED
1957

PHILCO

MODEL PICTURED
E-818

AC/DC operated AM
receiver

TUBES
6

POWER SUPPLY
105-125 volts AC

TUNING RANGE
540-1620KC

MFR/SUPPLIER
Philco Corp.

PHOTOFACT SET
383-9

PUBLISHED
1958

PHILCO

MODEL PICTURED
F-963

AC/DC operated AM
receiver

TUBES
6

POWER SUPPLY
105-120 volts AC/DC

TUNING RANGE
535-1620KC

MFR/SUPPLIER
Philco Corp.

PHOTOFACT SET
396-6

PUBLISHED
1958

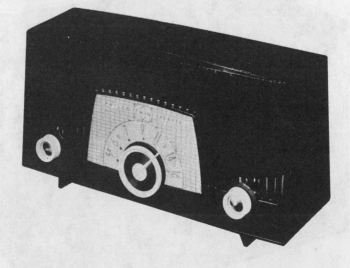

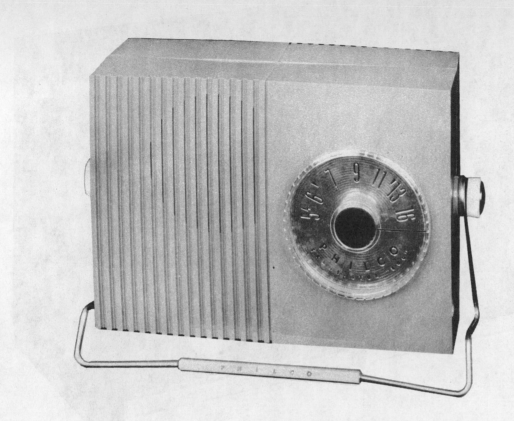

PHILCO

MODEL PICTURED
T-700

Battery operated portable
AM transistorized
receiver

TUBES
7 transistor

POWER SUPPLY
6 volts DC

TUNING RANGE
535-1620KC

MFR/SUPPLIER
Philco Corp.

PHOTOFACT SET
401-10

PUBLISHED
1958

PHILCO

MODEL PICTURED
F-809

AC operated AM receiver

TUBES
5

POWER SUPPLY
105-120 volts AC

TUNING RANGE
540-1620KC

MFR/SUPPLIER
Philco Corp.

PHOTOFACT SET
413-12

PUBLISHED
1958

PHILCO

MODEL PICTURED
F-974

AC/DC operated FM-AM
receiver

TUBES
7

POWER SUPPLY
105-120 volts AC/DC

TUNING RANGE
540-1620KC,
88-108MC

MFR/SUPPLIER
Philco Corp.

PHOTOFACT SET
426-12

PUBLISHED
1959

PHILCO

MODEL PICTURED
T-6

Battery operated portable
transistorized AM
receiver

TUBES
6 transistor

POWER SUPPLY
3 volts DC

TUNING RANGE
535-1620KC

MFR/SUPPLIER
Philco Corp.

PHOTOFACT SET
426-13

PUBLISHED
1959

PHILCO

MODEL PICTURED
F-1803

AC operated AM-FM
tuner with four-speed
automatic record changer

TUBES
8

POWER SUPPLY
110-120 volts AC

TUNING RANGE
540-1620KC,
88-108MC

MFR/SUPPLIER
Philco Corp.

PHOTOFACT SET
427-9

PUBLISHED
1959

PHILCO

MODEL PICTURED
T-9

Battery operated
multi-band transistorized
portable receiver

TUBES
9 transistor

POWER SUPPLY
1.5 or 6 volts DC

TUNING RANGE
7 bands

MFR/SUPPLIER
Philco Corp.

PHOTOFACT SET
429-10

PUBLISHED
1959

PHILCO

MODEL PICTURED
G-681

Three-power operated
portable AM receiver

TUBES
4

POWER SUPPLY
110-120 volts AC/DC or
7.5 volts A & 90 volts B
supply

TUNING RANGE
540-1620KC

MFR/SUPPLIER
Philco Corp.

PHOTOFACT SET
434-10

PUBLISHED
1959

PHILCO

MODEL PICTURED
T-4

Battery operated portable
transistorized AM
receiver

TUBES
4 transistor

POWER SUPPLY
6 volts DC

TUNING RANGE
535-1620KC

MFR/SUPPLIER
Philco Corp.

PHOTOFACT SET
437-10

PUBLISHED
1959

PHILCO

MODEL PICTURED
G-822

AC/DC operated AM
receiver

TUBES
5

POWER SUPPLY
105-120 volts AC/DC

TUNING RANGE
540-1620KC

MFR/SUPPLIER
Philco Corp.

PHOTOFACT SET
443-9

PUBLISHED
1959

PHILCO

MODEL PICTURED
G-751

AC operated AM receiver
with electric clock

TUBES
5

POWER SUPPLY
105-120 volts AC

TUNING RANGE
540-1620KC

MFR/SUPPLIER
Philco Corp.

PHOTOFACT SET
459-12

PUBLISHED
1959

PHILCO

MODEL PICTURED
G-1707S

AC operated FM-AM
tuner and amplifier with
four-speed auto record
changer

TUBES
6

POWER SUPPLY
105-120 volts AC

TUNING RANGE
540-1620KC,
88-108MC

MFR/SUPPLIER
Philco Corp.

PHOTOFACT SET
463-11

PUBLISHED
1959

PHILCO

MODEL PICTURED
G-1907S

AC operated FM-AM
tuner with amplifier and
four-speed auto record
changer

TUBES
8

POWER SUPPLY
105-120 volts AC

TUNING RANGE
540-1620KC,
88-108MC

MFR/SUPPLIER
Philco Corp.

PHOTOFACT SET
464-17

PUBLISHED
1959

226

PHILCO

MODEL PICTURED
T-75

Battery operated portable
transistorized AM
receiver

TUBES
7 transistor

POWER SUPPLY
3 volts DC

TUNING RANGE
535-1620KC

MFR/SUPPLIER
Philco Corp.

PHOTOFACT SET
465-11

PUBLISHED
1959

PHILCO

MODEL PICTURED
T-60

Battery operated portable
transistorized AM
receiver

TUBES
6 transistor

POWER SUPPLY
3 volts DC

TUNING RANGE
535-1620KC

MFR/SUPPLIER
Philco Corp.

PHOTOFACT SET
467-13

PUBLISHED
1959

227

PHILCO

MODEL PICTURED
T-65

Battery operated portable
transistorized AM
receiver

TUBES
6 transistor

POWER SUPPLY
3 volts DC

TUNING RANGE
535-1620KC

MFR/SUPPLIER
Philco Corp.

PHOTOFACT SET
468-10

PUBLISHED
1959

PHILCO

MODEL PICTURED
T-78

Battery operated portable
transistorized AM
receiver

TUBES
7 transistor

POWER SUPPLY
6 volts DC

TUNING RANGE
535-1620KC

MFR/SUPPLIER
Philco Corp.

PHOTOFACT SET
473-11

PUBLISHED
1960

PHILCO

MODEL PICTURED
G-747

AC operated AM receiver
with electric clock

TUBES
4

POWER SUPPLY
105-120 volts AC

TUNING RANGE
540-1620KC

MFR/SUPPLIER
Philco Corp.

PHOTOFACT SET
475-5

PUBLISHED
1960

PHILCO

MODEL PICTURED
T-1000

Battery operated
transistorized AM
receiver with clock

TUBES
6 transistor

POWER SUPPLY
3 volts DC (radio) & 1.5
volts DC (clock)

TUNING RANGE
535-1620KC

MFR/SUPPLIER
Philco Corp.

PHOTOFACT SET
483-12

PUBLISHED
1960

PHILCO

MODEL PICTURED
H-973

AC/DC operated AM
receiver

TUBES
6

POWER SUPPLY
105-125 volts AC

TUNING RANGE
535-1620KC

MFR/SUPPLIER
Philco Corp.

PHOTOFACT SET
486-20

PUBLISHED
1960

PHILCO

MODEL PICTURED
TC-47

Battery operated
transistorized portable
AM receiver with clock

TUBES
4 transistor

POWER SUPPLY
6 volts DC (radio) & 1.5
volts DC (clock)

TUNING RANGE
535-1620KC

MFR/SUPPLIER
Philco Corp.

PHOTOFACT SET
500-12

PUBLISHED
1960

PHILCO

MODEL PICTURED
H984AQ

AC/DC operated FM-AM
receiver

TUBES
8

POWER SUPPLY
105-125 volts AC/DC

TUNING RANGE
535-1620KC,
88-108MC

MFR/SUPPLIER
Philco Corp.

PHOTOFACT SET
507-13

PUBLISHED
1960

VOLUME CONTROL TONE SWITCH BAND PHONO TUNING CONTROL
ON-OFF SWITCH   BASS SWITCH

PHILHARMONIC

MODEL PICTURED
8712

AC operated phono-radio
combo AM-SW
superheterodyne receiver
with self-contained loop
antenna

TUBES
8

POWER SUPPLY
110-120 volts AC

TUNING RANGE
535-1680KC,
2.2-7.2MC,
6.9-23.5MC

MFR/SUPPLIER
Philharmonic Radio Corp.

PHOTOFACT SET
18-27

PUBLISHED
1947

PHILHARMONIC

MODEL PICTURED
100C

AC operated phono-radio
combo AM
superheterodyne receiver
with loop antenna

TUBES
7

POWER SUPPLY
110-120 volts AC

TUNING RANGE
540-1700KC

MFR/SUPPLIER
Philharmonic Radio Corp.

PHOTOFACT SET
38-16

PUBLISHED
1948

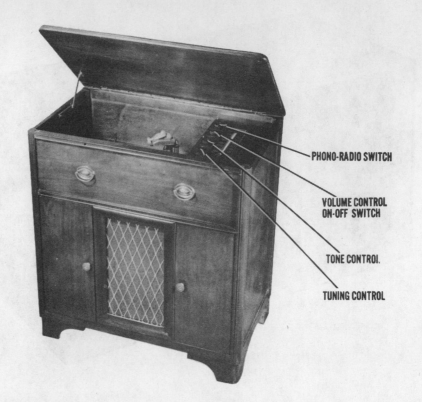

PHONO-RADIO SWITCH

VOLUME CONTROL
ON-OFF SWITCH

TONE CONTROL

TUNING CONTROL

PHILHARMONIC

MODEL PICTURED
349C

AC operated phono-radio
combo AM-FM
superheterodyne receiver
with loop antenna

TUBES
7

POWER SUPPLY
110-125 volts AC

TUNING RANGE
540-1620KC,
88-108MC

MFR/SUPPLIER
Philharmonic Radio Corp.

PHOTOFACT SET
58-17

PUBLISHED
1949

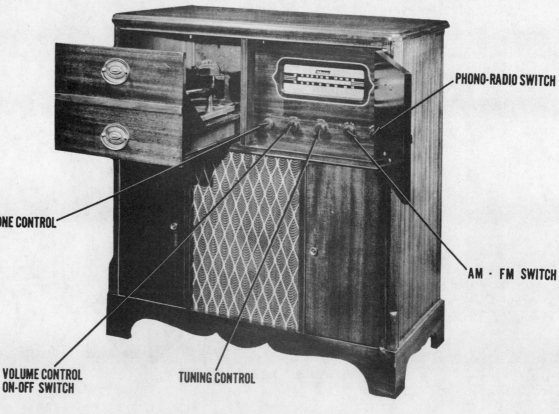

PHONO-RADIO SWITCH

TONE CONTROL

AM - FM SWITCH

VOLUME CONTROL
ON-OFF SWITCH

TUNING CONTROL

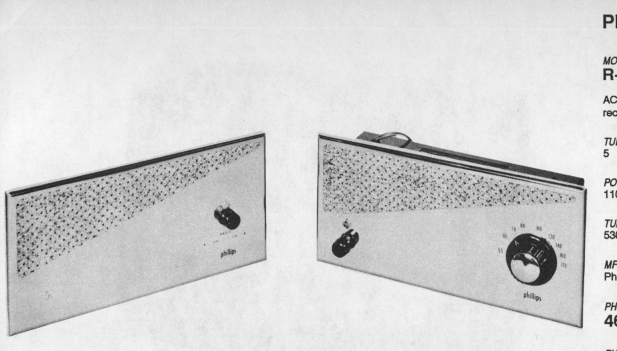

PHILLIPS

MODEL PICTURED
R-100

AC/DC operated AM
receiver

TUBES
5

POWER SUPPLY
110-120 volts AC/DC

TUNING RANGE
530-1700KC

MFR/SUPPLIER
Phillips Radio

PHOTOFACT SET
467-14

PUBLISHED
1959

PHILLIPS 66

MODEL PICTURED
3-81A

AC operated phono-radio
combo AM-FM-SW
superheterodyne receiver
with loop antenna

TUBES
14

POWER SUPPLY
110-120 volts AC

TUNING RANGE
535-1720KC,
87.4-108.7MC,
5.6-18.5MC

MFR/SUPPLIER
Phillips Petroleum Co.

PHOTOFACT SET
48-20

PUBLISHED
1948

VOLUME CONTROL
ON-OFF SWITCH

TONE SWITCH

PHONO-RADIO SWITCH

TUNING CONTROL

PILOT

MODEL PICTURED
T-511

AC/DC operated
two-band
superheterodyne receiver
with self-contained loop
antenna

TUBES
6

POWER SUPPLY
110-125 volts AC/DC

TUNING RANGE
535-1720KC,
5.67-24.0MC

MFR/SUPPLIER
Pilot Radio

PHOTOFACT SET
5-24

PUBLISHED
1946

VOLUME CONTROL
ON-OFF SWITCH

TONE CONTROL

BAND SWITCH

TUNING CONTROL

PILOT

MODEL PICTURED
T-500U

AC/DC operated
two-band
superheterodyne receiver

TUBES
5

POWER SUPPLY
105-120 volts AC/DC

TUNING RANGE
535-1720KC,
5.6-24.0MC

MFR/SUPPLIER
Pilot Radio

PHOTOFACT SET
12-23

PUBLISHED
1947

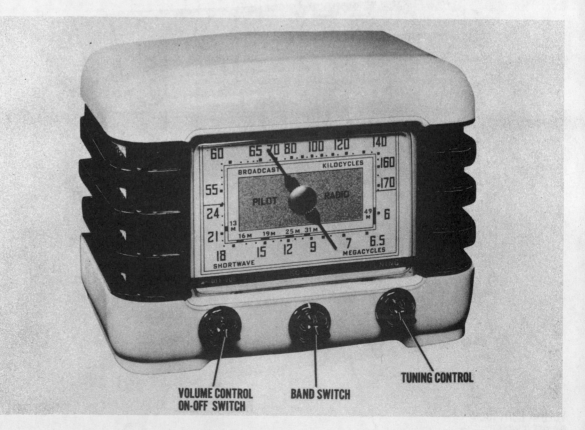

VOLUME CONTROL
ON-OFF SWITCH

BAND SWITCH

TUNING CONTROL

234

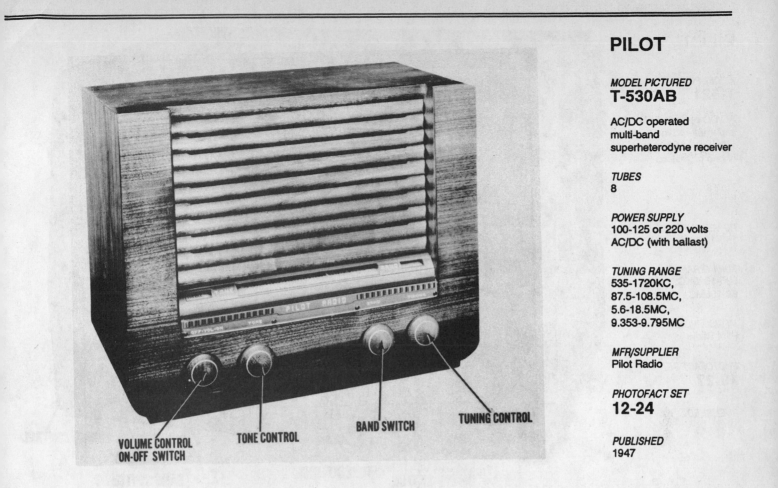

VOLUME CONTROL
ON-OFF SWITCH

TONE CONTROL

BAND SWITCH

TUNING CONTROL

PILOT

MODEL PICTURED
T-530AB

AC/DC operated
multi-band
superheterodyne receiver

TUBES
8

POWER SUPPLY
100-125 or 220 volts
AC/DC (with ballast)

TUNING RANGE
535-1720KC,
87.5-108.5MC,
5.6-18.5MC,
9.353-9.795MC

MFR/SUPPLIER
Pilot Radio

PHOTOFACT SET
12-24

PUBLISHED
1947

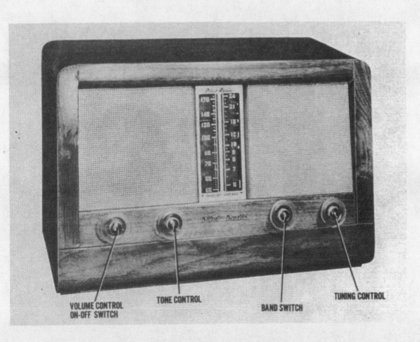

VOLUME CONTROL
ON-OFF SWITCH

TONE CONTROL

BAND SWITCH

TUNING CONTROL

PILOT

MODEL PICTURED
T-411-U

AC operated two-band
superheterodyne receiver
with self-contained loop
antenna

TUBES
6

POWER SUPPLY
105-125 volts AC

TUNING RANGE
535-1720KC,
5.67-24MC

MFR/SUPPLIER
Pilot Radio

PHOTOFACT SET
15-25

PUBLISHED
1947

PILOT

MODEL PICTURED
T-521

AC/DC operated FM-AM
superheterodyne receiver
with loop antenna and
phono provisions

TUBES
8

POWER SUPPLY
110-120 volts AC/DC

TUNING RANGE
535-1620KC,
88-108MC

MFR/SUPPLIER
Pilot Radio

PHOTOFACT SET
19-27

PUBLISHED
1947

**VOLUME CONTROL
ON-OFF SWITCH**   **TONE CONTROL**   **PHONO-
BAND SWITCH**   **TUNING CONTROL**

PILOT

MODEL PICTURED
T-601 Pilotuner

AC operated FM
superheterodyne receiver

TUBES
5

POWER SUPPLY
110-120 volts AC

TUNING RANGE
88-108MC

MFR/SUPPLIER
Pilot Radio

PHOTOFACT SET
28-26

PUBLISHED
1947

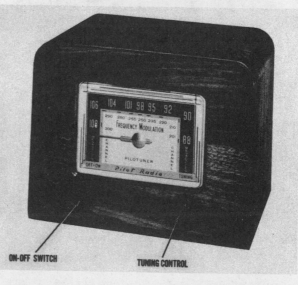

ON-OFF SWITCH   **TUNING CONTROL**

PILOT

VOLUME CONTROL
ON-OFF SWITCH   TONE CONTROL   PHONO- BAND SWITCH   TUNING CONTROL

MODEL PICTURED
T-741

AC/DC operated
multi-band
superheterodyne receiver

TUBES
7

POWER SUPPLY
105-125 or 220-240 volts
AC/DC

TUNING RANGE
5 bands

MFR/SUPPLIER
Pilot Radio

PHOTOFACT SET
37-18

PUBLISHED
1948

PILOT

MODEL PICTURED
PT-1036

AC operated FM-AM
receiver with four-speed
automatic record changer

TUBES
14

POWER SUPPLY
105-120 volts AC

TUNING RANGE
535-1720KC,
88-108MC

MFR/SUPPLIER
Pilot Radio

PHOTOFACT SET
421-12

PUBLISHED
1958

PILOT

MODEL PICTURED
PT-1031

AC operated FM-AM
receiver with four-speed
automatic record changer

TUBES
11

POWER SUPPLY
105-125 volts AC

TUNING RANGE
535-1700KC,
88-108MC

MFR/SUPPLIER
Pilot Radio

PHOTOFACT SET
428-11

PUBLISHED
1959

POLICALARM

MODEL PICTURED
PR-31

AC/DC operated FM
superheterodyne receiver

TUBES
6

POWER SUPPLY
110-120 volts AC

TUNING RANGE
30-50MC

MFR/SUPPLIER
Radio Apparatus Corp.

PHOTOFACT SET
105-8

PUBLISHED
1950

VOLUME CONTROL
ON-OFF SWITCH

TUNING CONTROL

POLICALARM

MODEL PICTURED
PR-31A

AC/DC operated FM
superheterodyne receiver

TUBES
6

POWER SUPPLY
110-120 volts AC/DC

TUNING RANGE
30-50MC

MFR/SUPPLIER
Monitoradio, Div. I.D.E.A.

PHOTOFACT SET
295-10

PUBLISHED
1955

**ON-OFF
SWITCH
VOLUME
CONTROL**

TUNING

POLICALARM

MODEL PICTURED
PR-9

AC/DC operated FM
receiver

TUBES
6

POWER SUPPLY
110-120 volts AC/DC

TUNING RANGE
152-174MC

MFR/SUPPLIER
Monitoradio, Div. I.D.E.A.

PHOTOFACT SET
359-12

PUBLISHED
1957

PORTO
BARADIO

MODEL PICTURED
PB-520

AC/DC operated AM
superheterodyne receiver
with loop antenna

TUBES
5

POWER SUPPLY
110-120 volts AC/DC

TUNING RANGE
540-1780KC

MFR/SUPPLIER
Porto Products

PHOTOFACT SET
33-16

PUBLISHED
1948

**VOLUME CONTROL
ON-OFF SWITCH**　　　　**TUNING CONTROL**

PORTO
BARADIO

MODEL PICTURED
**PB-520
(Revised)**

AC/DC operated AM
superheterodyne receiver
with loop antenna

TUBES
5

POWER SUPPLY
110-120 volts AC/DC

TUNING RANGE
540-1670KC

MFR/SUPPLIER
Porto Products

PHOTOFACT SET
48-21

PUBLISHED
1948

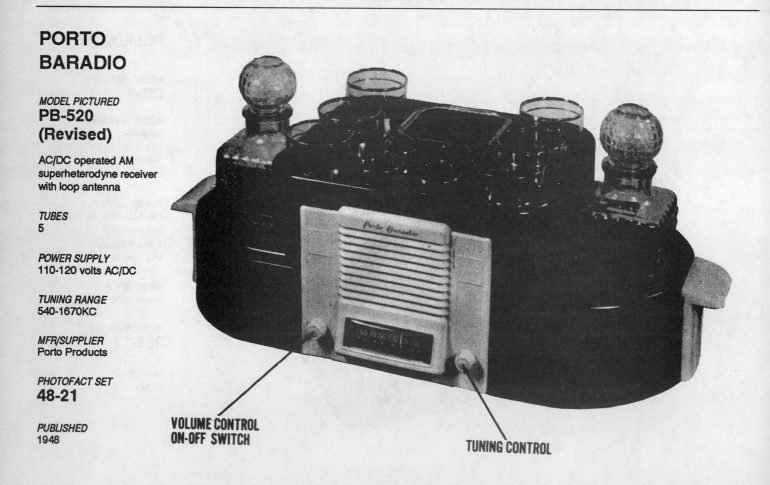

**VOLUME CONTROL
ON-OFF SWITCH**　　　　**TUNING CONTROL**

PREMIER

MODEL PICTURED
15LW

AC/DC operated AM
superheterodyne receiver
with self-contained loop
antenna

TUBES
5

POWER SUPPLY
117 volts AC/DC

TUNING RANGE
535-1720KC

MFR/SUPPLIER
Premier Crystal
Laboratories Inc.

PHOTOFACT SET
6-24

PUBLISHED
1946

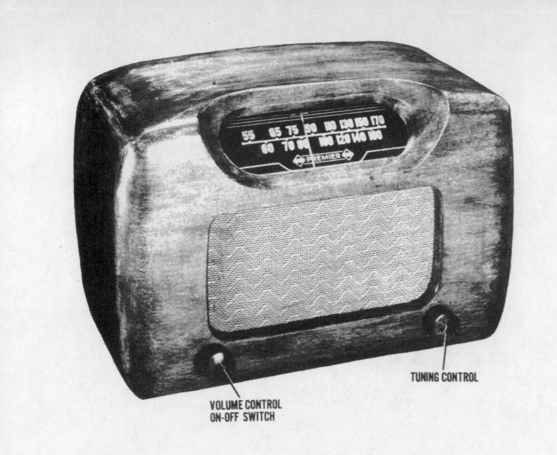

TUNING CONTROL

VOLUME CONTROL
ON-OFF SWITCH

PURITAN

MODEL PICTURED
501

AC/DC operated AM
superheterodyne receiver
with self-contained loop
antenna

TUBES
5

POWER SUPPLY
117 volts AC/DC

TUNING RANGE
540-1600KC

MFR/SUPPLIER
Pure Oil Co.

PHOTOFACT SET
4-5

PUBLISHED
1946

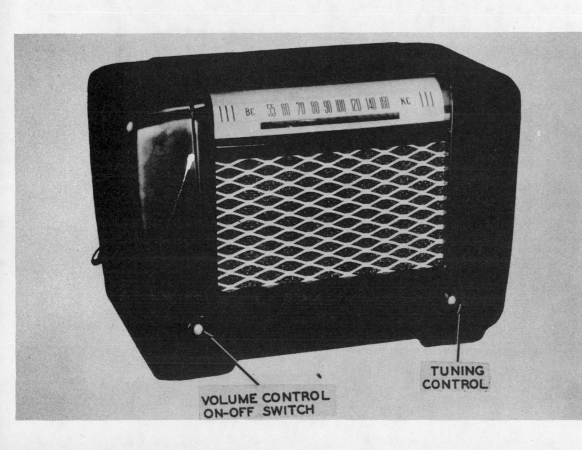

TUNING
CONTROL

VOLUME CONTROL
ON-OFF SWITCH

PURITAN

MODEL PICTURED
502X

AC/DC operated AM
superheterodyne receiver
with self-contained loop
antenna

TUBES
5

POWER SUPPLY
117 volts AC/DC

TUNING RANGE
540-1610KC

MFR/SUPPLIER
Pure Oil Co.

PHOTOFACT SET
4-26

PUBLISHED
1946

VOLUME CONTROL
ON-OFF SWITCH

TUNING
CONTROL

PURITAN

MODEL PICTURED
508

AC operated three-band
superheterodyne receiver
with loop antenna and
five-button auto tuning

TUBES
7

POWER SUPPLY
105-125 volts AC

TUNING RANGE
540-1725KC,
11.4-15.5MC,
5.9-10MC

MFR/SUPPLIER
Pure Oil Co.

PHOTOFACT SET
4-31

PUBLISHED
1946

TONE
CONTROL

VOLUME CONTROL
ON-OFF SWITCH

TUNING
CONTROL

BAND
SWITCH

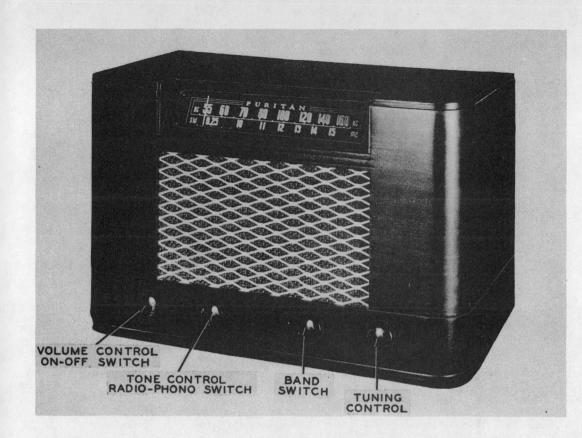

VOLUME CONTROL
ON-OFF SWITCH

TONE CONTROL
RADIO-PHONO SWITCH

BAND
SWITCH

TUNING
CONTROL

PURITAN

MODEL PICTURED
504

AC operated two-band
superheterodyne receiver
with self-contained loop
antenna

TUBES
6

POWER SUPPLY
117 volts AC

TUNING RANGE
540-1600KC,
9-15.5MC

MFR/SUPPLIER
Pure Oil Co.

PHOTOFACT SET
5-39

PUBLISHED
1946

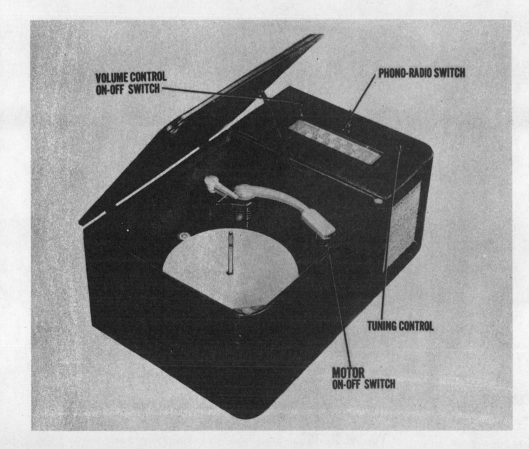

VOLUME CONTROL
ON-OFF SWITCH

PHONO-RADIO SWITCH

TUNING CONTROL

MOTOR
ON-OFF SWITCH

PURITAN

MODEL PICTURED
503

AC operated phono-radio
combo AM
superheterodyne receiver
with self-contained loop
antenna

TUBES
5

POWER SUPPLY
105-125 volts AC

TUNING RANGE
540-1600KC

MFR/SUPPLIER
Pure Oil Co.

PHOTOFACT SET
10-25

PUBLISHED
1946

PURITAN

MODEL PICTURED
509

Three-power operated
portable AM
superheterodyne receiver
with loop antenna

TUBES
6

POWER SUPPLY
105-125 volts AC/DC or 9
volts A & 90 volts B
supply

TUNING RANGE
532-1700KC

MFR/SUPPLIER
Pure Oil Co.

PHOTOFACT SET
26-21

PUBLISHED
1947

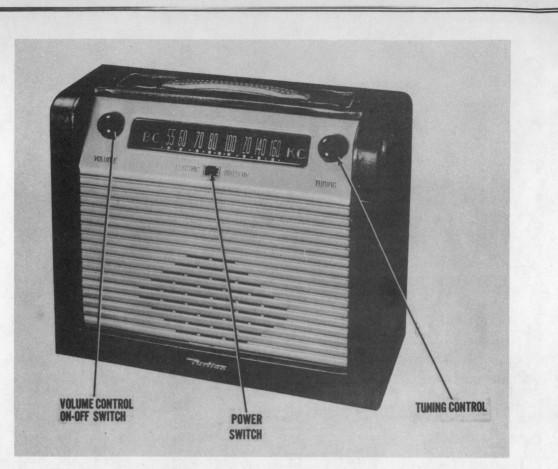

VOLUME CONTROL
ON-OFF SWITCH

POWER
SWITCH

TUNING CONTROL

RADIOETTE

MODEL PICTURED
PR-2

Three-power operated
portable AM
superheterodyne receiver
with loop antenna

TUBES
4

POWER SUPPLY
105-120 volts AC/DC or
4.5 volts A & 67.5 volts B
supply

TUNING RANGE
540-1650KC

MFR/SUPPLIER
Alamo Electronics

PHOTOFACT SET
50-15

PUBLISHED
1948

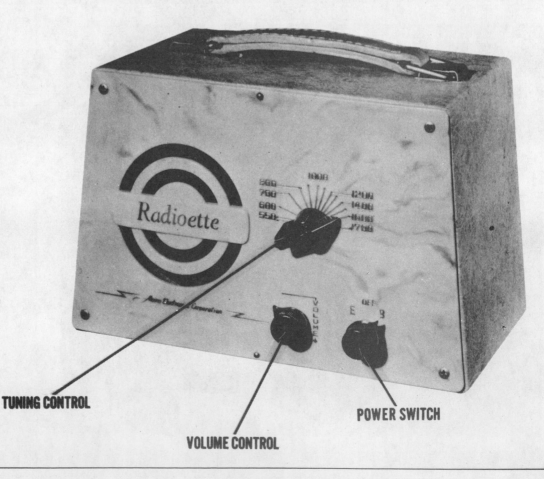

TUNING CONTROL

VOLUME CONTROL

POWER SWITCH

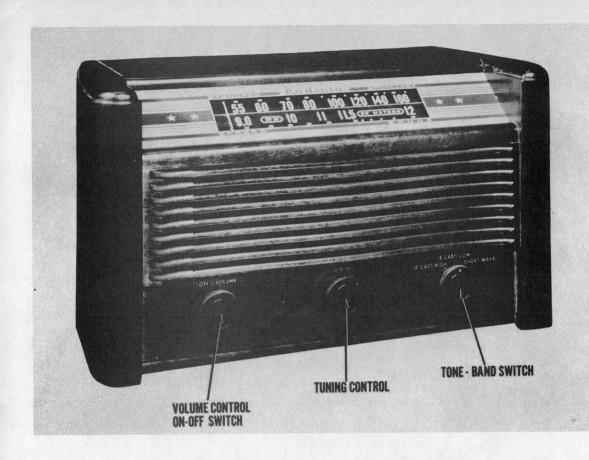

VOLUME CONTROL
ON-OFF SWITCH

TUNING CONTROL

TONE - BAND SWITCH

RADIOLA

MODEL PICTURED
61-5

AC/DC operated
two-band
superheterodyne receiver
with self-contained loop
antenna

TUBES
6

POWER SUPPLY
105-125 volts AC/DC

TUNING RANGE
540-1600KC,
9-11.8MC

MFR/SUPPLIER
RCA Home Instrument
Division

PHOTOFACT SET
12-25

PUBLISHED
1947

VOLUME CONTROL
ON-OFF SWITCH

TUNING CONTROL

TONE SWITCH

RADIOLA

MODEL PICTURED
61-1

AC/DC operated AM
superheterodyne receiver
with self-contained loop
antenna

TUBES
6

POWER SUPPLY
105-125 volts AC/DC

TUNING RANGE
540-1600KC

MFR/SUPPLIER
RCA Home Instrument
Division

PHOTOFACT SET
14-25

PUBLISHED
1947

RADIOLA

MODEL PICTURED
61-8, 61-9

AC/DC operated AM
superheterodyne receiver
with loop antenna

TUBES
5

POWER SUPPLY
105-125 volts AC/DC

TUNING RANGE
540-1600KC

MFR/SUPPLIER
RCA Home Instrument
Division

PHOTOFACT SET
27-21

PUBLISHED
1947

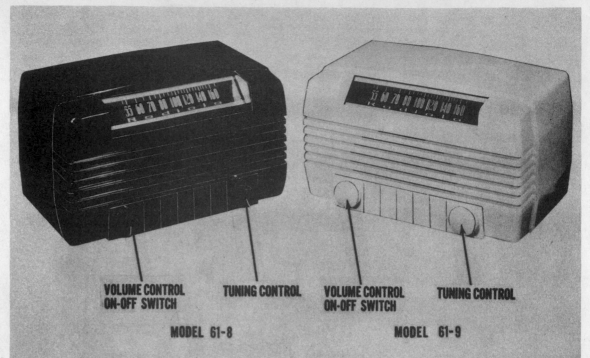

VOLUME CONTROL
ON-OFF SWITCH   TUNING CONTROL   VOLUME CONTROL
ON-OFF SWITCH   TUNING CONTROL

MODEL 61-8   MODEL 61-9

RADIOLA

MODEL PICTURED
75ZU

AC operated phono-radio
combo AM
superheterodyne receiver
with loop antenna

TUBES
5

POWER SUPPLY
105-125 volts AC

TUNING RANGE
540-1600KC

MFR/SUPPLIER
RCA Home Instrument
Division

PHOTOFACT SET
36-19

PUBLISHED
1948

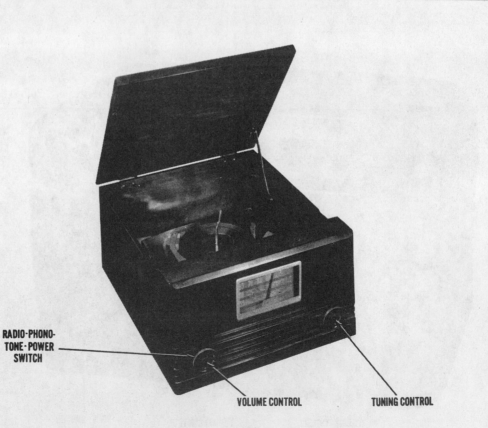

RADIO-PHONO-
TONE-POWER
SWITCH

VOLUME CONTROL   TUNING CONTROL

RADIOLA

MODEL PICTURED
76ZX11

AC/DC operated AM
superheterodyne receiver
with loop antenna

TUBES
6

POWER SUPPLY
105-125 volts AC/DC

TUNING RANGE
540-1600KC

MFR/SUPPLIER
RCA Home Instrument
Division

PHOTOFACT SET
36-20

PUBLISHED
1948

VOLUME CONTROL
ON-OFF SWITCH

TUNING CONTROL

VOLUME CONTROL
ON-OFF SWITCH

TONE
SWITCH

BAND SWITCH

TUNING CONTROL

RADIONIC

MODEL PICTURED
Y62W, Y728

AC/DC operated
two-band
superheterodyne receiver
with self-contained loop
antenna

TUBES
6

POWER SUPPLY
105-125 volts AC/DC

TUNING RANGE
535-1700KC,
5.3-18MC

MFR/SUPPLIER
Radionic

PHOTOFACT SET
26-22

PUBLISHED
1947

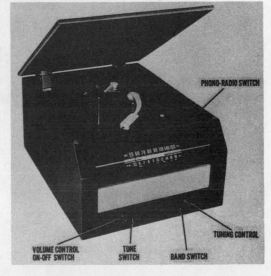

PHONO-RADIO SWITCH

VOLUME CONTROL
ON-OFF SWITCH

TONE
SWITCH

BAND SWITCH

TUNING CONTROL

RANGER

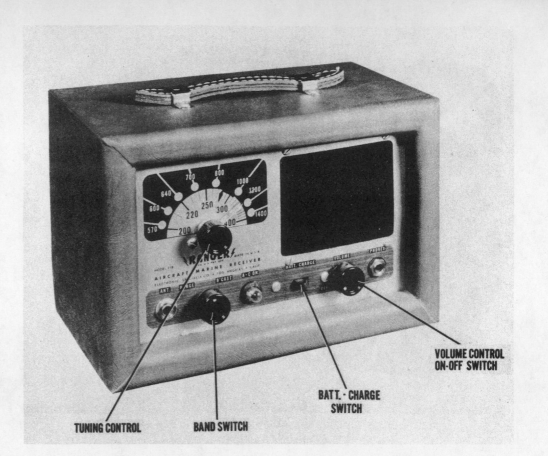

MODEL PICTURED
118

Three-power operated portable two-band superheterodyne receiver with loop antenna

TUBES
5

POWER SUPPLY
110-120 volts AC/DC or 4.5 volts A & 67.5 volts B supply

TUNING RANGE
550-1550KC, 195-410KC

MFR/SUPPLIER
Electronic Specialty

PHOTOFACT SET
28-27

PUBLISHED
1947

VOLUME CONTROL
ON-OFF SWITCH

BATT. - CHARGE
SWITCH

TUNING CONTROL

BAND SWITCH

RAYENERGY

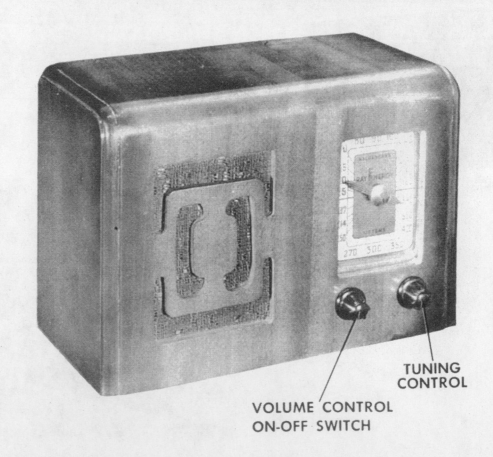

MODEL PICTURED
AD

AC/DC operated superheterodyne receiver with self-contained loop antenna

TUBES
5

POWER SUPPLY
117 volts AC

TUNING RANGE
530-1670KC

MFR/SUPPLIER
RayEnergy Radio & Television Corp.

PHOTOFACT SET
7-24

PUBLISHED
1946

TUNING
CONTROL

VOLUME CONTROL
ON-OFF SWITCH

RAYENERGY

MODEL PICTURED
AD4

AC/DC operated AM
superheterodyne receiver

TUBES
5

POWER SUPPLY
117 volts AC/DC

TUNING RANGE
550-1650KC

MFR/SUPPLIER
RayEnergy Radio &
Television Corp.

PHOTOFACT SET
7-25

PUBLISHED
1946

VOLUME CONTROL
ON-OFF SWITCH

TUNING CONTROL

RAYENERGY

MODEL PICTURED
SRB-1X

AC/DC operated AM
superheterodyne receiver
with self-contained loop
antenna

TUBES
5

POWER SUPPLY
110-120 volts AC/DC

TUNING RANGE
540-1650KC

MFR/SUPPLIER
RayEnergy Radio &
Television Corp.

PHOTOFACT SET
13-26

PUBLISHED
1947

VOLUME CONTROL
ON-OFF SWITCH

TUNING CONTROL

RAYTHEON

MODEL PICTURED
CR-41

AC operated AM
superheterodyne receiver
with electric clock

TUBES
4

POWER SUPPLY
115 volts AC

TUNING RANGE
540-1600KC

MFR/SUPPLIER
Raytheon Television &
Radio Corp.

PHOTOFACT SET
212-5

PUBLISHED
1953

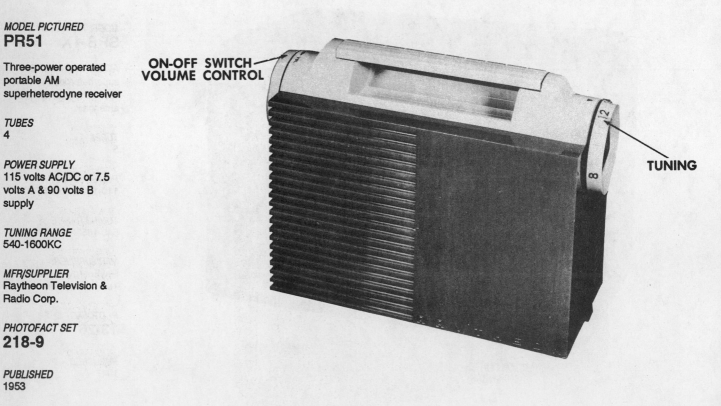

AUTO
ON-OFF
SWITCH

SLEEP
SWITCH

ALARM
ON-OFF

TUNING CONTROL

VOLUME CONTROL

RAYTHEON

MODEL PICTURED
PR51

Three-power operated
portable AM
superheterodyne receiver

TUBES
4

POWER SUPPLY
115 volts AC/DC or 7.5
volts A & 90 volts B
supply

TUNING RANGE
540-1600KC

MFR/SUPPLIER
Raytheon Television &
Radio Corp.

PHOTOFACT SET
218-9

PUBLISHED
1953

ON-OFF SWITCH
VOLUME CONTROL

TUNING

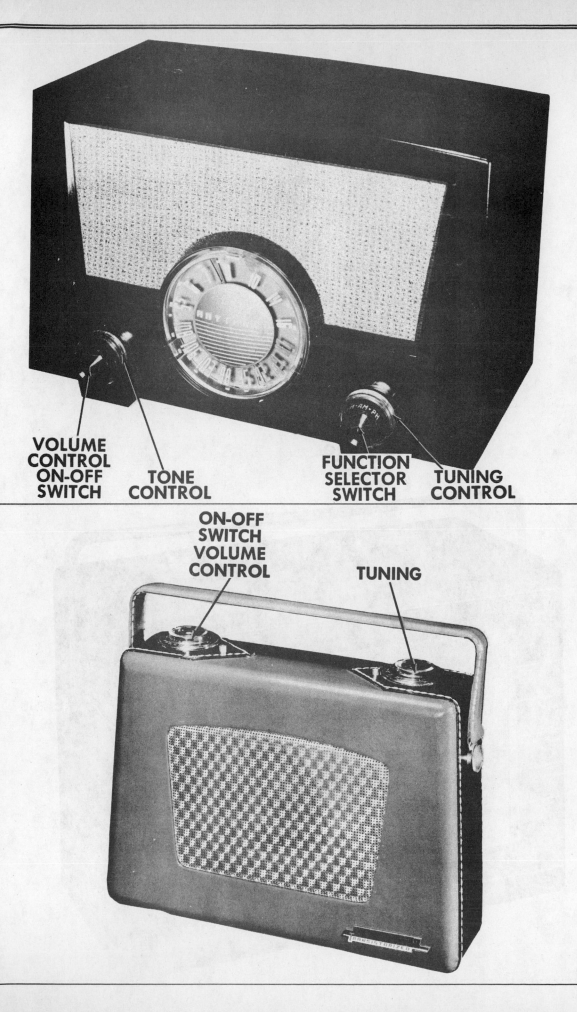

VOLUME
CONTROL
ON-OFF
SWITCH

TONE
CONTROL

FUNCTION
SELECTOR
SWITCH

TUNING
CONTROL

ON-OFF
SWITCH
VOLUME
CONTROL

TUNING

RAYTHEON

MODEL PICTURED
FR81A

AC operated AM-FM
superheterodyne receiver

TUBES
9

POWER SUPPLY
115 volts AC

TUNING RANGE
540-1600KC,
88-108MC

MFR/SUPPLIER
Raytheon Television &
Radio Corp.

PHOTOFACT SET
232-6

PUBLISHED
1954

RAYTHEON

MODEL PICTURED
8TP-1

Battery operated portable
transistorized AM
superheterodyne receiver

TUBES
8 transistor

POWER SUPPLY
6 volts DC

TUNING RANGE
540-1600KC

MFR/SUPPLIER
Raytheon Television &
Radio Corp.

PHOTOFACT SET
292-9

PUBLISHED
1955

RAYTHEON

MODEL PICTURED
5R-10B

AC/DC operated AM
superheterodyne receiver

TUBES
5

POWER SUPPLY
115 volts AC/DC

TUNING RANGE
540-1600KC

MFR/SUPPLIER
Raytheon Television &
Radio Corp.

PHOTOFACT SET
303-11

PUBLISHED
1956

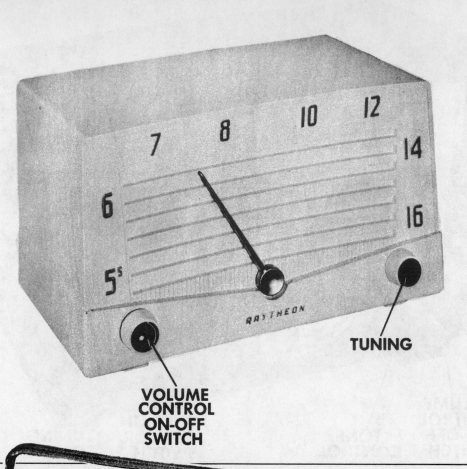

**VOLUME
CONTROL
ON-OFF
SWITCH**

TUNING

RAYTHEON

MODEL PICTURED
T2500

Battery operated portable
transistorized AM
superheterodyne receiver

TUBES
7 transistor

POWER SUPPLY
6 volts A battery or four
1.5 volt batteries

TUNING RANGE
540-1600KC

MFR/SUPPLIER
Raytheon Television &
Radio Corp.

PHOTOFACT SET
329-11

PUBLISHED
1956

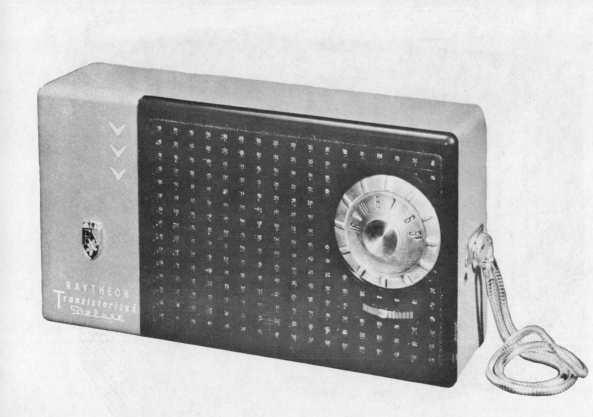

RAYTHEON

MODEL PICTURED
T-100-1

Battery operated portable
transistorized AM
superheterodyne receiver

TUBES
4 transistor

POWER SUPPLY
9 volts DC

TUNING RANGE
540-1600KC

MFR/SUPPLIER
Raytheon Television &
Radio Corp.

PHOTOFACT SET
331-11

PUBLISHED
1956

RAYTHEON

MODEL PICTURED
T-150-1

Battery operated portable
transistorized AM
superheterodyne receiver

TUBES
6 transistor

POWER SUPPLY
9 volts DC

TUNING RANGE
540-1600KC

MFR/SUPPLIER
Raytheon Television &
Radio Corp.

PHOTOFACT SET
335-13

PUBLISHED
1956

RCA VICTOR

MODEL PICTURED
55F

Battery operated AM
superheterodyne receiver

TUBES
5

POWER SUPPLY
1.5 volts A & 90 volts B
supply

TUNING RANGE
540-1720KC

MFR/SUPPLIER
RCA Victor

PHOTOFACT SET
4-6

PUBLISHED
1946

POWER
SWITCH

VOLUME CONTROL
ON-OFF SWITCH

TUNING
CONTROL

RCA VICTOR

MODEL PICTURED
64F3

Battery operated AM
superheterodyne receiver

TUBES
4

POWER SUPPLY
1.5 volts A & 90 volts B
supply

TUNING RANGE
540-1600KC

MFR/SUPPLIER
RCA Victor

PHOTOFACT SET
4-16

PUBLISHED
1946

VOLUME CONTROL
ON-OFF SWITCH

TUNING
CONTROL

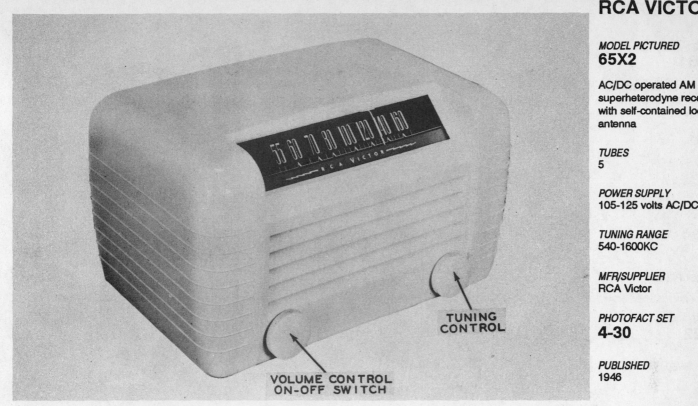

VOLUME CONTROL
ON-OFF SWITCH

TUNING
CONTROL

RCA VICTOR

MODEL PICTURED
65X2

AC/DC operated AM
superheterodyne receiver
with self-contained loop
antenna

TUBES
5

POWER SUPPLY
105-125 volts AC/DC

TUNING RANGE
540-1600KC

MFR/SUPPLIER
RCA Victor

PHOTOFACT SET
4-30

PUBLISHED
1946

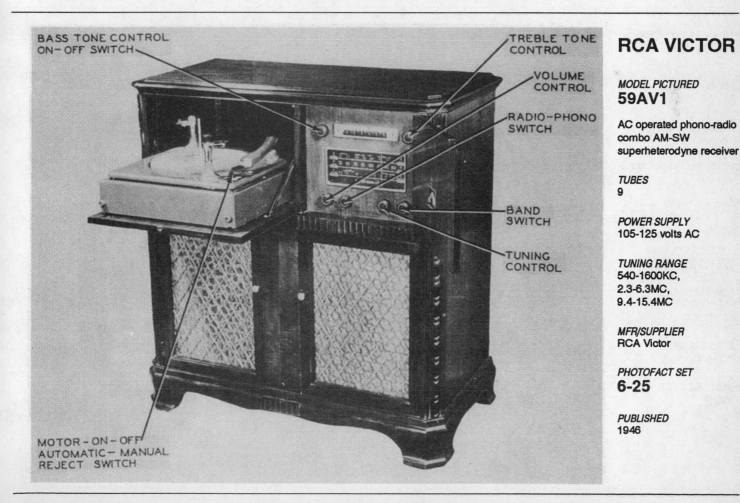

BASS TONE CONTROL
ON-OFF SWITCH

TREBLE TONE
CONTROL

VOLUME
CONTROL

RADIO-PHONO
SWITCH

BAND
SWITCH

TUNING
CONTROL

MOTOR-ON-OFF
AUTOMATIC-MANUAL
REJECT SWITCH

RCA VICTOR

MODEL PICTURED
59AV1

AC operated phono-radio
combo AM-SW
superheterodyne receiver

TUBES
9

POWER SUPPLY
105-125 volts AC

TUNING RANGE
540-1600KC,
2.3-6.3MC,
9.4-15.4MC

MFR/SUPPLIER
RCA Victor

PHOTOFACT SET
6-25

PUBLISHED
1946

RCA VICTOR

MODEL PICTURED
54B1

Battery operated portable AM superheterodyne receiver

TUBES
4

POWER SUPPLY
1.5 volts A & 67.5 volts B supply

TUNING RANGE
550-1600KC

MFR/SUPPLIER
RCA Victor

PHOTOFACT SET
7-22

PUBLISHED
1946

ON-OFF SWITCH   VOLUME CONTROL

LOOP ANT.(28)   TUNING CONTROL

RCA VICTOR

MODEL PICTURED
66X1

AC/DC operated two-band superheterodyne receiver with self-contained loop antenna

TUBES
6

POWER SUPPLY
105-125 volts AC/DC

TUNING RANGE
540-1600KC, 9.0-12.0MC

MFR/SUPPLIER
RCA Victor

PHOTOFACT SET
7-23

PUBLISHED
1946

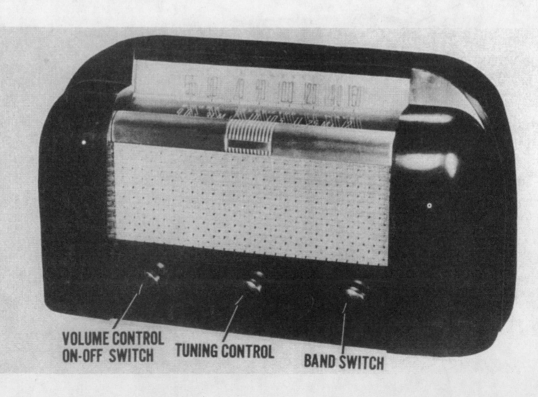

VOLUME CONTROL
ON-OFF SWITCH   TUNING CONTROL   BAND SWITCH

RCA VICTOR

MODEL PICTURED
67AV1

AC operated phono-radio combo two-band superheterodyne receiver with self-contained loop antenna

TUBES
7

POWER SUPPLY
105-125 volts AC

TUNING RANGE
540-1600KC, 9.2-16.0MC

MFR/SUPPLIER
RCA Victor

PHOTOFACT SET
9-27

PUBLISHED
1946

TUNING CONTROL

BAND-PHONO SWITCH

TONE CONTROL

VOLUME CONTROL ON-OFF SWITCH

RCA VICTOR

MODEL PICTURED
65U

AC operated phono-radio combo AM superheterodyne receiver with loop antenna

TUBES
5

POWER SUPPLY
105-125 volts AC

TUNING RANGE
540-1600KC

MFR/SUPPLIER
RCA Victor

PHOTOFACT SET
14-23

PUBLISHED
1947

ON-OFF PHONO-RADIO-TONE SWITCH

VOLUME CONTROL

TUNING CONTROL

RCA VICTOR

MODEL PICTURED
66BX

Three-power operated portable AM superheterodyne receiver with self-contained loop antenna

TUBES
6

POWER SUPPLY
110-125 volts AC/DC or 9 volts A & 90 volts B supply

TUNING RANGE
540-1600KC

MFR/SUPPLIER
RCA Victor

PHOTOFACT SET
14-24

PUBLISHED
1947

VOLUME CONTROL   TUNING CONTROL

RCA VICTOR

MODEL PICTURED
54B5

Battery operated portable AM superheterodyne receiver with self-contained loop antenna

TUBES
4

POWER SUPPLY
1.5 volts A & 67.5 volts B supply

TUNING RANGE
550-1600KC

MFR/SUPPLIER
RCA Victor

PHOTOFACT SET
17-25

PUBLISHED
1947

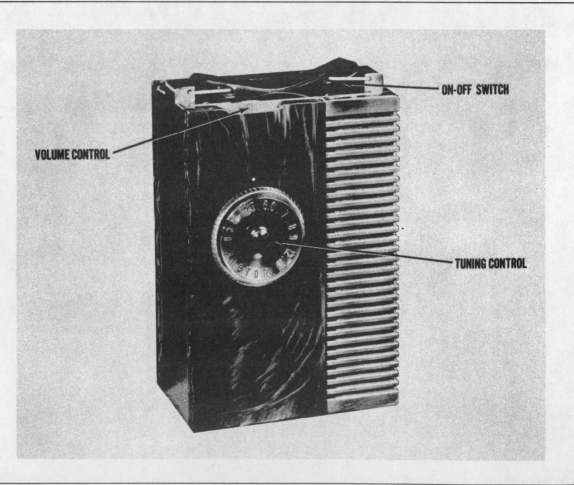

ON-OFF SWITCH

VOLUME CONTROL

TUNING CONTROL

PHONO MOTOR
ON-OFF SWITCH

TUNING
CONTROL

AUTO.- MAN.
START- REJECT
SWITCH

VOLUME
CONTROL

BAND- PHONO
SWITCH

ON - OFF
SWITCH

BASS TONE
CONTROL

TREBLE TONE
CONTROL

RCA VICTOR

MODEL PICTURED
612V3

AC operated phono-radio combo AM-FM superheterodyne receiver with self-contained loop antenna

TUBES
12

POWER SUPPLY
110-120 volts AC

TUNING RANGE
540-1600KC,
88-108MC,
9.2-16MC

MFR/SUPPLIER
RCA Victor

PHOTOFACT SET
17-27

PUBLISHED
1947

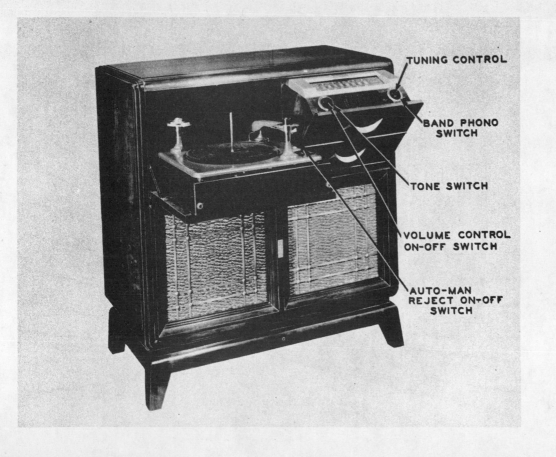

TUNING CONTROL

BAND PHONO
SWITCH

TONE SWITCH

VOLUME CONTROL
ON-OFF SWITCH

AUTO-MAN
REJECT ON-OFF
SWITCH

RCA VICTOR

MODEL PICTURED
711V2

AC operated phono-radio combo FM-AM superheterodyne receiver with loop antenna

TUBES
11

POWER SUPPLY
110-120 volts AC

TUNING RANGE
540-1600KC,
88-108MC,
9.2-16.0MC

MFR/SUPPLIER
RCA Victor

PHOTOFACT SET
22-24

PUBLISHED
1947

RCA VICTOR

MODEL PICTURED
65BR9

AC/Battery operated portable AM superheterodyne receiver with loop antenna and self-contained battery

TUBES
5

POWER SUPPLY
110-120 volts AC or two volt battery

TUNING RANGE
540-1600KC

MFR/SUPPLIER
RCA Victor

PHOTOFACT SET
23-16

PUBLISHED
1947

VOLUME CONTROL   TUNING CONTROL   POWER SWITCH

RCA VICTOR

MODEL PICTURED
68R1

AC operated AM-FM superheterodyne receiver with loop antenna

TUBES
8

POWER SUPPLY
105-125 volts AC

TUNING RANGE
540-1600KC,
88-108MC

MFR/SUPPLIER
RCA Victor

PHOTOFACT SET
23-17

PUBLISHED
1947

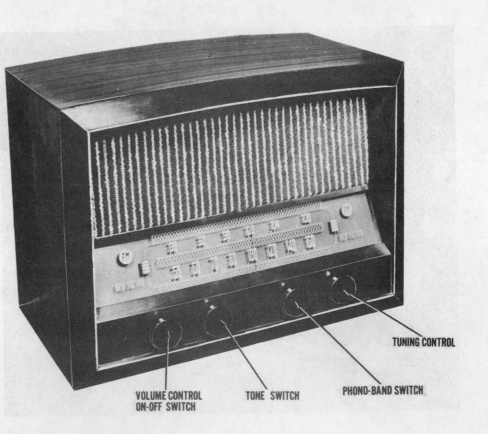

VOLUME CONTROL ON-OFF SWITCH   TONE SWITCH   PHONO-BAND SWITCH   TUNING CONTROL

RCA VICTOR

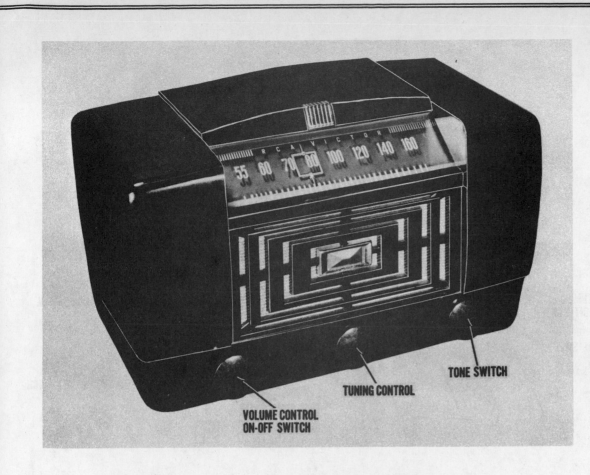

TONE SWITCH

TUNING CONTROL

**VOLUME CONTROL
ON-OFF SWITCH**

MODEL PICTURED
66X11

AC/DC operated AM
superheterodyne receiver
with loop antenna

TUBES
6

POWER SUPPLY
105-125 volts AC/DC

TUNING RANGE
540-1600KC

MFR/SUPPLIER
RCA Victor

PHOTOFACT SET
27-20

PUBLISHED
1947

RCA VICTOR

**VOLUME CONTROL
ON-OFF SWITCH**

TUNING CONTROL

MODEL PICTURED
65X1

AC/DC operated AM
superheterodyne receiver
with loop antenna

TUBES
5

POWER SUPPLY
105-125 volts AC/DC

TUNING RANGE
540-1600KC

MFR/SUPPLIER
RCA Victor

PHOTOFACT SET
31-26

PUBLISHED
1948

RCA VICTOR

MODEL PICTURED
610V2

AC operated phono-radio combo FM-AM superheterodyne receiver with loop antenna

TUBES
10

POWER SUPPLY
105-125 volts AC

TUNING RANGE
540-1600KC, 88-108MC

MFR/SUPPLIER
RCA Victor

PHOTOFACT SET
31-27

PUBLISHED
1948

VOLUME CONTROL
ON-OFF SWITCH

TONE SWITCH

TUNING CONTROL

PHONO- BAND SWITCH

RCA VICTOR

MODEL PICTURED
75X11

AC/DC operated AM superheterodyne receiver with loop antenna

TUBES
5

POWER SUPPLY
105-125 volts AC/DC

TUNING RANGE
540-1600KC

MFR/SUPPLIER
RCA Victor

PHOTOFACT SET
33-21

PUBLISHED
1948

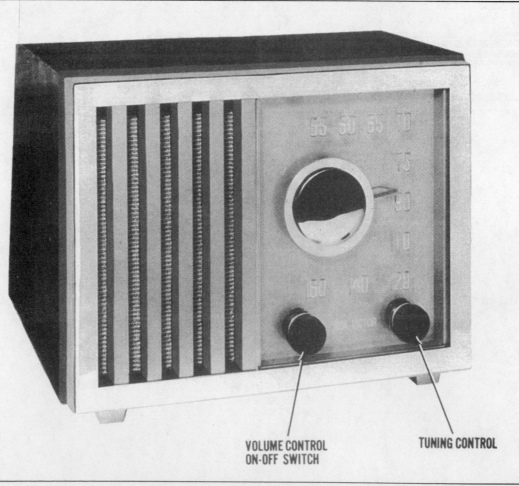

VOLUME CONTROL
ON-OFF SWITCH

TUNING CONTROL

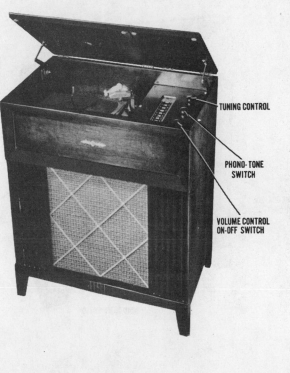

RCA VICTOR

MODEL PICTURED
77U

AC operated phono-radio combo AM superheterodyne receiver with loop antenna

TUBES
6

POWER SUPPLY
105-125 volts AC

TUNING RANGE
540-1600KC

MFR/SUPPLIER
RCA Victor

PHOTOFACT SET
38-17

PUBLISHED
1948

VOLUME CONTROL

PHONO-TONE-POWER SWITCH

TUNING CONTROL

RCA VICTOR

MODEL PICTURED
77V1

AC operated phono-radio combo AM superheterodyne receiver with loop antenna

TUBES
7

POWER SUPPLY
105-125 volts AC

TUNING RANGE
540-1600KC

MFR/SUPPLIER
RCA Victor

PHOTOFACT SET
38-18

PUBLISHED
1948

TUNING CONTROL

PHONO-TONE SWITCH

VOLUME CONTROL ON-OFF SWITCH

RCA VICTOR

MODEL PICTURED
8X53

AC/DC operated AM
superheterodyne receiver
with loop antenna

TUBES
5

POWER SUPPLY
105-125 volts AC/DC

TUNING RANGE
540-1600KC

MFR/SUPPLIER
RCA Victor

PHOTOFACT SET
39-17

PUBLISHED
1948

TUNING CONTROL

**VOLUME CONTROL
ON-OFF SWITCH**

RCA VICTOR

MODEL PICTURED
77V2

AC operated phono-radio
combo two-band
superheterodyne receiver
with loop antenna

TUBES
7

POWER SUPPLY
105-125 volts AC

TUNING RANGE
540-1600KC,
9.2-16MC

MFR/SUPPLIER
RCA Victor

PHOTOFACT SET
39-18

PUBLISHED
1948

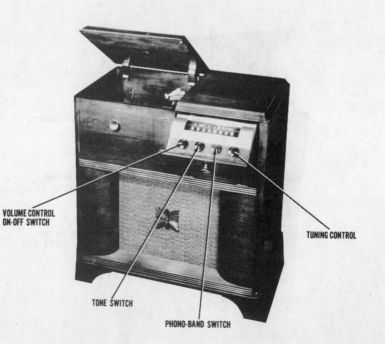

**VOLUME CONTROL
ON-OFF SWITCH**

TUNING CONTROL

TONE SWITCH

PHONO-BAND SWITCH

RCA VICTOR

MODEL PICTURED
710V2

AC operated phono-radio
combo AM-FM
superheterodyne receiver

TUBES
10

POWER SUPPLY
105-125 volts AC

TUNING RANGE
540-1600KC,
88-108MC

MFR/SUPPLIER
RCA Victor

PHOTOFACT SET
40-15

PUBLISHED
1948

VOLUME CONTROL
ON-OFF SWITCH

TONE SWITCH

BAND SWITCH

TUNING CONTROL

RCA VICTOR

MODEL PICTURED
8BX5

Three-power operated
portable AM
superheterodyne receiver
with loop antenna

TUBES
5

POWER SUPPLY
110-125 volts AC/DC or
7.5 volts A & 75 volts B
supply

TUNING RANGE
540-1600KC

MFR/SUPPLIER
RCA Victor

PHOTOFACT SET
46-20

PUBLISHED
1948

TUNING CONTROL

VOLUME CONTROL
ON-OFF SWITCH

RCA VICTOR

MODEL PICTURED
8X521

AC/DC operated AM
superheterodyne receiver
with loop antenna

TUBES
5

POWER SUPPLY
110-120 volts AC/DC

TUNING RANGE
540-1600KC

MFR/SUPPLIER
RCA Victor

PHOTOFACT SET
52-17

PUBLISHED
1948

TUNING CONTROL

VOLUME CONTROL
ON-OFF SWITCH

RCA VICTOR

MODEL PICTURED
8R71

AC operated AM-FM
superheterodyne receiver
with loop antenna

TUBES
7

POWER SUPPLY
110-120 volts AC

TUNING RANGE
540-1600KC,
88-108MC

MFR/SUPPLIER
RCA Victor

PHOTOFACT SET
53-20

PUBLISHED
1949

VOLUME CONTROL
ON-OFF SWITCH

TONE SWITCH   TUNING CONTROL   PHONO-BAND SWITCH

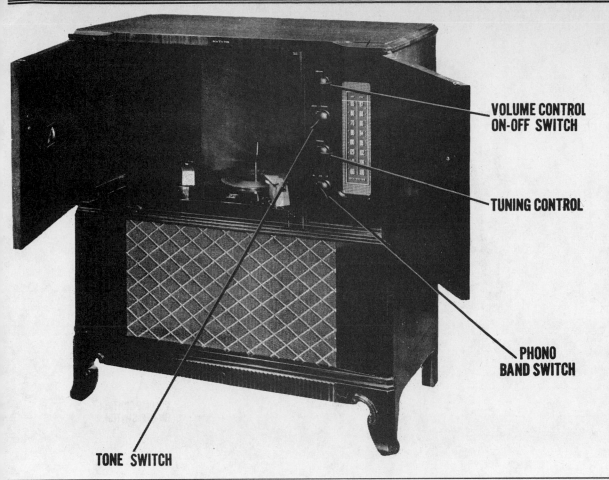

**VOLUME CONTROL
ON-OFF SWITCH**

TUNING CONTROL

**PHONO
BAND SWITCH**

TONE SWITCH

RCA VICTOR

MODEL PICTURED
8V90

AC operated phono-radio
combo AM-FM
superheterodyne receiver
with loop antenna

TUBES
9

POWER SUPPLY
110-120 volts AC

TUNING RANGE
540-1600KC,
88-108MC

MFR/SUPPLIER
RCA Victor

PHOTOFACT SET
56-20

PUBLISHED
1949

RCA VICTOR

MODEL PICTURED
8V111

AC operated phono-radio
combo AM-FM
superheterodyne receiver
with loop antenna

TUBES
11

POWER SUPPLY
110-120 volts AC

TUNING RANGE
540-1600KC,
88-108MC

MFR/SUPPLIER
RCA Victor

PHOTOFACT SET
58-18

PUBLISHED
1949

**VOLUME CONTROL
ON-OFF SWITCH**

TONE CONTROL

TUNING CONTROL

PHONO-BAND SWITCH

RCA VICTOR

MODEL PICTURED
8X542

AC/DC operated AM
superheterodyne receiver
with loop antenna

TUBES
5

POWER SUPPLY
110-120 volts AC/DC

TUNING RANGE
540-1600KC

MFR/SUPPLIER
RCA Victor

PHOTOFACT SET
59-16

PUBLISHED
1949

TUNING CONTROL

VOLUME CONTROL
ON-OFF SWITCH

RCA VICTOR

MODEL PICTURED
8X71

AC/DC operated AM-FM
superheterodyne receiver
with loop antenna

TUBES
7

POWER SUPPLY
110-120 volts AC/DC

TUNING RANGE
540-1600KC,
88-108MC

MFR/SUPPLIER
RCA Victor

PHOTOFACT SET
63-15

PUBLISHED
1949

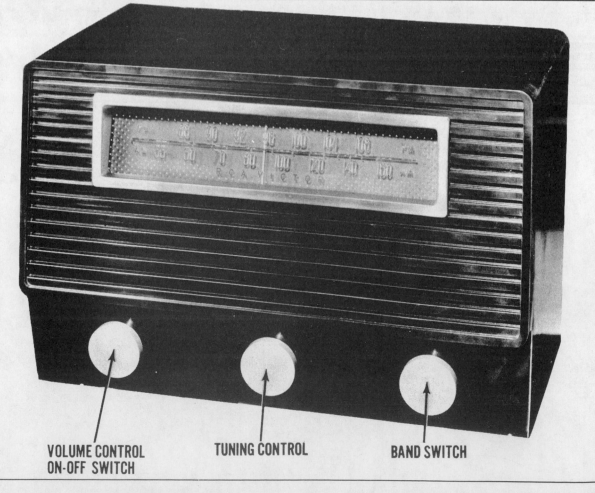

VOLUME CONTROL
ON-OFF SWITCH

TUNING CONTROL

BAND SWITCH

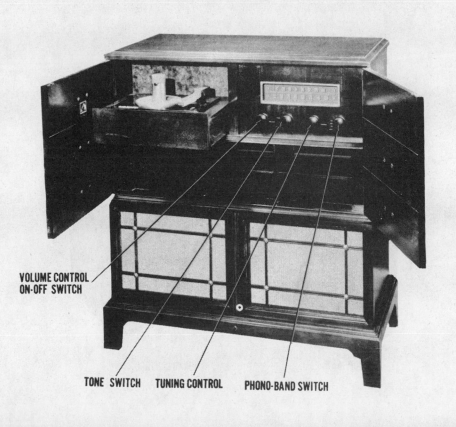

RCA VICTOR

MODEL PICTURED
8X682

AC/DC operated
two-band
superheterodyne receiver
with loop antenna

TUBES
6

POWER SUPPLY
110-120 volts AC/DC

TUNING RANGE
540-1600KC,
9.4-12MC

MFR/SUPPLIER
RCA Victor

PHOTOFACT SET
65-10

PUBLISHED
1949

VOLUME CONTROL
ON-OFF SWITCH

TONE CONTROL

BAND SWITCH

TUNING CONTROL

RCA VICTOR

MODEL PICTURED
9W101

AC operated phono-radio
combo AM-FM
superheterodyne receiver
with loop antenna

TUBES
10

POWER SUPPLY
110-120 volts AC

TUNING RANGE
540-1600KC,
88-108MC

MFR/SUPPLIER
RCA Victor

PHOTOFACT SET
73-10

PUBLISHED
1949

VOLUME CONTROL
ON-OFF SWITCH

TONE SWITCH   TUNING CONTROL   PHONO-BAND SWITCH

RCA VICTOR

MODEL PICTURED
9Y7

AC operated phono-radio
combo AM
superheterodyne receiver
with loop antenna

TUBES
6

POWER SUPPLY
110-120 volts AC

TUNING RANGE
540-1600KC

MFR/SUPPLIER
RCA Victor

PHOTOFACT SET
75-13

PUBLISHED
1949

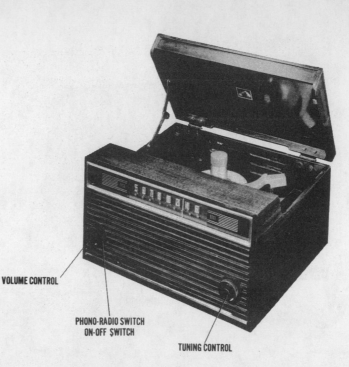

VOLUME CONTROL

PHONO-RADIO SWITCH
ON-OFF SWITCH

TUNING CONTROL

RCA VICTOR

MODEL PICTURED
8B42

Battery operated portable
AM superheterodyne
receiver with loop
antenna

TUBES
4

POWER SUPPLY
1.5 volts A & 67.5 volts B
supply

TUNING RANGE
540-1600KC

MFR/SUPPLIER
RCA Victor

PHOTOFACT SET
76-16

PUBLISHED
1949

VOLUME CONTROL

LOOP ANT

TUNING CONTROL

RCA VICTOR

MODEL PICTURED
9BX56

Three-power operated portable AM superheterodyne receiver with loop antenna

TUBES
4

POWER SUPPLY
110-120 volts AC/DC or 7.5 volts A & 67.5 volts B supply

TUNING RANGE
540-1600KC

MFR/SUPPLIER
RCA Victor

PHOTOFACT SET
79-13

PUBLISHED
1949

VOLUME CONTROL
ON-OFF SWITCH

TUNING CONTROL

RCA VICTOR

MODEL PICTURED
9X642

AC/DC operated AM superheterodyne receiver with loop antenna

TUBES
6

POWER SUPPLY
110-120 volts AC/DC

TUNING RANGE
540-1600KC

MFR/SUPPLIER
RCA Victor

PHOTOFACT SET
87-9

PUBLISHED
1950

VOLUME CONTROL
ON-OFF SWITCH

TUNING CONTROL

271

RCA VICTOR

MODEL PICTURED
9W106

AC operated phono-radio combo AM-FM superheterodyne receiver with loop antenna

TUBES
10

POWER SUPPLY
110-120 volts AC

TUNING RANGE
540-1600KC,
88-108MC

MFR/SUPPLIER
RCA Victor

PHOTOFACT SET
97-12

PUBLISHED
1950

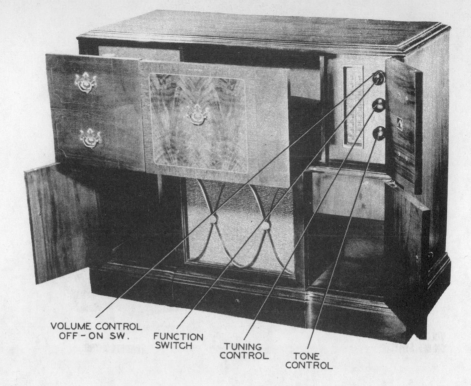

VOLUME CONTROL OFF-ON SW.   FUNCTION SWITCH   TUNING CONTROL   TONE CONTROL

RCA VICTOR

MODEL PICTURED
8F43

Battery operated AM superheterodyne receiver

TUBES
4

POWER SUPPLY
1.5 volts A & 90 volts B supply

TUNING RANGE
540-1600KC

MFR/SUPPLIER
RCA Victor

PHOTOFACT SET
97-13

PUBLISHED
1950

VOLUME CONTROL ON-OFF SWITCH

TUNING CONTROL

RCA VICTOR

MODEL PICTURED
9Y51

AC operated phono-radio
combo AM
superheterodyne receiver
with loop antenna

TUBES
5

POWER SUPPLY
110-120 volts AC

TUNING RANGE
540-1600KC

MFR/SUPPLIER
RCA Victor

PHOTOFACT SET
98-11

PUBLISHED
1950

VOLUME CONTROL

RADIO-PHONO-TONE
ON-OFF SWITCH

TUNING CONTROL

RCA VICTOR

MODEL PICTURED
9X561

AC/DC operated AM
superheterodyne receiver
with loop antenna

TUBES
5

POWER SUPPLY
110-120 volts AC/DC

TUNING RANGE
540-1600KC

MFR/SUPPLIER
RCA Victor

PHOTOFACT SET
101-9

PUBLISHED
1950

VOLUME CONTROL
ON-OFF SWITCH

TUNING CONTROL

RCA VICTOR

MODEL PICTURED
BX55

Three-power operated
portable AM
superheterodyne receiver
with loop antenna

TUBES
4

POWER SUPPLY
110-120 volts AC/DC or
7.5 volts A & 75 volts B
supply

TUNING RANGE
540-1600KC

MFR/SUPPLIER
RCA Victor

PHOTOFACT SET
102-11

PUBLISHED
1950

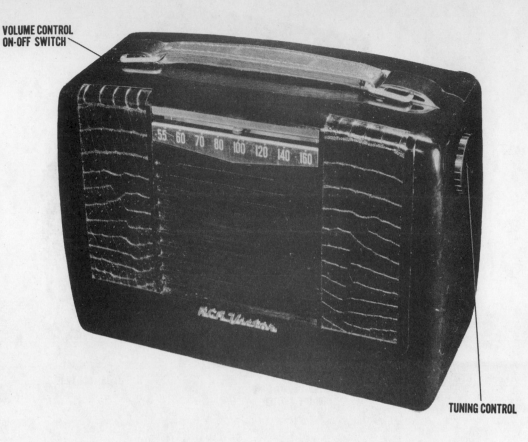

VOLUME CONTROL
ON-OFF SWITCH

TUNING CONTROL

RCA VICTOR

MODEL PICTURED
BX6

Three-power operated
portable AM
superheterodyne receiver
with loop antenna

TUBES
5

POWER SUPPLY
110-120 volts AC/DC or 9
volts A & 90 volts B
supply

TUNING RANGE
540-1600KC

MFR/SUPPLIER
RCA Victor

PHOTOFACT SET
103-13

PUBLISHED
1950

VOLUME CONTROL

TUNING CONTROL

RCA VICTOR

MODEL PICTURED
9X652

AC/DC operated
two-band
superheterodyne receiver
with loop antenna

TUBES
6

POWER SUPPLY
110-120 volts AC/DC

TUNING RANGE
540-1600KC,
5.9-17.9MC

MFR/SUPPLIER
RCA Victor

PHOTOFACT SET
104-9

PUBLISHED
1950

VOLUME CONTROL
ON-OFF SWITCH

PHONO-RADIO SWITCH

TUNING CONTROL

RCA VICTOR

MODEL PICTURED
9X571

AC/DC operated AM
superheterodyne receiver
with loop antenna

TUBES
5

POWER SUPPLY
110-120 volts AC/DC

TUNING RANGE
540-1600KC

MFR/SUPPLIER
RCA Victor

PHOTOFACT SET
107-7

PUBLISHED
1950

OFF-ON
RADIO-
PHONO
TONE
CONT.

VOLUME
CONTROL

TUNING
CONTROL

RCA VICTOR

AC operated phono-radio
combo AM
superheterodyne receiver
with loop antenna

TUBES
5

POWER SUPPLY
110-120 volts AC

TUNING RANGE
540-1600KC

MFR/SUPPLIER
RCA Victor

PHOTOFACT SET
109-10

PUBLISHED
1950

TONE CONTROL ON-OFF SWITCH VOLUME CONTROL PHONO-RADIO SWITCH TUNING CONTROL

RCA VICTOR

MODEL PICTURED
X552

AC/DC operated AM
superheterodyne receiver
with loop antenna

TUBES
5

POWER SUPPLY
110-120 volts AC/DC

TUNING RANGE
540-1600KC

MFR/SUPPLIER
RCA Victor

PHOTOFACT SET
129-9

PUBLISHED
1951

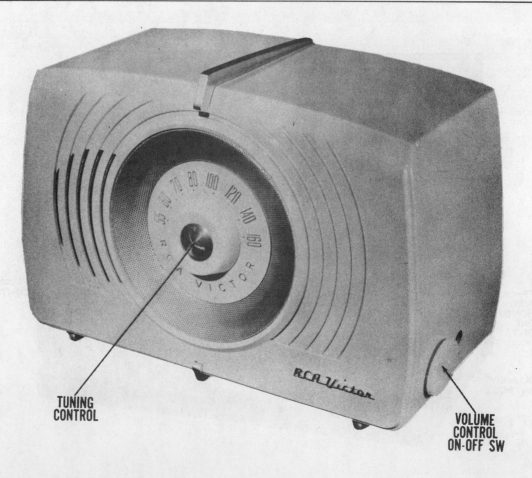

TUNING CONTROL

VOLUME CONTROL ON-OFF SW

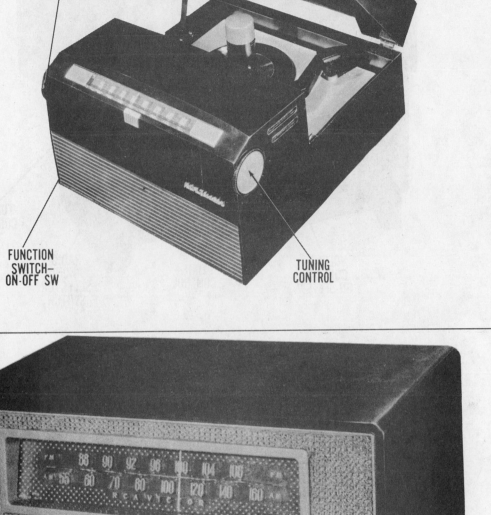

VOLUME
CONTROL

FUNCTION
SWITCH—
ON-OFF SW

TUNING
CONTROL

RCA VICTOR

MODEL PICTURED
9Y510

AC operated phono-radio
combo AM
superheterodyne receiver
with loop antenna

TUBES
5

POWER SUPPLY
110-120 volts AC

TUNING RANGE
540-1600KC

MFR/SUPPLIER
RCA Victor

PHOTOFACT SET
131-13

PUBLISHED
1951

VOLUME
CONTROL
ON-OFF SW

TUNING
CONTROL

PHONO
AM-FM
SELECTOR
SWITCH

RCA VICTOR

MODEL PICTURED
X711

AC/DC operated AM-FM
superheterodyne receiver
with loop antenna

TUBES
7

POWER SUPPLY
110-120 volts AC/DC

TUNING RANGE
540-1600KC,
88-108MC

MFR/SUPPLIER
RCA Victor

PHOTOFACT SET
133-11

PUBLISHED
1951

RCA VICTOR

MODEL PICTURED
A-82

AC operated phono-radio
combo AM
superheterodyne receiver
with loop antenna

TUBES
8

POWER SUPPLY
110-120 volts AC

TUNING RANGE
540-1600KC

MFR/SUPPLIER
RCA Victor

PHOTOFACT SET
137-10

PUBLISHED
1951

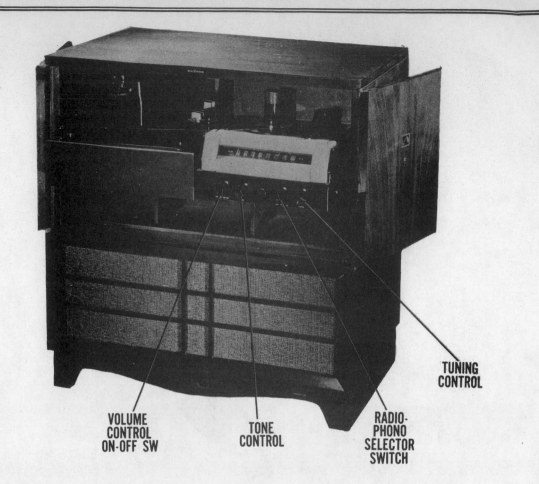

VOLUME
CONTROL
ON-OFF SW

TONE
CONTROL

RADIO-
PHONO
SELECTOR
SWITCH

TUNING
CONTROL

RCA VICTOR

MODEL PICTURED
45-W-10

AC operated phono-radio
combo AM-FM
superheterodyne receiver
with loop antenna

TUBES
10

POWER SUPPLY
110-120 volts AC

TUNING RANGE
540-1600KC,
88-108MC

MFR/SUPPLIER
RCA Victor

PHOTOFACT SET
138-8

PUBLISHED
1951

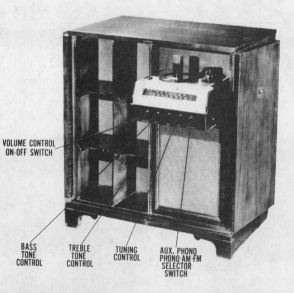

VOLUME CONTROL
ON-OFF SWITCH

BASS
TONE
CONTROL

TREBLE
TONE
CONTROL

TUNING
CONTROL

AUX. PHONO
PHONO-AM-FM
SELECTOR
SWITCH

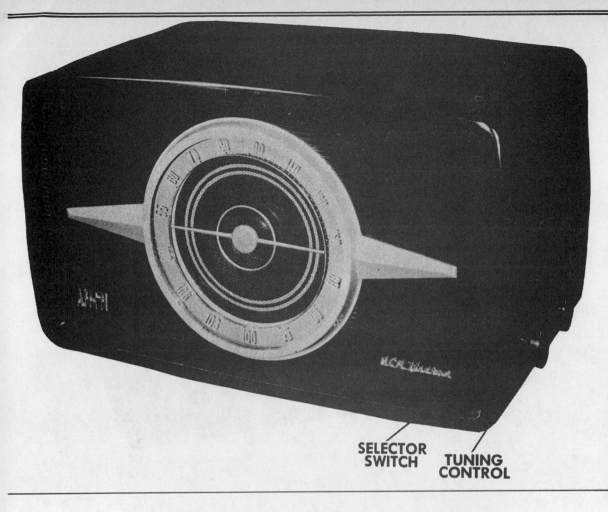

RCA VICTOR

MODEL PICTURED
1R81

AC operated AM-FM
superheterodyne receiver
with loop antenna

TUBES
8

POWER SUPPLY
110-120 volts AC

TUNING RANGE
540-1600KC,
88-108MC

MFR/SUPPLIER
RCA Victor

PHOTOFACT SET
156-10

PUBLISHED
1952

SELECTOR
SWITCH

TUNING
CONTROL

RCA VICTOR

MODEL PICTURED
1X592

AC/DC operated AM
superheterodyne receiver
with loop antenna

TUBES
5

POWER SUPPLY
110-120 volts AC/DC

TUNING RANGE
540-1600KC

MFR/SUPPLIER
RCA Victor

PHOTOFACT SET
159-12

PUBLISHED
1952

TUNING
CONTROL

VOLUME
CONTROL
ON-OFF SW.

RCA VICTOR

MODEL PICTURED
PX600

Three-power operated
portable AM
superheterodyne receiver

TUBES
5

POWER SUPPLY
110-120 volts AC/DC or 9
volts A & 90 volts B
supply

TUNING RANGE
540-1600KC

MFR/SUPPLIER
RCA Victor

PHOTOFACT SET
168-12

PUBLISHED
1952

VOLUME
CONTROL
ON-OFF SW

TUNING
CONTROL

RCA VICTOR

MODEL PICTURED
1X51

AC/DC operated AM
superheterodyne receiver

TUBES
5

POWER SUPPLY
110-120 volts AC/DC

TUNING RANGE
540-1600KC

MFR/SUPPLIER
RCA Victor

PHOTOFACT SET
172-8

PUBLISHED
1952

TUNING
CONTROL

VOLUME
CONTROL
ON-OFF SW.

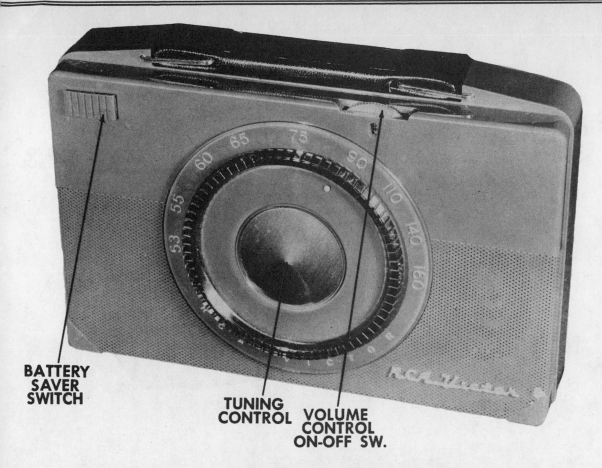

RCA VICTOR

MODEL PICTURED
2B400

Battery operated portable
AM superheterodyne
receiver

TUBES
4

POWER SUPPLY
1.5 volts A & 67.5 volts B
supply

TUNING RANGE
540-1600KC

MFR/SUPPLIER
RCA Victor

PHOTOFACT SET
181-10

PUBLISHED
1952

**BATTERY
SAVER
SWITCH**

**TUNING
CONTROL**

**VOLUME
CONTROL
ON-OFF SW.**

RCA VICTOR

MODEL PICTURED
2US7

AC operated phono-radio
combo AM
superheterodyne receiver

TUBES
6

POWER SUPPLY
110-120 volts AC

TUNING RANGE
535-1600KC

MFR/SUPPLIER
RCA Victor

PHOTOFACT SET
182-8

PUBLISHED
1952

**VOLUME
CONTROL**

**RADIO-
PHONO
TONE SW.
ON-OFF SW.**

**TUNING
CONTROL**

RCA VICTOR

MODEL PICTURED
2BX63

Three-power operated
portable AM
superheterodyne receiver

TUBES
5

POWER SUPPLY
110-120 volts AC/DC or 9
volts A & 90 volts B
supply

TUNING RANGE
540-1600KC

MFR/SUPPLIER
RCA Victor

PHOTOFACT SET
193-7

PUBLISHED
1953

VOLUME
CONTROL
ON-OFF
SWITCH

TUNING
CONTROL

RCA VICTOR

MODEL PICTURED
2C521

AC operated AM
superheterodyne receiver
with electric clock

TUBES
5

POWER SUPPLY
110-120 volts AC

TUNING RANGE
540-1600KC

MFR/SUPPLIER
RCA Victor

PHOTOFACT SET
194-11

PUBLISHED
1953

AUTO
ON-OFF
SWITCH

SLEEP
SWITCH

ALARM
SET

VOLUME
CONTROL

TUNING
CONTROL

RCA VICTOR

MODEL PICTURED
2-R-51

AC operated AM
superheterodyne receiver

TUBES
4

POWER SUPPLY
110-120 volts AC

TUNING RANGE
540-1600KC

MFR/SUPPLIER
RCA Victor

PHOTOFACT SET
196-13

PUBLISHED
1953

ON-OFF
SWITCH
VOLUME
CONTROL

TUNING
CONTROL

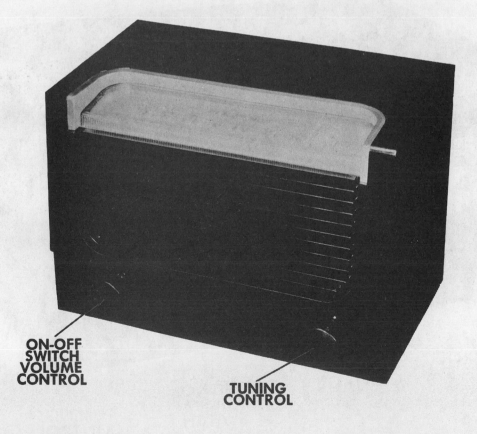

RCA VICTOR

MODEL PICTURED
2X61

AC/DC operated AM
superheterodyne receiver

TUBES
6

POWER SUPPLY
110-120 volts AC/DC

TUNING RANGE
540-1600KC

MFR/SUPPLIER
RCA Victor

PHOTOFACT SET
197-8

PUBLISHED
1953

ON-OFF
SWITCH
VOLUME
CONTROL

TUNING
CONTROL

RCA VICTOR

MODEL PICTURED
2-X-621

AC/DC operated
two-band
superheterodyne receiver

TUBES
6

POWER SUPPLY
110-120 volts AC/DC

TUNING RANGE
540-1600KC,
5.8-18.0MC

MFR/SUPPLIER
RCA Victor

PHOTOFACT SET
199-9

PUBLISHED
1953

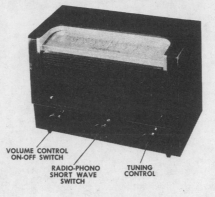

VOLUME CONTROL
ON-OFF SWITCH

RADIO-PHONO
SHORT WAVE
SWITCH

TUNING
CONTROL

RCA VICTOR

MODEL PICTURED
2-XF-91

AC/DC operated AM-FM
superheterodyne receiver

TUBES
8

POWER SUPPLY
110-120 volts AC/DC

TUNING RANGE
540-1600KC,
88-108MC

MFR/SUPPLIER
RCA Victor

PHOTOFACT SET
206-9

PUBLISHED
1953

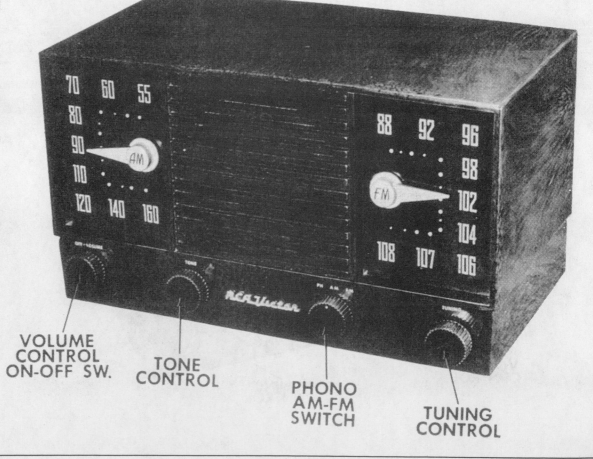

VOLUME
CONTROL
ON-OFF SW.

TONE
CONTROL

PHONO
AM-FM
SWITCH

TUNING
CONTROL

284

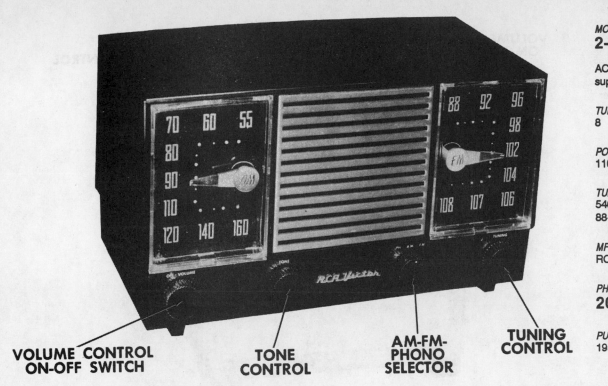

RCA VICTOR

MODEL PICTURED
2-XF-931

AC operated AM-FM
superheterodyne receiver

TUBES
8

POWER SUPPLY
110-120 volts AC

TUNING RANGE
540-1600KC,
88-108MC

MFR/SUPPLIER
RCA Victor

PHOTOFACT SET
209-9

PUBLISHED
1953

**VOLUME CONTROL
ON-OFF SWITCH**

**TONE
CONTROL**

**AM-FM-
PHONO
SELECTOR**

**TUNING
CONTROL**

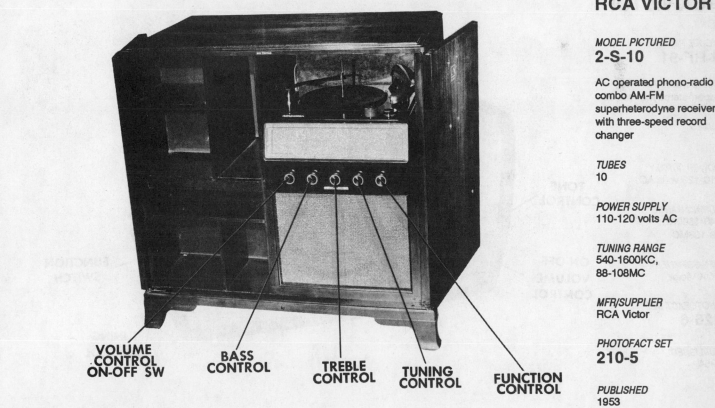

RCA VICTOR

MODEL PICTURED
2-S-10

AC operated phono-radio
combo AM-FM
superheterodyne receiver
with three-speed record
changer

TUBES
10

POWER SUPPLY
110-120 volts AC

TUNING RANGE
540-1600KC,
88-108MC

MFR/SUPPLIER
RCA Victor

PHOTOFACT SET
210-5

PUBLISHED
1953

**VOLUME
CONTROL
ON-OFF SW**

**BASS
CONTROL**

**TREBLE
CONTROL**

**TUNING
CONTROL**

**FUNCTION
CONTROL**

RCA VICTOR

MODEL PICTURED
2-S-7

AC operated phono-radio
combo AM
superheterodyne receiver

TUBES
6

POWER SUPPLY
110-120 volts AC

TUNING RANGE
540-1600KC

MFR/SUPPLIER
RCA Victor

PHOTOFACT SET
222-11

PUBLISHED
1953

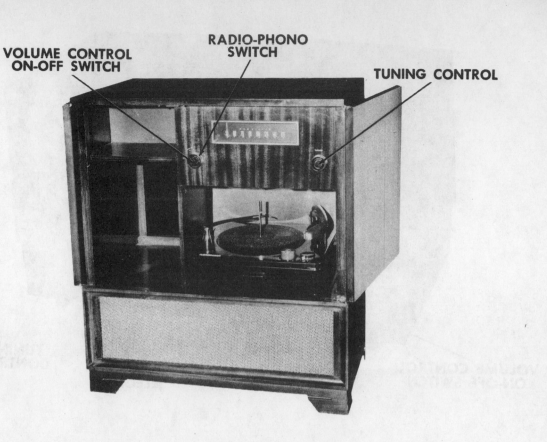

VOLUME CONTROL
ON-OFF SWITCH

RADIO-PHONO
SWITCH

TUNING CONTROL

RCA VICTOR

MODEL PICTURED
3-RF-91

AC operated AM-FM
superheterodyne receiver

TUBES
9

POWER SUPPLY
110-120 volts AC

TUNING RANGE
540-1600KC,
88-108MC

MFR/SUPPLIER
RCA Victor

PHOTOFACT SET
226-6

PUBLISHED
1954

TONE
CONTROL

ON-OFF
VOLUME
CONTROL

FUNCTION
SWITCH

TUNING
CONTROL

RCA VICTOR

MODEL PICTURED
3-X-521

AC/DC operated AM
superheterodyne receiver

TUBES
5

POWER SUPPLY
110-120 volts AC/DC

TUNING RANGE
540-1600KC

MFR/SUPPLIER
RCA Victor

PHOTOFACT SET
226-7

PUBLISHED
1954

**TUNING
CONTROL**

**ON-OFF
VOLUME
CONTROL**

RCA VICTOR

MODEL PICTURED
3-BX-51

Three-power operated
portable AM
superheterodyne receiver

TUBES
4

POWER SUPPLY
115 volts AC/DC or 7.5
volts A & 75 volts B
supply

TUNING RANGE
540-1600KC

MFR/SUPPLIER
RCA Victor

PHOTOFACT SET
227-11

PUBLISHED
1954

**ON-OFF SWITCH
VOLUME CONTROL**

TUNING CONTROL

RCA VICTOR

MODEL PICTURED
3-BX-671

Three-power operated
portable multi-band
superheterodyne receiver

TUBES
5

POWER SUPPLY
110-120 volts AC/DC or 9
volts A & 90 volts B
supply

TUNING RANGE
7 bands

MFR/SUPPLIER
RCA Victor

PHOTOFACT SET
228-14

PUBLISHED
1954

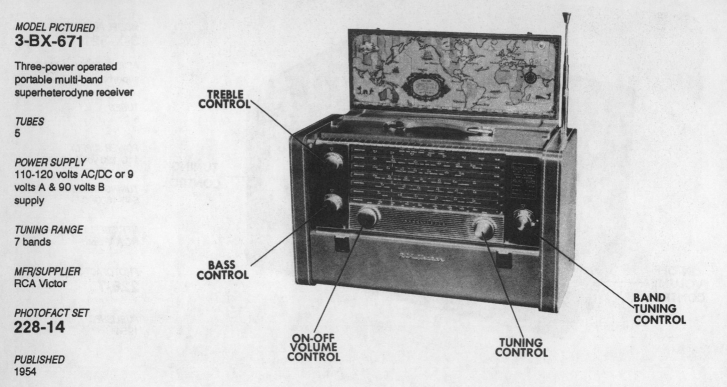

TREBLE
CONTROL

BASS
CONTROL

ON-OFF
VOLUME
CONTROL

TUNING
CONTROL

BAND
TUNING
CONTROL

RCA VICTOR

MODEL PICTURED
4X641

AC/DC operated AM
superheterodyne receiver

TUBES
6

POWER SUPPLY
110-120 volts AC/DC

TUNING RANGE
540-1620KC

MFR/SUPPLIER
RCA Victor

PHOTOFACT SET
259-13

PUBLISHED
1954

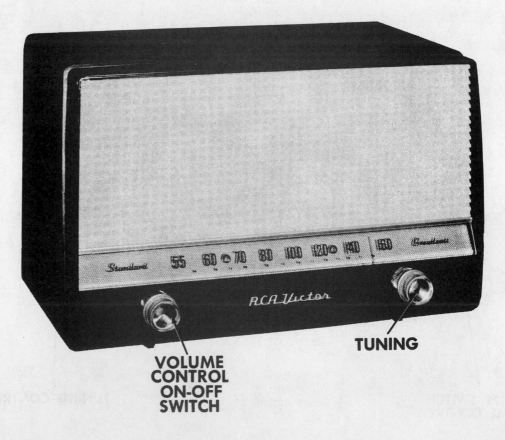

VOLUME
CONTROL
ON-OFF
SWITCH

TUNING

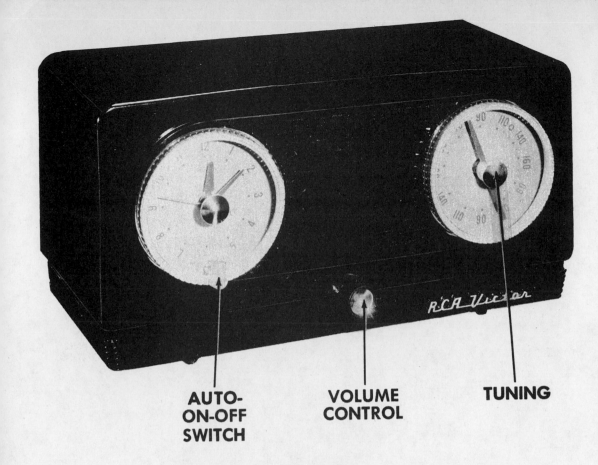

RCA VICTOR

MODEL PICTURED
4-C-531

AC operated AM
superheterodyne receiver
with electric clock

TUBES
5

POWER SUPPLY
110-120 volts AC

TUNING RANGE
540-1620KC

MFR/SUPPLIER
RCA Victor

PHOTOFACT SET
260-13

PUBLISHED
1954

**AUTO-
ON-OFF
SWITCH**　　**VOLUME
CONTROL**　　**TUNING**

RCA VICTOR

MODEL PICTURED
4-Y-511

AC operated phono-radio
combo AM
superheterodyne receiver
with 45RPM auto record
changer

TUBES
5

POWER SUPPLY
110-120 volts AC

TUNING RANGE
540-1620KC

MFR/SUPPLIER
RCA Victor

PHOTOFACT SET
261-12

PUBLISHED
1954

**VOLUME
CONTROL
ON-OFF
SWITCH**　　**RADIO-PHONO
SWITCH**　　**TUNING**

RCA VICTOR

MODEL PICTURED
4-X-661

AC/DC operated
two-band
superheterodyne receiver

TUBES
6

POWER SUPPLY
110-120 volts AC/DC

TUNING RANGE
540-1600KC,
5.95-17.9MC

MFR/SUPPLIER
RCA Victor

PHOTOFACT SET
265-9

PUBLISHED
1955

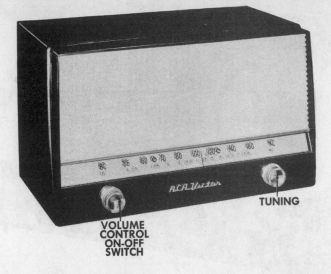

VOLUME
CONTROL
ON-OFF
SWITCH

TUNING

RCA VICTOR

MODEL PICTURED
3-US-5

AC operated phono-radio
combo AM
superheterodyne receiver
with three-speed auto
record changer

TUBES
5

POWER SUPPLY
110-120 volts AC

TUNING RANGE
540-1600KC

MFR/SUPPLIER
RCA Victor

PHOTOFACT SET
267-10

PUBLISHED
1955

VICTROLA

VOLUME
CONTROL

RADIO
PHONO
ON-OFF
SWITCH

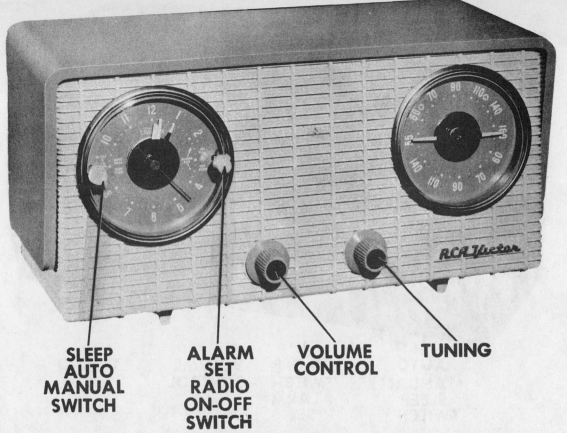

RCA VICTOR

MODEL PICTURED
4-C-671

AC operated AM
superheterodyne receiver
with electric clock

TUBES
6

POWER SUPPLY
110-120 volts AC

TUNING RANGE
540-1600KC

MFR/SUPPLIER
RCA Victor

PHOTOFACT SET
269-11

PUBLISHED
1955

SLEEP AUTO MANUAL SWITCH

ALARM SET RADIO ON-OFF SWITCH

VOLUME CONTROL

TUNING

RCA VICTOR

MODEL PICTURED
4-X-551

AC/DC operated AM
superheterodyne receiver
with phono provisions

TUBES
5

POWER SUPPLY
110-120 volts AC/DC

TUNING RANGE
540-1600KC

MFR/SUPPLIER
RCA Victor

PHOTOFACT SET
271-10

PUBLISHED
1955

TUNING

VOLUME CONTROL ON-OFF SWITCH

RCA VICTOR

MODEL PICTURED
4-C-541

AC operated AM
superheterodyne receiver
with electric clock

TUBES
5

POWER SUPPLY
110-120 volts AC

TUNING RANGE
540-1600KC

MFR/SUPPLIER
RCA Victor

PHOTOFACT SET
273-11

PUBLISHED
1955

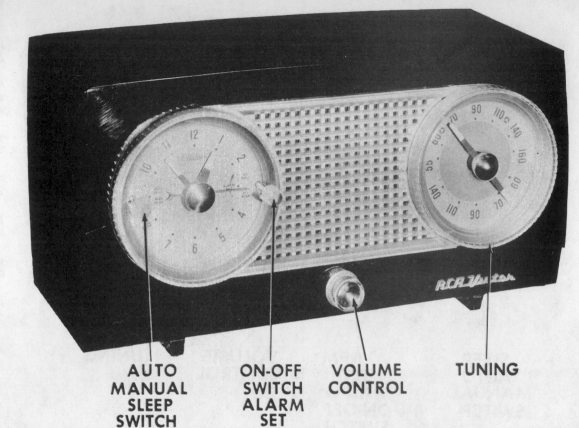

AUTO   ON-OFF   VOLUME   TUNING
MANUAL   SWITCH   CONTROL
SLEEP   ALARM
SWITCH   SET

RCA VICTOR

MODEL PICTURED
5-BX-41

Three-power operated
portable AM
superheterodyne receiver

TUBES
4

POWER SUPPLY
110-120 volts AC/DC or
1.5 volts A & 67.5 volts B
supply

TUNING RANGE
540-1600KC

MFR/SUPPLIER
RCA Victor

PHOTOFACT SET
278-10

PUBLISHED
1955

VOLUME
CONTROL
ON-OFF
SWITCH

TUNING

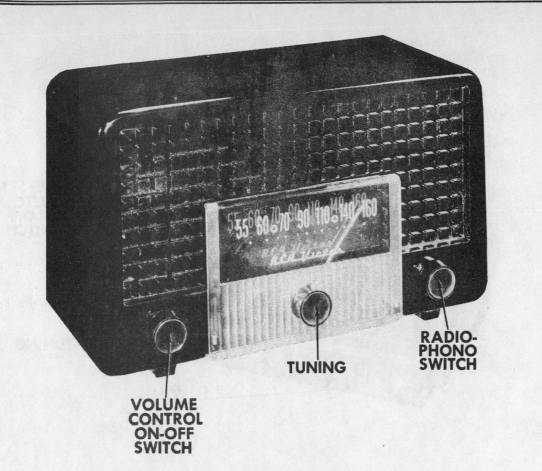

VOLUME CONTROL ON-OFF SWITCH

TUNING

RADIO-PHONO SWITCH

RCA VICTOR

MODEL PICTURED
5-X-560

AC/DC operated AM superheterodyne receiver

TUBES
5

POWER SUPPLY
110-120 volts AC/DC

TUNING RANGE
540-1600KC

MFR/SUPPLIER
RCA Victor

PHOTOFACT SET
279-12

PUBLISHED
1955

SLEEP TIME ADJ.

AUTO ON-OFF SWITCH

TUNING

VOLUME CONTROL

RCA VICTOR

MODEL PICTURED
5-C-581

AC operated AM superheterodyne receiver with electric clock

TUBES
5

POWER SUPPLY
110-120 volts AC

TUNING RANGE
540-1600KC

MFR/SUPPLIER
RCA Victor

PHOTOFACT SET
284-11

PUBLISHED
1955

RCA VICTOR

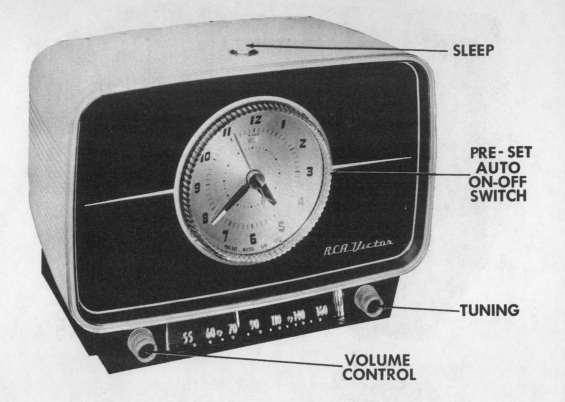

SLEEP

PRE-SET
AUTO
ON-OFF
SWITCH

TUNING

VOLUME
CONTROL

RCA VICTOR

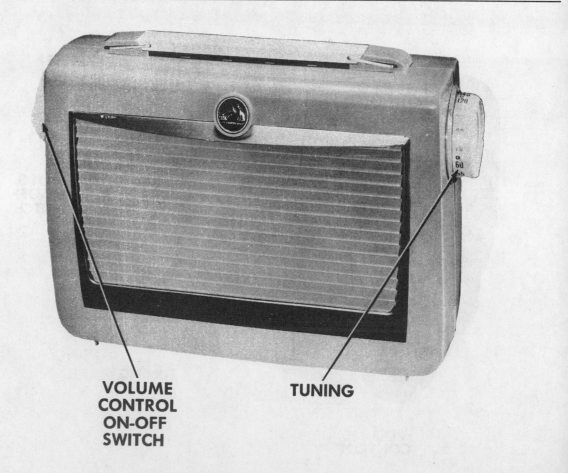

VOLUME
CONTROL
ON-OFF
SWITCH

TUNING

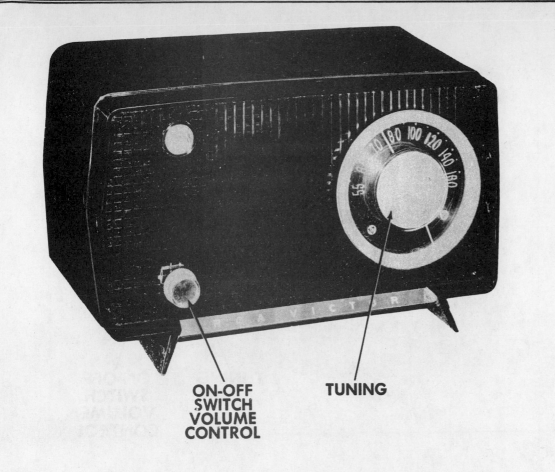

ON-OFF
SWITCH
VOLUME
CONTROL

TUNING

RCA VICTOR

MODEL PICTURED
6-X-7A

AC/DC operated AM
superheterodyne receiver

TUBES
5

POWER SUPPLY
110-120 volts AC/DC

TUNING RANGE
540-1620KC

MFR/SUPPLIER
RCA Victor

PHOTOFACT SET
303-9

PUBLISHED
1956

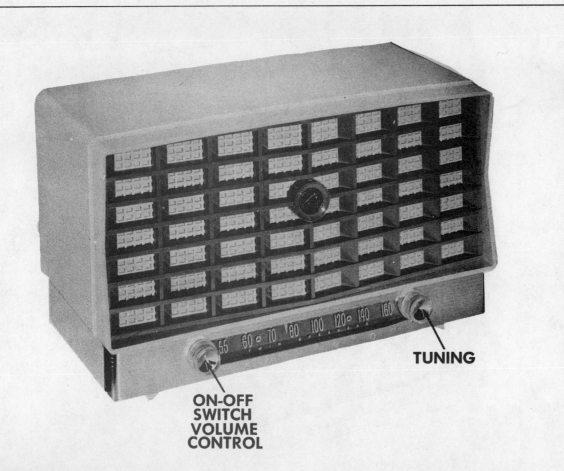

ON-OFF
SWITCH
VOLUME
CONTROL

TUNING

RCA VICTOR

MODEL PICTURED
6-XD-5

AC/DC operated AM
superheterodyne receiver

TUBES
5

POWER SUPPLY
105-125 volts AC/DC

TUNING RANGE
540-1600KC

MFR/SUPPLIER
RCA Victor

PHOTOFACT SET
309-14

PUBLISHED
1956

RCA VICTOR

MODEL PICTURED
6-X-5A

AC/DC operated AM
superheterodyne receiver

TUBES
5

POWER SUPPLY
105-125 volts AC/DC

TUNING RANGE
540-1600KC

MFR/SUPPLIER
RCA Victor

PHOTOFACT SET
316-12

PUBLISHED
1956

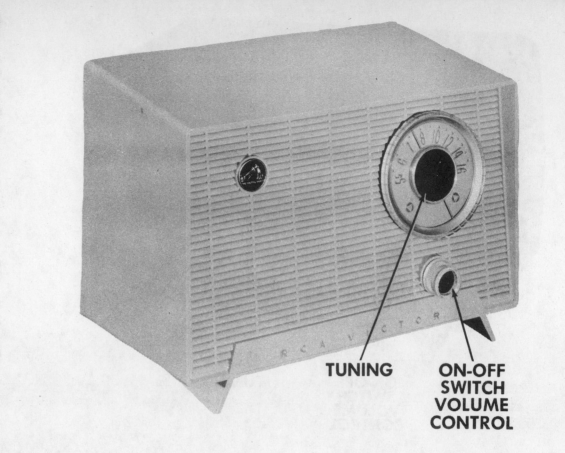

TUNING

**ON-OFF
SWITCH
VOLUME
CONTROL**

RCA VICTOR

MODEL PICTURED
6-BY-4A

Battery operated portable
AM superheterodyne
receiver with 45RPM
record changer

TUBES
4

POWER SUPPLY
1.5 volts A & 90 volts B
supply

TUNING RANGE
540-1600KC

MFR/SUPPLIER
RCA Victor

PHOTOFACT SET
321-12

PUBLISHED
1956

RCA VICTOR

MODEL PICTURED
6HF3

AC operated AM-FM
superheterodyne receiver
with three-speed auto
record changer

TUBES
13

POWER SUPPLY
110-120 volts AC

TUNING RANGE
540-1600KC,
88-108MC

MFR/SUPPLIER
RCA Victor

PHOTOFACT SET
323-11

PUBLISHED
1956

RCA VICTOR

MODEL PICTURED
7-BT-9J

Battery operated portable
transistorized AM
superheterodyne receiver

TUBES
6 transistor

POWER SUPPLY
9 volts DC

TUNING RANGE
540-1600KC

MFR/SUPPLIER
RCA Victor

PHOTOFACT SET
327-10

PUBLISHED
1956

RCA VICTOR

MODEL PICTURED
7-BT-10K

Battery operated portable
transistorized AM
superheterodyne receiver

TUBES
7 transistor

POWER SUPPLY
9 volts DC

TUNING RANGE
540-1600KC

MFR/SUPPLIER
RCA Victor

PHOTOFACT SET
329-10

PUBLISHED
1956

RCA VICTOR

MODEL PICTURED
6-XY-5A

AC operated AM
superheterodyne receiver

TUBES
5

POWER SUPPLY
110-120 volts AC

TUNING RANGE
540-1600KC

MFR/SUPPLIER
RCA Victor

PHOTOFACT SET
333-11

PUBLISHED
1956

RCA VICTOR

MODEL PICTURED
6-HF-1

AC operated AM-FM
superheterodyne receiver
with three-speed auto
record changer

TUBES
17

POWER SUPPLY
110-120 volts AC

TUNING RANGE
540-1600KC,
88-108MC

MFR/SUPPLIER
RCA Victor

PHOTOFACT SET
334-8

PUBLISHED
1956

RCA VICTOR

MODEL PICTURED
6-C-5A

AC operated AM
superheterodyne receiver
with electric clock

TUBES
5

POWER SUPPLY
110-120 volts AC

TUNING RANGE
540-1600KC

MFR/SUPPLIER
RCA Victor

PHOTOFACT SET
340-15

PUBLISHED
1956

RCA VICTOR

MODEL PICTURED
7-BX-6

Three-power operated
portable AM receiver

TUBES
4

POWER SUPPLY
110-120 volts AC/DC or
7.5 volts A & 90 volts B
supply

TUNING RANGE
540-1600KC

MFR/SUPPLIER
RCA Victor

PHOTOFACT SET
344-11

PUBLISHED
1957

RCA VICTOR

MODEL PICTURED
7-BX-8J

Three-power operated
portable AM receiver

TUBES
5

POWER SUPPLY
110-120 volts AC/DC or 9
volts A & 90 volts B
supply

TUNING RANGE
540-1600KC

MFR/SUPPLIER
RCA Victor

PHOTOFACT SET
345-12

PUBLISHED
1957

RCA VICTOR

MODEL PICTURED
7-BX-9H

Three-power operated
portable AM-SW receiver

TUBES
5

POWER SUPPLY
110-120 volts AC/DC or 9
volts A & 90 volts B
supply

TUNING RANGE
540-1600KC,
2-5MC

MFR/SUPPLIER
RCA Victor

PHOTOFACT SET
349-10

PUBLISHED
1957

RCA VICTOR

MODEL PICTURED
8-BT-10K

Battery operated portable
AM transistorized
receiver

TUBES
7 transistor

POWER SUPPLY
9 volts DC

TUNING RANGE
540-1600KC

MFR/SUPPLIER
RCA Victor

PHOTOFACT SET
350-13

PUBLISHED
1957

RCA VICTOR

MODEL PICTURED
8-C-7EE

AC operated AM receiver
with electric clock

TUBES
5

POWER SUPPLY
110-120 volts AC

TUNING RANGE
540-1600KC

MFR/SUPPLIER
RCA Victor

PHOTOFACT SET
359-13

PUBLISHED
1957

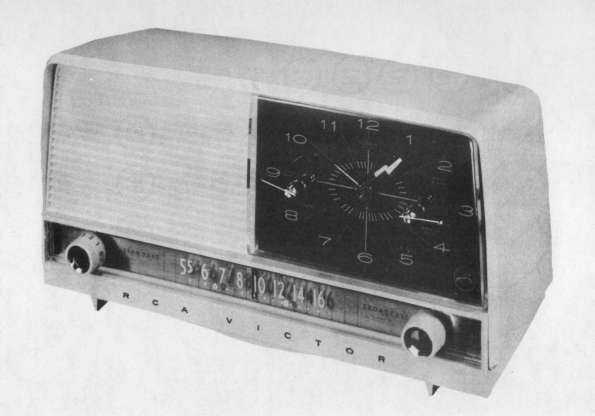

RCA VICTOR

MODEL PICTURED
8-X-8D

AC/DC operated AM
receiver

TUBES
5

POWER SUPPLY
110-120 volts AC/DC

TUNING RANGE
540-1600KC

MFR/SUPPLIER
RCA Victor

PHOTOFACT SET
359-14

PUBLISHED
1957

RCA VICTOR

MODEL PICTURED
7-HFR-1

AC operated AM-FM
receiver with tape
recorder and four-speed
auto record changer

TUBES
18

POWER SUPPLY
110-120 volts AC

TUNING RANGE
540-1600KC,
88-108MC

MFR/SUPPLIER
RCA Victor

PHOTOFACT SET
365-11

PUBLISHED
1957

RCA VICTOR

MODEL PICTURED
8-X-5D

AC/DC operated AM
receiver

TUBES
5

POWER SUPPLY
110-120 volts AC/DC

TUNING RANGE
540-1600KC

MFR/SUPPLIER
RCA Victor

PHOTOFACT SET
368-11

PUBLISHED
1957

RCA VICTOR

MODEL PICTURED
8-C-5E

AC operated AM receiver
with electric clock

TUBES
5

POWER SUPPLY
110-120 volts AC

TUNING RANGE
540-1600KC

MFR/SUPPLIER
RCA Victor

PHOTOFACT SET
369-18

PUBLISHED
1957

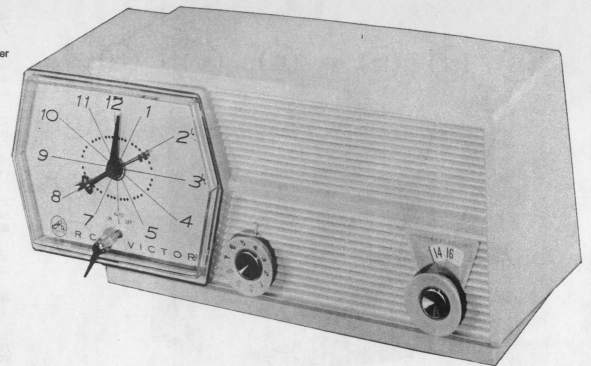

RCA VICTOR

MODEL PICTURED
9-BT-9H

Battery operated portable
AM transistorized
receiver

TUBES
6 transistor

POWER SUPPLY
9 volts DC

TUNING RANGE
540-1600KC

MFR/SUPPLIER
RCA Victor

PHOTOFACT SET
371-7

PUBLISHED
1957

RCA VICTOR

MODEL PICTURED
8-RF-13

AC operated AM-FM
receiver

TUBES
13

POWER SUPPLY
110-120 volts AC

TUNING RANGE
540-1600KC,
88-108MC

MFR/SUPPLIER
RCA Victor

PHOTOFACT SET
390-9

PUBLISHED
1958

RCA VICTOR

MODEL PICTURED
9X10FE

AC/DC operated AM
receiver

TUBES
5

POWER SUPPLY
110-120 volts AC/DC

TUNING RANGE
540-1600KC

MFR/SUPPLIER
RCA Victor

PHOTOFACT SET
394-14

PUBLISHED
1958

RCA VICTOR

AC operated AM receiver
with four-speed auto
record changer

TUBES
5

POWER SUPPLY
110-120 volts AC

TUNING RANGE
540-1600KC

MFR/SUPPLIER
RCA Victor

PHOTOFACT SET
395-12

PUBLISHED
1958

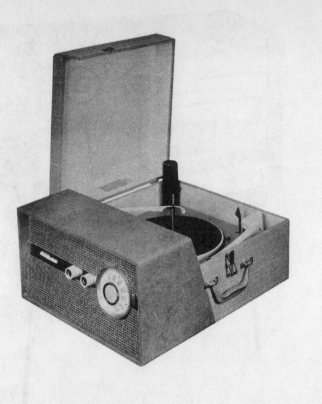

RCA VICTOR

AC operated AM-FM
receiver with four-speed
auto record changer

TUBES
11

POWER SUPPLY
110-120 volts AC

TUNING RANGE
540-1600KC,
88-108MC

MFR/SUPPLIER
RCA Victor

PHOTOFACT SET
396-7

PUBLISHED
1958

RCA VICTOR

MODEL PICTURED
9INT1

AC operated FM-AM-SW
receiver

TUBES
7

POWER SUPPLY
110-120 volts AC

TUNING RANGE
510-1650KC,
88-108MC,
2.2-7MC,
5.8-18.5MC

MFR/SUPPLIER
RCA Victor

PHOTOFACT SET
399-11

PUBLISHED
1958

RCA VICTOR

MODEL PICTURED
1BT46

Battery operated portable
AM transistorized
receiver

TUBES
6 transistor

POWER SUPPLY
4.5 volts DC

TUNING RANGE
540-1600KC

MFR/SUPPLIER
RCA Victor

PHOTOFACT SET
410-14

PUBLISHED
1958

RCA VICTOR

MODEL PICTURED
SHF-1

AC operated FM-AM
receiver with four-speed
automatic record changer

TUBES
33

POWER SUPPLY
110-120 volts AC

TUNING RANGE
540-1600KC,
88-108MC

MFR/SUPPLIER
RCA Victor

PHOTOFACT SET
418-11

PUBLISHED
1958

RCA VICTOR

MODEL PICTURED
1BX57

Three-power operated
portable AM receiver

TUBES
4

POWER SUPPLY
110-120 volts AC or 7.5
volts A & 90 volts B
supply

TUNING RANGE
540-1600KC

MFR/SUPPLIER
RCA Victor

PHOTOFACT SET
422-12

PUBLISHED
1958

RCA VICTOR

MODEL PICTURED
SHC4

AC operated AM-FM
receiver with four-speed
auto record changer

TUBES
14

POWER SUPPLY
110-120 volts AC

TUNING RANGE
540-1600KC,
88-108MC

MFR/SUPPLIER
RCA Victor

PHOTOFACT SET
435-10

PUBLISHED
1959

RCA VICTOR

MODEL PICTURED
1MBT6

Battery operated portable
transistorized multi-band
receiver

TUBES
9 transistor

POWER SUPPLY
13.5 volts DC

TUNING RANGE
7 bands

MFR/SUPPLIER
RCA Victor

PHOTOFACT SET
436-13

PUBLISHED
1959

RCA VICTOR

Battery operated portable
AM transistorized
receiver

TUBES
7 transistor

POWER SUPPLY
6 volts DC

TUNING RANGE
540-1600KC

MFR/SUPPLIER
RCA Victor

PHOTOFACT SET
442-10

PUBLISHED
1959

RCA VICTOR

MODEL PICTURED
1BT21

Battery operated portable
transistorized AM
receiver

TUBES
6 transistor

POWER SUPPLY
6 volts DC

TUNING RANGE
540-1600KC

MFR/SUPPLIER
RCA Victor

PHOTOFACT SET
447-10

PUBLISHED
1959

RCA VICTOR

MODEL PICTURED
C1E

AC operated AM receiver
with electric clock

TUBES
5

POWER SUPPLY
110-120 volts AC

TUNING RANGE
535-1620KC

MFR/SUPPLIER
RCA Victor

PHOTOFACT SET
465-14

PUBLISHED
1959

RCA VICTOR

MODEL PICTURED
T1JE

Battery operated
transistorized portable
AM receiver

TUBES

6 transistor

POWER SUPPLY
4.5 volts DC

TUNING RANGE
535-1620KC

MFR/SUPPLIER
RCA Victor

PHOTOFACT SET
478-15

PUBLISHED
1960

RCA VICTOR

MODEL PICTURED
XF3EM

AC/DC operated FM-AM
receiver

TUBES
7

POWER SUPPLY
110-120 volts AC/DC

TUNING RANGE
535-1620KC,
88-108MC

MFR/SUPPLIER
RCA Victor

PHOTOFACT SET
479-14

PUBLISHED
1960

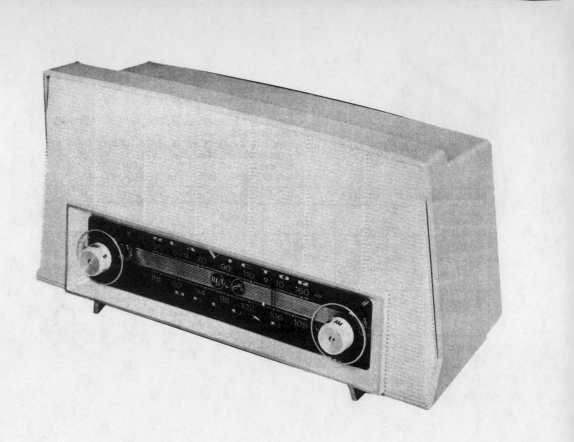

RCA VICTOR

MODEL PICTURED
C-4EM

AC operated AM receiver

TUBES
5

POWER SUPPLY
110-120 volts AC

TUNING RANGE
535-1620KC

MFR/SUPPLIER
RCA Victor

PHOTOFACT SET
480-10

PUBLISHED
1960

RCA VICTOR

MODEL PICTURED
TX1JE

Battery operated
transistorized AM
receiver

TUBES
6 transistor

POWER SUPPLY
4.5 volts DC

TUNING RANGE
540-1600KC

MFR/SUPPLIER
RCA Victor

PHOTOFACT SET
489-17

PUBLISHED
1960

RCA VICTOR

MODEL PICTURED
TPM11, KS11

AC operated FM-AM
tuner, stereo preamp,
stereo amplifier and
four-speed record
changer

TUBES
8

POWER SUPPLY
110-120 volts AC

TUNING RANGE
535-1620KC,
88-108MC

MFR/SUPPLIER
RCA Victor

PHOTOFACT SET
491-12

PUBLISHED
1960

MODEL TPM11   MODEL KS11

RCA VICTOR

MODEL PICTURED
TPR-8

AC operated FM-AM
tuner, stereo preamp, and
stereo amplifier

TUBES
20

POWER SUPPLY
110-120 volts AC

TUNING RANGE
540-1600KC,
88-108MC

MFR/SUPPLIER
RCA Victor

PHOTOFACT SET
496-14

PUBLISHED
1960

RCA VICTOR

MODEL PICTURED
1T4J

Battery operated
transistorized portable
AM receiver

TUBES
8 transistor

POWER SUPPLY
6 volts DC

TUNING RANGE
535-1620KC

MFR/SUPPLIER
RCA Victor

PHOTOFACT SET
499-11

PUBLISHED
1960

INDEXES

Use these indexes to learn more about the radios pictured in this book and other models with similar configurations and similar tube usage

- To locate the picture or PHOTOFACT set number of a model number you know, go to page 318

- To find out which tubes a particular radio holds, go to page 329

- To determine which tubes can substitute for the tubes you need to replace, go to page 339

- To find out what brands various manufacturers produced during the baby boom years, go to page 342

Pictured Radios and Similar Models, by Model Number

BRAND/MODEL	PF SET	SIMILAR TO	PAGE	BRAND/MODEL	PF SET	SIMILAR TO	PAGE	BRAND/MODEL	PF SET	SIMILAR TO	PAGE
MOTOROLA				107F31B	33-14	107F31	13	52L3	190-11	52L1	39
A1B	491-9	A1W	80	10KT12B	471-10	10KT12M	77	52L3A	190-11	52L1	39
A1N	491-9	A1W	80	10KT12M	471-10	-	77	52M1U	188-10	-	37
A1R	491-9	A1W	80	10KT12W	471-10	10KT12M	77	52M2U	188-10	52M1U	37
A1W	491-9	-	80	10T28B	402-11	-	67	52M3U	188-10	52M1U	37
A2G	492-13	C3S-1	81	10T28M	402-11	10T28B	67	52R11	188-11	52R14	38
A2N	492-13	C3S-1	81	10T28MC	402-11	10T28B	67	52R11A	178-7	52R12A	36
A2P	492-13	C3S-1	81	13KT15B	451-10	13KT15M	71	52R11U	177-11	52R12U	36
A2W	492-13	C3S-1	81	13KT15CW	451-10	13KT15M	71	52R12	188-11	52R14	38
C1N	492-13	C3S-1	81	13KT15M	451-10	-	71	52R12A	178-7	-	36
C1W	492-13	C3S-1	81	15KT25B-1S	408-12	15KT25MC-1	68	52R12U	177-11	-	36
C2B	492-13	C3S-1	81	15KT25B1	408-12	15KT25MC-1	68	52R13	188-11	52R14	38
C2P	492-13	C3S-1	81	15KT25M-1	408-12	15KT25MC-1	68	52R13A	178-7	52R12A	36
C2W	492-13	C3S-1	81	15KT25M-1S	408-12	15KT25MC-1	68	52R14	188-11	-	38
C3G	492-13	C3S-1	81	15KT25MC-1	408-12	-	68	52R14A	178-7	52R12A	36
C3G-1	492-13	C3S-1	81	15KT25MC-1S	408-12	15KT25MC-1	68	52R14U	177-11	52R12U	36
C3S	492-13	C3S-1	81	15KT25MCH-1	408-12	15KT25MC-1	68	52R15	188-11	52R14	38
C3S-1	492-13	-	81	15KT25MCH-1S	408-12	15KT25MC-1	68	52R15A	178-7	52R12A	36
C3W	492-13	C3S-1	81	21B1	191-14	42B1	40	52R15U	177-11	52R12U	36
C3W-1	492-13	C3S-1	81	45B12	9-23	-	6	52R16	188-11	52R14	38
C45	494-11	C4B	81	45CD4	319-8	56CC1	55	52R16A	178-7	52R12A	36
C4B	494-11	-	81	45P1	308-8	-	53	52R16U	177-11	52R12U	36
HK-18B	408-12	15KT25MC-1	68	45P2	308-8	45P1	53	53C1	236-7	-	44
HK-18C	408-12	15KT25MC-1	68	47B11	29-17	-	9	53C2	236-7	53C1	44
HK-18M	408-12	15KT25MC-1	68	48L11	47-13	-	15	53C3	236-7	53C1	44
HK-18W	408-12	15KT25MC-1	68	49L11Q	77-7	-	22	53C4	236-7	53C1	44
HK-19	408-12	15KT25MC-1	68	49L13Q	77-7	49L11Q	22	53C6	235-7	-	44
L12G	471-9	L12N	76	51M1U	149-8	-	33	53C7	235-7	53C6	44
L12N	471-9	-	76	51M2U	149-8	51M1U	33	53C8	235-7	53C6	44
L14E	470-16	-	76	52B1U	190-10	-	39	53C9	235-7	53C6	44
SK25B	472-6	SK25MC	77	52C1	191-15	-	40	53D1	253-9	-	48
SK25M	472-6	SK25MC	77	52C6	177-10	-	35	53F2	234-9	-	43
SK25MC	472-6	-	77	52C7	177-10	52C6	35	53H1	250-16	-	47
SK25MCH	472-6	SK25MC	77	52C8	177-10	52C6	35	53H2	250-16	53H1	47
SK32W	490-10	-	80	52CW1	198-10	-	41	53H3	250-16	53H1	47
SK33W	490-10	SK32W	80	52CW2	198-10	52CW1	41	53H4	250-16	53H1	47
SK35W	490-10	SK32W	80	52CW3	198-10	52CW1	41	53LC1	217-10	-	42
X11B	481-9	-	79	52CW4	198-10	52CW1	41	53LC2	217-10	53LC1	42
X11E	473-9	-	78	52H11U	176-6	-	34	53LC3	217-10	53LC1	42
X11G	481-9	X11B	79	52H12U	176-6	52H11U	34	53R1,U	247-8	-	46
X11R	481-9	X11B	79	52H13U	176-6	52H11U	34	53R1A	273-8	-	49
X12A	472-7	-	78	52H14U	176-6	52H11U	34	53R2,U	247-8	53R1,U	46
X12A-1	486-18	-	79	52L1	190-11	-	39	53R2A	273-8	53R1A	49
X12E	472-7	X12A	78	52L1A	190-11	52L1	39	53R3,U	247-8	53R1,U	46
X12E-1	486-18	X12A-1	79	52L2	190-11	52L1	39	53R3A	273-8	53R1A	49
107F31	33-14	-	13	52L2A	190-11	52L1	39	53R4,U	247-8	53R1,U	46

BRAND/MODEL	PF SET	SIMILAR TO	PAGE	BRAND/MODEL	PF SET	SIMILAR TO	PAGE	BRAND/MODEL	PF SET	SIMILAR TO	PAGE
53R4A	273-8	53R1A	49	56CS3	317-10	56CE1	54	57R1	347-9	-	60
53R5,U	247-8	53R1,U	46	56CS4	317-10	56CE1	54	57R2	347-9	57R1	60
53R5A	273-8	53R1A	49	56L1A	341-9	56B1A	59	57R3	347-9	57R1	60
53R6,U	247-8	53R1,U	46	56L1U	341-9	56B1A	59	57R4	347-9	57R1	60
53R6A	273-8	53R1A	49	56L2A	341-9	56B1A	59	57RF1	352-10	-	61
53X1	236-8	-	45	56L2U	341-9	56B1A	59	57RF2	352-10	57RF1	61
53X2	236-8	53X1	45	56R1	320-10	-	56	57W1	353-9	-	62
53X3	236-8	53X1	45	56R2	320-10	56R1	56	57W1MC	353-9	57W1	62
53X4	236-8	53X1	45	56R3	320-10	56R1	56	57X1	351-13	-	60
54L1	266-9	-	48	56R4	320-10	56R1	56	57X11	28-25	-	8
54L2	266-9	54L1	48	56RF1	324-10	-	57	57X12	28-25	57X11	8
54L3	266-9	54L1	48	56RF2	324-10	56RF1	57	57X2	351-13	57X1	60
54L4	266-9	54L1	48	56T1	339-12	-	59	582R13U	177-11	52R12U	36
54L5	266-9	54L1	48	56W1	316-11	-	54	58A11	52-13	-	16
54L6	266-9	54L1	48	56W1B	316-11	56W1	54	58A12	52-13	58A11	16
54X1	282-9	-	51	56X1	318-8	-	55	58G11	64-8	-	20
54X2	282-9	54X1	51	56X11	28-24	-	8	58G12	64-8	58G11	20
54X3	282-9	54X1	51	56X2	318-8	56X1	55	58L11	45-17	-	14
55A1	299-5	-	51	56X3	318-8	56X1	55	58R11	49-14	-	15
55A2	299-5	55A1	51	578W1B	353-9	57W1	62	58R11A	69-11	-	21
55A3	299-5	55A1	51	57A1	355-7	-	63	58R12	49-14	58R11	15
55B1	303-6	-	52	57A2	355-7	57A1	63	58R12A	69-11	58R11A	21
55C1	280-7	-	50	57A3	355-7	57A1	63	58R13	49-14	58R11	15
55C2	280-7	55C1	50	57CD1	359-9	-	64	58R13A	69-11	58R11A	21
55C3	280-7	55C1	50	57CD1A	359-9	57CD1	64	58R14	49-14	58R11	15
55C4	280-7	55C1	50	57CD2	359-9	57CD1	64	58R14A	69-11	58R11A	21
55F11	4-14	-	4	57CD2A	359-9	57CD1	64	58R15	49-14	58R11	15
55J1	301-7	-	52	57CD3	359-9	57CD1	64	58R15A	69-11	58R11A	21
55J2	301-7	55J1	52	57CD3A	359-9	57CD1	64	58R16	49-14	58R11	15
55L1	303-6	55B1	52	57CD4	359-9	57CD1	64	58R16A	69-11	58R11A	21
55L2	303-6	55B1	52	57CD4A	359-9	57CD1	64	58X11	53-15	-	16
55L3	303-6	55B1	52	57CE	353-8	-	62	58X12	53-15	58X11	16
55L4	303-6	55B1	52	57CS1	361-7	-	65	59F11	68-12	-	20
55M1	303-6	55B1	52	57CS1A	361-7	57CS1	65	59H11U	97-9	-	26
55M2	303-6	55B1	52	57CS2	361-7	57CS1	65	59H121U	97-9	59H11U	26
55M3	303-6	55B1	52	57CS2A	361-7	57CS1	65	59L11Q	78-10	59L12Q	22
56B1A	341-9	-	59	57CS3	361-7	57CS1	65	59L12Q	78-10	-	22
56B1U	341-9	56B1A	59	57CS3A	361-7	57CS1	65	59L14Q	78-10	59L12Q	22
56CC1	319-8	-	55	57CS4	361-7	57CS1	65	59R11	79-10	-	23
56CC2	319-8	56CC1	55	57CS4A	361-7	57CS1	65	59R121	79-10	59R11	23
56CD1	319-8	56CC1	55	57H1	358-8	-	64	59R13M	79-10	59R11	23
56CD2	319-8	56CC1	55	57H1A	358-8	57H1	64	59R14E	79-10	59R11	23
56CD3	319-8	56CC1	55	57H2	358-8	57H1	64	59R15G	79-10	59R11	23
56CE1	317-10	-	54	57H2A	358-8	57H1	64	59R16Y	79-10	59R11	23
56CJ1	322-9	-	57	57H3	358-8	57H1	64	59X11	81-11	-	24
56CJ2	322-9	56CJ1	57	57H3A	358-8	57H1	64	59X121	81-11	59X11	24
56CS1	317-10	56CE1	54	57H4	358-8	57H1	64	59X21U	98-6	-	26
56CS2	317-10	56CE1	54	57H4A	358-8	57H1	64	59X221U	98-6	59X21U	26

MOTOROLA (cont.)

BRAND/MODEL	PF SET	SIMILAR TO	PAGE	BRAND/MODEL	PF SET	SIMILAR TO	PAGE	BRAND/MODEL	PF SET	SIMILAR TO	PAGE
5A7	29-16	-	9	5P32R	363-14	5P31A	66	63C2	266-10	63C1	49
5A7A	29-16	5A7	9	5P32Y	363-14	5P31A	66	63C3	266-10	63C1	49
5C1	116-9	-	31	5P33EW	363-14	5P31A	66	63L1	222-8	-	42
5C11E	454-13	-	72	5R11U	115-6	-	30	63L2	222-8	63L1	42
5C12M	454-13	5C11E	72	5R12U	115-6	5R11U	30	63L3	222-8	63L1	42
5C12P	454-13	5C11E	72	5R13U	115-6	5R11U	30	63LSS	251-13	-	47
5C12W	454-13	5C11E	72	5R14U	115-6	5R11U	30	63X1	238-9	-	45
5C13B	460-13	5C13M	74	5R15U	115-6	5R11U	30	63X1A	238-9	63X1	45
5C13M	460-13	-	74	5R16U	115-6	5R11U	30	63X2	238-9	63X1	45
5C13P	460-13	5C13M	74	5R23G	409-13	-	68	63X21	249-11	-	46
5C13W	460-13	5C13M	74	5R23G-1	409-13	5R23G	68	63X3	238-9	63X1	45
5C14CW	460-13	5C13M	74	5R23N	409-13	5R23G	68	64X1	277-9	-	50
5C14GW	460-13	5C13M	74	5R23N-1	409-13	5R23G	68	64X2	277-9	64X1	50
5C14PW	460-13	5C13M	74	5T11G	444-13	5T11M	70	65F11	6-19	-	5
5C15BW	457-12	-	73	5T11M	444-13	-	70	65F21	4-12	-	3
5C15GW	457-12	5C15BW	73	5T11R	444-13	5T11M	70	65L1	305-13	-	53
5C15VW	457-12	5C15BW	73	5T11W	444-13	5T11M	70	65L11	8-22	-	6
5C16NW	460-13	5C13M	74	5T12B	462-11	5T13P	74	65L12	8-22	65L11	6
5C16W	460-13	5C13M	74	5T12M	462-11	5T13P	74	65L2	305-13	65L1	53
5C2	116-9	5C1	31	5T12P	462-11	5T13P	74	65X11A	4-8	65X14A	3
5C22M	410-11	-	69	5T12W	462-11	5T13P	74	65X12A	4-8	65X14A	3
5C22N	410-11	5C22M	69	5T13P	462-11	-	74	65X13A	4-8	65X14A	3
5C22P	410-11	5C22M	69	5T13S	462-11	5T13P	74	65X14A	4-8	-	3
5C22W	410-11	5C22M	69	5T14GW	462-11	5T13P	74	65X14B	4-8	65X14A	3
5C22Y	410-11	5C22M	69	5T14W	462-11	5T13P	74	66C1	325-9	-	58
5C23CW	410-11	5C22M	69	5T24GW-1	407-14	-	67	66C2	325-9	66C1	58
5C23GW	410-11	5C22M	69	5T24WN-1	407-14	5T24GW-1	67	66L1	338-7	-	58
5C23PW	410-11	5C22M	69	5T25B-1	407-14	5T24GW-1	67	66L2	338-7	66L1	58
5C3	116-9	5C1	31	5T25M-1	407-14	5T24GW-1	67	66T1	366-8	-	66
5C4	116-9	5C1	31	5T25MC-1	407-14	5T24GW-1	67	66X1	321-8	-	56
5H11U	117-9	-	31	5X11U	114-7	-	29	66X2	321-8	66X1	56
5H12U	117-9	5H11U	31	5X12U	114-7	5X11U	29	67C1	357-6	-	63
5H13U	117-9	5H11U	31	5X13U	114-7	5X11U	29	67C2	357-6	67C1	63
5J1	100-7	5L1	27	5X21U	120-9	-	32	67F11	31-20	-	11
5J1U	100-7	5L1	27	5X22U	120-9	5X21U	32	67F12	31-20	67F11	11
5L1	100-7	-	27	5X23U	120-9	5X21U	32	67F12B	31-20	67F11	11
5L1U	100-7	5L1	27	62C1	189-12	-	38	67L11	31-21	-	11
5M1	101-7	-	27	62C2	189-12	62C1	38	67X1	352-11	-	61
5M1U	101-7	5M1	27	62C3	189-12	62C1	38	67X11	30-20	-	10
5M2	101-7	5M1	27	62CW1	196-7	-	41	67X12	30-20	67X11	10
5M2U	101-7	5M1	27	62L1U	183-10	-	37	67X13	30-20	67X11	10
5P21B	431-15	5P21N	70	62L2U	183-10	62L1U	37	67X2	352-11	67X1	61
5P21N	431-15	-	70	62L3U	183-10	62L1U	37	67XM21	32-14	-	12
5P21R	431-15	5P21N	70	62X11U	175-14	-	34	68F11	58-13	-	19
5P31A	363-14	-	66	62X12U	175-14	62X11U	34	68F12	58-13	68F11	19
5P32C	363-14	5P31A	66	62X13U	175-14	62X11U	34	68F14	58-13	68F11	19
5P32E	363-14	5P31A	66	62X21	228-12	-	43	68F14B	58-13	68F11	19
				63C1	266-10	-	49	68F14M	58-13	68F11	19

BRAND/MODEL	PF SET	SIMILAR TO	PAGE	BRAND/MODEL	PF SET	SIMILAR TO	PAGE	BRAND/MODEL	PF SET	SIMILAR TO	PAGE
68L11	45-18	-	14	79FM21R	88-7	-	25	NC-183R	49-15	-	88
68T11	54-14	-	17	79MX21	85-9		25	NC-183T	49-15	NC-183R	88
68X11	56-16	-	18	79XM22	85-9	79MX21	25	NC-2-40DR	41-16	-	86
68X11A	56-16	68X11	18	7F11	113-5	-	29	NC-2-40DT	41-16	NC-2-40DR	86
68X12	56-16	68X11	18	7F11B	113-5	7F11	29	NC-33	47-14	-	86
68X12A	56-16	68X11	18	7X23E	462-12	7X24S	75	NC-46	9-26	-	85
69L11	76-15	-	21	7X24S	462-12	-	75	NC-57	48-14	-	87
69X11	82-9	-	24	7X24W	462-12	7X24S	75	NC-88	233-7	-	91
69X121	82-9	69X11	24	7X25P	467-10	-	75	NC-98	264-14	-	92
6F11	117-10	-	32	7X25W	467-10	7X25P	75				
6F11B	117-10	6F11	32	85F21	6-20	-	5	**NATIONAL UNION**			
6L1	102-7	-	28	85K21	5-3	-	4	G-613	19-23	-	93
6L2	102-7	6L1	28	88FM21	54-15	-	17	571	17-22	-	92
6T15N	456-15	6T15S	72	8FM21	121-9	-	33	571A	17-22	571	92
6T15S	456-15	-	72	8FM21B	121-9	8FM21	33	571B	17-22	571	92
6X11U	112-5	-	28	8K26E	459-8		73				
6X12U	112-5	6X11U	28	8K26S	459-8	8K26E	73	**NEC**			
6X28B	444-14	6X28N	71	95F31	19-22	-	7	NT-61	497-13	-	93
6X28N	444-14	-	71	95F31B	19-22	95F31	7	NT-620	498-13	-	94
6X28P	444-14	6X28N	71	95F33	19-22	95F31	7				
6X28W	444-14	6X28N	71	99FM21R	80-10	-	23	**NORELCO**			
6X39A	419-10	6X39A-2	69	9FM21	114-8	-	30	B2X98A/70R	511-10	-	98
6X39A-1	419-10	6X39A-2	69	9FM21B	114-8	9FM21	30	B3X88AU/70	508-15	B3X88U/71	97
6X39A-2	419-10	-	69					B3X88U/71	508-15	-	97
72XM21	176-7	-	35	**MUNTZ**				B3X88U/72	508-15	B3X88U/71	97
75F21	19-21	-	7	804	421-11		82	B5X88A	501-11	-	94
75F31	29-18	-	10	806A	434-8	-	83	L1X75T/64R	510-13	-	97
75F31A	29-18	75F31	10	R-10	428-9	-	82	L2X97T	503-14	-	95
75F31B	29-18	75F31	10	R-10 Revised	456-16	R-12	83	L3X86T	504-17	-	95
76F31	29-18	75F31	10	R-11	457-13	R-13	84	L3X88T	506-16	-	96
76T1	360-7	-	65	R-12	456-16	-	83	L4X95T	507-12	-	96
76T2	360-7	76T1	65	R-13	457-13	-	84				
77FM21	33-13	-	12					**OLSON**			
77FM22	33-13	77FM21	12	**NANOLA (NANAO)**				RA-315	483-11	-	98
77FM22M	33-13	77FM21	12	6TP-106	476-8	-	84	RA-323	486-19	-	99
77FM22WM	33-13	77FM21	12								
77FM23	33-13	77FM21	12	**NATIONAL**				**OLYMPIC**			
77XM21	34-12	-	13	HFS	62-14	-	89	A590	456-17	689M	123
77XM22	34-12	77XM21	13	HR0-7R	50-12	HRO-7T	88	HF500	256-11	-	109
77XM22B	34-12	77XM21	13	HR0-7T	50-12		88	SE822P	501-12	7511	128
78F11	56-17	-	18	HRO-50	112-7	-	89	SE823	501-12	7511	128
78F11M	56-17	78F11	18	HRO-50R1	169-11	-	90	402	286-8	-	111
78F12	56-17	78F11	18	HRO-50T1	169-11	HRO-50R1	90	403	315-8	-	112
78FM21	59-13	-	19	HRO-60	202-4	-	91	412	454-15	-	122
78FM21M	59-13	78FM21	19	NC-108R	47-15	NC-108T	87	441	430-10	-	117
78FM22M	59-13	78FM21	19	NC-108T	47-15	-	87	442W-1	502-14	-	129
79FM21	88-7	79FM21R	25	NC-125	139-10	-	90	445	264-15	-	111
79FM21B	88-7	79FM21R	25	NC-173R	40-13	NC-173T	85	447	349-7	-	114
				NC-173T	40-13	-	85	450-V	363-16	-	114
								455	429-9	-	116

BRAND/MODEL	PF SET	SIMILAR TO	PAGE
OLYMPIC (cont.)			
460	455-14	461	122
461	455-14	-	122
465	464-14	-	124
489	154-9	-	109
505	259-10	-	110
505B	259-10	505	110
509	431-16	-	117
51-421W	151-9	-	108
541	452-14	551	121
544	428-10	-	116
550	452-14	551	121
551	452-14	-	121
552	477-8	-	126
555	464-15	-	125
557	510-14	-	129
572B	257-11	-	110
572M	257-11	572B	110
5720W	390-8	-	115
574	319-9	-	113
575	317-11	-	112
576	321-9	-	113
577	321-9	576	113
5781W	392-9	5783W	115
5783W	392-9	-	115
589	456-17	689M	123
592	435-7	593	118
593	435-7	-	118
594	435-7	593	118
6-501	4-10	6-502P	100
6-502	4-10	6-502P	100
6-502P	4-10	-	100
6-503	4-10	6-502P	100
6-601V	8-24	6-601W	101
6-601W	8-24	-	101
6-602	8-24	6-601W	101
6-604 Series	22-21	6-604W	101
6-606	4-36	-	100
6-617	4-7	-	99
666	434-9	-	118
682	437-9	683	119
683	437-9	-	119
688	453-10	-	121
689M	456-17	-	123
694	459-10	-	123
695	472-8	697	125
696	472-8	697	125
697	472-8	-	125

BRAND/MODEL	PF SET	SIMILAR TO	PAGE
7-421V	57-13	7-421W	107
7-421W	57-13	-	107
7-421X	57-13	7-421W	107
7-435V	34-13	-	104
7-435W	34-13	7-435V	104
7-526	30-21	-	102
7-532V	32-15	7-532W	103
7-532W	32-15	-	103
7-537	37-13	-	105
7-622	34-14	-	104
7-638	34-14	7-622	104
7-724	29-19	-	102
7-925	31-22	-	103
7-934	31-22	7-925	103
7-936	31-22	7-925	103
7-939	31-22	7-925	103
700	472-8	697	125
730	479-13	-	126
7501	501-12	7511	128
7502	501-12	7511	128
7511	501-12	-	128
766	436-11	-	119
768	449-16	-	120
770	460-15	-	124
771	450-11	-	120
777	500-11	-	128
8-451	48-15	-	106
8-533V	57-14	8-533W	107
8-533W	57-14	-	107
8-618	35-16	-	105
8-925	45-19	8-934	106
8-934	45-19	-	106
8-936	45-19	8-934	106
808	480-8	-	127
859	497-15	-	127
9-435V	152-11	-	108
9-435W	152-11	9-435V	108

OLYMPIC-CONTINENTAL

BRAND/MODEL	PF SET	SIMILAR TO	PAGE
GB374	451-13	-	130
GB375	474-10	-	131
GB376	465-9	-	130
GBS388	508-16	-	132
205	495-14	-	132
300	494-13	-	131

OLYMPIC-OPTA

BRAND/MODEL	PF SET	SIMILAR TO	PAGE
52804	420-6	-	134
52805	420-6	52804	134

BRAND/MODEL	PF SET	SIMILAR TO	PAGE
5711W	424-11	-	134
5735W	389-12	5804T/W	133
5804T/W	389-12		133
5805T/W	389-12	5804T/W	133
5806T/W	398-9		133
5920	441-10	-	135

PACKARD-BELL

BRAND/MODEL	PF SET	SIMILAR TO	PAGE
PT-1	480-9	RPC-3	153
RMS-1	480-9	RPC-3	153
RPC-1	459-11	11RP7S	152
RPC-2	480-9	RPC-3	153
RPC-3	480-9	-	153
RPT-1	480-9	RPC-3	153
SAC-3	480-9	RPC-3	153
100	53-16	-	143
1052	8-26	1052A	136
1052A	8-26	-	136
1054-B	13-23	-	137
1063	18-25	-	139
10RP1	327-7	-	148
10RP2	377-13	-	150
1181	75-12	-	145
1181A	75-12	1181	145
11RP2	377-13	10RP2	150
11RP6S	459-11	11RP7S	152
11RP7S	459-11	-	152
11RP8S	459-11	11RP7S	152
11RP9S	459-11	11RP7S	152
1273	46-19	-	142
1472	48-17	-	143
471	30-22	-	140
4RB1	369-17	-	149
4RC1	496-13	-	154
531	231-11	-	146
532	232-4	-	146
541	270-12	543	148
543	270-12	-	148
568	19-24	-	139
572	22-22	-	140
5DA	16-29	-	138
5R1	343-9	-	149
5R5	457-16	-	151
5R6	473-10	-	153
5RC1	372-11	-	150
5RC3	372-11	5RC1	150
5RC4	372-11	5RC1	150
5RC7	473-10	5R6	153

BRAND/MODEL	PF SET	SIMILAR TO	PAGE
621	181-8	-	145
631	256-12	-	147
632	266-11	-	147
651	4-42	-	135
661	8-25	-	136
662	13-22	-	137
673A	46-18	-	141
673B	46-18	673A	141
682	54-16	-	144
6R1	417-8	6RC1	151
6RC1	417-8	-	151
7R2	463-10	7R3	152
7R3	463-10	-	152
861	17-23	-	138
872	31-23	-	141
880	46-18	673A	141
880A	46-18	673A	141
881-A	47-17	-	142
881-B	47-17	881-A	142
884	74-6	-	144
892	74-6	884	144

PHILCO

BRAND/MODEL	PF SET	SIMILAR TO	PAGE
B-1352	235-9	-	210
B-1752	240-6	-	210
B-1753	240-6	B-1752	210
B-956	218-8	-	206
B1349	259-11	-	212
B1756	241-10	-	211
B569	261-11	-	212
B570	228-13	-	208
B570	257-12	-	211
B572	257-12	B570	211
B574	229-9	-	208
B650	226-5	-	207
B652	234-10	-	209
B710	223-8	-	207
B714	229-10	-	209
B714X	229-10	B714	209
C-570	272-9	-	214
C-583	272-9	C-570	214
C-584	272-9	C-570	214
C-587	272-9	C-570	214
C-660	271-8	-	213
C-663	271-9	-	213
C-666	294-9	-	215
C-667	279-9	-	214
C-710	272-9	C-570	214

BRAND/MODEL	PF SET	SIMILAR TO	PAGE
C-721	272-9	C-570	214
C-722	272-9	C-570	214
C-723	272-9	C-570	214
C1348	259-11	B1349	212
D-1345	347-10	-	217
D-579	328-8	-	216
D-590	328-8	D-579	216
D-591	328-8	D-579	216
D-592	321-10	-	215
D-593	321-10	D-592	215
D-595	321-10	D-592	215
D-598	321-10	D-592	215
D-664	324-11	-	216
D-665	324-11	D-664	216
D-717	328-8	D-579	216
D-726	328-8	D-579	216
D-727	321-10	D-592	215
D-728	321-10	D-592	215
D-730	321-10	D-592	215
D-736	321-10	D-592	215
D719	328-8	D-579	216
E-1370	359-11	-	219
E-670	346-8	-	217
E-672	346-8	E-670	217
E-675	346-8	E-670	217
E-676	346-8	E-670	217
E-740	351-14	-	218
E-742	351-14	E-740	218
E-748	383-9	E-818	220
E-808	351-14	E-740	218
E-810	351-14	E-740	218
E-812	351-14	E-740	218
E-814	351-14	E-740	218
E-818	383-9	-	220
E-976	378-11	-	219
F-1600	427-9	F-1803	223
F-1700	427-9	F-1803	223
F-1702	427-9	F-1803	223
F-1802	427-9	F-1803	223
F-1803	427-9	-	223
F-1805	427-9	F-1803	223
F-743	413-12	F-809	221
F-750	413-12	F-809	221
F-752	413-12	F-809	221
F-754	413-12	F-809	221
F-758	413-12	F-809	221
F-760	396-6	F-963	220
F-809	413-12	-	221

BRAND/MODEL	PF SET	SIMILAR TO	PAGE
F-813	413-12	F-809	221
F-815	413-12	F-809	221
F-817	413-12	F-809	221
F-963	396-6	-	220
F-974	426-12	-	222
G-1706	463-11	G-1707S	226
G-1706S	463-11	G-1707S	226
G-1707	463-11	G-1707S	226
G-1707S	463-11	-	226
G-1906S	464-17	G-1907S	226
G-1907S	464-17	-	226
G-681	434-10	-	224
G-747	475-5	-	229
G-749	459-12	G-751	225
G-749X	459-12	G-751	225
G-751	459-12	-	225
G-753	459-12	G-751	225
G-755	459-12	G-751	225
G-820	443-9	G-822	225
G-822	443-9	-	225
G-824	443-9	G-822	225
G-826	443-9	G-822	225
G-828	443-9	G-822	225
H-973	486-20	-	230
H984AQ	507-13	-	231
H984E	507-13	H984AQ	231
HFT-1	427-9	F-1803	223
RSB-10	464-17	G-1907S	226
RT-150	463-11	G-1707S	226
RT-202 Series	464-17	G-1907S	226
SA-1500	463-11	G-1707S	226
SA-3000	464-17	G-1907S	226
T-1000	483-12	-	229
T-6	426-13	-	222
T-60	467-13	-	227
T-600	426-13	T-6	222
T-65	468-10	-	228
T-7	347-11	-	218
T-700	401-10	-	221
T-75	465-11	-	227
T-78	473-11	-	228
T-7X	465-11	T-75	227
T-800	401-10	T-700	221
T-9	429-10	-	223
T4	437-10	-	224
T4J	437-10	T4	224
TC-47	500-12	-	230
46-1201	4-35	-	155

BRAND/MODEL	PF SET	SIMILAR TO	PAGE	BRAND/MODEL	PF SET	SIMILAR TO	PAGE	BRAND/MODEL	PF SET	SIMILAR TO	PAGE
PHILCO (cont.)				48-472	43-15	48-472-I	173	50-522-I	78-11	50-522	185
46-1201 Revised	29-21		160	48-472 Revised	48-18	48-472	175	50-522-I	96-8	50-526	189
46-1209	13-24	-	158	48-472-I	43-15	-	173	50-524	78-11	50-522	185
46-1213	12-33	-	157	48-475	40-14	-	170	50-524	96-8	50-526	189
46-1226	15-24	-	158	48-482	30-24	-	160	50-526	96-8	-	189
46-131 Revised	32-16	46-131	161	48-485	49-19	49-1100	176	50-527	80-11	-	185
46-131	5-13	-	156	49-101	87-8	-	187	50-527-I	80-11	50-527	185
46-132	4-20	-	154	49-1100	49-19	-	176	50-620	85-11	-	186
46-142	36-16		166	49-1401	45-21	-	174	50-621	89-11	-	188
46-350	10-24	-	157	49-1405	54-24	-	180	50-920	88-8	50-921	187
46-420	6-22	-	156	49-1600	50-13	-	177	50-921	88-8	-	187
46-420-1	6-22	46-420	156	49-1613	91-9	-	188	50-922	88-8	50-921	187
46-421	5-12	-	155	49-1615	64-9	-	184	50-925	99-12	-	191
46-421-1	5-12	46-421	155	49-500	48-19	-	175	50-926	99-12	50-925	191
46-480	19-25	-	159	49-500I	48-19	49-500	175	51-1330	130-11	-	194
46-901	56-19		181	49-501	56-18	-	181	51-1730	140-8	-	195
47-204	33-18	-	163	49-501-I	56-18	49-501	181	51-1730L	140-8	51-1730	195
47-205	33-18	47-204	163	49-503	52-15	-	178	51-1731	124-7	51-1732	193
48-1201	31-25	48-1260	161	49-504	54-17	-	180	51-1732	124-7	-	193
48-1253	36-17	-	166	49-504-I	54-17	49-504	180	51-1733	137-9	-	195
48-1256	34-18	-	165	49-505	53-18	-	179	51-1733(L)	137-9	51-1733	195
48-1260	31-25	-	161	49-506	48-19	49-500	175	51-1734	137-9	51-1733	195
48-1262	35-18	-	165	49-601	42-21	-	172	51-530	122-7	-	192
48-1263	32-18	-	162	49-602	41-18	-	171	51-532	122-7	51-530	192
48-1264	36-18	-	167	49-603	59-15	-	183	51-534	122-7	51-530	192
48-1266	39-15	-	170	49-605	58-15	49-607	182	51-537	126-10	-	193
48-1270	42-20	-	172	49-607	58-15	-	182	51-537-I	126-10	51-537	193
48-1274	41-17	48-1276	171	49-900-E	49-16	-	176	51-629	136-13	-	194
48-1276	41-17	-	171	49-902	51-16	-	178	51-631	106-12	-	192
48-1282	35-18	48-1262	165	49-904	58-16	-	183	51-632	136-13	51-629	194
48-1284	45-20	-	173	49-905	52-16	-	179	51-930	153-11	-	196
48-1286	51-15	-	177	49-906	57-16	-	182	51-931	153-11	51-930	196
48-1290	47-18	-	174	49-990-I	49-16	49-900-E	176	51-932	153-11	51-930	196
48-150	34-16	-	164	50-1420	97-11	50-1421	190	51-934	102-10	-	191
48-200	33-19	48-200-I	163	50-1421	97-11	-	190	52-1340	160-8	-	197
48-200-I	33-19	-	163	50-1422	97-11	50-1421	190	52-544	163-9	52-544-I	198
48-206	37-16	-	168	50-1423	97-11	50-1421	190	52-544-I	163-9	-	198
48-214	33-19	48-200-I	163	50-1720	93-8	-	189	52-544-W	163-9	52-544-I	198
48-225	37-15	48-230	167	50-1721	98-9	50-1724	190	52-640	153-12	-	196
48-230	37-15	-	167	50-1723	98-9	50-1724	190	52-641	153-12	52-640	196
48-250	32-17	48-250-I	162	50-1724	98-9	-	190	52-643	161-7	-	198
48-250-I	32-17	-	162	50-1725	93-8	50-1720	189	52-940	156-9	-	197
48-300	37-17	-	168	50-1726	91-9	49-1613	188	52-941	156-9	52-940	197
48-360	38-14	-	169	50-1727	86-7	-	186	52-942	156-9	52-940	197
48-460	34-17	-	164	50-520	73-9	-	184	52-944	169-12	-	199
48-460-I	34-17	48-460	164	50-5201	73-9	50-520	184	53-1350	203-7	53-1750	204
48-461	38-15	-	169	50-522	78-11	-	185	53-1750	203-7	-	204
48-464	26-20	-	159	50-522	96-8	50-526	189	53-1754	214-8	-	206

BRAND/MODEL	PF SET	SIMILAR TO	PAGE
53-559	213-6	-	205
53-560	189-13	-	201
53-561	188-12	53-562	200
53-562	188-12	-	200
53-563	196-12	-	202
53-564	188-12	53-562	200
53-566	185-11	-	199
53-652	234-10		209
53-656	187-10	-	200
53-658	187-10	53-656	200
53-700	193-6	53-701	201
53-700-I	193-6	53-701	201
53-701	193-6	-	201
53-701-I	193-6	53-701	201
53-702	202-5	-	204
53-706	202-5	53-702	204
53-707	202-5	53-702	204
53-800	210-4	53-804	205
53-804	210-4	-	205
53-950	200-6	-	203
53-952	200-6	53-950	203
53-954	200-6	53-950	203
53-956	218-8		206
53-958	200-7	-	203
53-960	199-7	-	202

PHILHARMONIC

100C	38-16	-	232
349C	58-17	-	232
6810	18-27	8712	231
8701	18-27	8712	231
8702	18-27	8712	231
8703	18-27	8712	231
8710	18-27	8712	231
8711	18-27	8712	231
8712	18-27	-	231

PHILLIPS

R-100	467-14	-	233

PHILLIPS 66

3-81A	48-20	-	233

PILOT

HF-30	421-12	PT-1036	237
PT-1031	428-11	-	238
PT-1036	421-12	-	237
T-411-U	15-25	-	235
T-500U Series	12-23	T-500U	234

BRAND/MODEL	PF SET	SIMILAR TO	PAGE
T-510	5-24	T-511	234
T-511	5-24	-	234
T-521	19-27	-	236
T-530 Series	12-24	T-530AB	235
T-601	28-26	-	236
T-741	37-18	-	237

POLICALARM

PR-31	105-8	-	238
PR-31A	295-10	-	239
PR-9	359-12	-	239

PORTO BARADIO

PA-510	33-16	PB-520	240
PA-510	48-21	PB-520 Revised	240
PB-520	33-16	-	240
PB-520 Revised	48-21	-	240

PREMIER

15LW	6-24	-	241

PURITAN

501 4-5		-	241
501X	4-26	502X	242
502 4-5		501	241
502X	4-26	-	242
503	10-25	-	243
504	5-39	-	243
508	4-31	-	242
509	26-21	-	244

RADIOETTE

PR-2	50-15	-	244

RADIOLA

61-1	14-25	-	245
61-10	12-25	61-5	245
61-2	14-25	61-1	245
61-3	14-25	61-1	245
61-5	12-25	-	245
61-8	27-21	-	246
61-9	27-21	61-8	246
75ZU	36-19	-	246
76ZX11	36-20	-	247
76ZX12	36-20	76ZX11	247

RADIONIC

Y62W	26-22	-	247
Y728	26-22	Y62W	247

BRAND/MODEL	PF SET	SIMILAR TO	PAGE
RANGE			
118	28-27	-	248
RAYENERGY			
AD	7-24	-	248
AD4	7-25	-	249
SRB-1X	13-26	-	249
RAYTHEON			
CR-41	212-5	-	250
CR-41A	212-5	CR-41	250
CR-42	212-5	CR-41	250
CR-42A	212-5	CR-41	250
CR-43	212-5	CR-41	250
CR-43A	212-5	CR-41	250
FR81A	232-6	-	251
FR82A	232-6	FR81A	251
PR51	218-9	-	250
PR51A	218-9	PR51	250
T-100-1	331-11	-	253
T-100-2	331-11	T-100-1	253
T-100-3	331-11	T-100-1	253
T-100-4	331-11	T-100-1	253
T-100-5	331-11	T-100-1	253
T-150-1	335-13	-	253
T-150-2	335-13	T-150-1	253
T-150-3	335-13	T-150-1	253
T-150-4	335-13	T-150-1	253
T-150-5	335-13	T-150-1	253
T2500	329-11	-	252
5R-10B	303-11	-	252
5R-11W	303-11	5R-10B	252
5R-12R	303-11	5R-10B	252
8TP-1	292-9	-	251
8TP-2	292-9	8TP-1	251
8TP-3	292-9	8TP-1	251
8TP-4	292-9	8TP-1	251
RCA VICTOR			
A-82	137-10	-	278
A106	97-12	9W106	272
A55	109-10	-	276
BC3	442-10	1BT32	310
BC3	447-10	1BT21	310
BCS4	447-10	1BT21	310
BCS4	442-10	1BT32	310
BK2	496-14	TPR-8	314
BX55	102-11	-	274
BX57	102-11	BX55	274

BRAND/MODEL	PF SET	SIMILAR TO	PAGE	BRAND/MODEL	PF SET	SIMILAR TO	PAGE	BRAND/MODEL	PF SET	SIMILAR TO	PAGE
RCA VICTOR (cont.)				TX1HE	489-17	TX1JE	313	1X55	172-8	1X51	280
BX6	103-13	-	274	TX1JE	489-17	-	313	1X56	172-8	1X51	280
C-4E	480-10	C-4EM	312	TX1K	489-17	TX1JE	313	1X57	172-8	1X51	280
C-4EM	480-10	-	312	X-4EF	480-10	C-4EM	312	1X591	159-12	1X592	279
C-4FE	480-10	C-4EM	312	X-4HE	480-10	C-4EM	312	1X592	159-12	-	279
C1D	465-14	C1E	311	X-4JE	480-10	C-4EM	312	2-R-51	196-13	-	283
C1E	465-14	-	311	X1	465-14	C1E	311	2-R-52	196-13	2-R-51	283
C1F	465-14	C1E	311	X2EF	465-14	C1E	311	2-S-10	210-5	-	285
C1L	465-14	C1E	311	X2HE	465-14	C1E	311	2-S-7	222-11	-	286
C2E	465-14	C1E	311	X2JE	465-14	C1E	311	2-X-621	199-9	-	284
C2FE	465-14	C1E	311	X3EM	465-14	C1E	311	2-XF-91	206-9	-	284
C2J	465-14	C1E	311	X3EN	465-14	C1E	311	2-XF-931	209-9	-	285
C3E	465-14	C1E	311	X3H3	465-14	C1E	311	2-XF-932	209-9	2-XF-931	285
C3EK	465-14	C1E	311	X551	129-9	X552	276	2-XF-933	209-9	2-XF-931	285
C3HE	465-14	C1E	311	X552	129-9	-	276	2-XF-934	209-9	2-XF-931	285
C3PC1	465-14	C1E	311	X711	133-11	-	277	2B400	181-10	-	281
DK109	496-14	TPR-8	314	XF2	479-14	XF3EM	312	2B401	181-10	2B400	281
DK111	496-14	TPR-8	314	XF3EH	479-14	XF3EM	312	2B402	181-10	2B400	281
KS11	491-12	-	313	XF3EM	479-14	-	312	2B403	181-10	2B400	281
KS12	491-12	KS11	313	XF3J	479-14	XF3EM	312	2B404	181-10	2B400	281
KS4	496-14	TPR-8	314	XF4	479-14	XF3EM	312	2B405	181-10	2B400	281
PM17	491-12	KS11	313	1BT21	447-10	-	310	2BX63	193-7	-	282
PM18	491-12	KS11	313	1BT24	447-10	1BT21	310	2C521	194-11	-	282
PT1	478-15	T1JE	311	1BT29	447-10	1BT21	310	2C522	194-11	2C521	282
PX600	168-12	-	280	1BT32	442-10	-	310	2C527	194-11	2C521	282
RK222	447-10	1BT21	310	1BT34	442-10	1BT32	310	2US7	182-8	-	281
RK222	442-10	1BT32	310	1BT36	442-10	1BT32	310	2X61	197-8	-	283
RK249	499-11	1T4J	314	1BT41	410-14	1BT46	307	2X62	197-8	2X61	283
SHC2	435-10	SHC4	309	1BT46	410-14	-	307	3-BX-51	227-11	-	287
SHC3	435-10	SHC4	309	1BT48	410-14	1BT46	307	3-BX-52	227-11	3-BX-51	287
SHC4	435-10	-	309	1BX57	422-12	-	308	3-BX-53	227-11	3-BX-51	287
SHC6	435-10	SHC4	309	1BX59	422-12	1BX57	308	3-BX-54	227-11	3-BX-51	287
SHF1	418-11	SHF-1	308	1BX62	422-12	1BX57	308	3-BX-671	228-14	-	288
SHF3	396-7	-	306	1BX64	422-12	1BX57	308	3-RF-91	226-6	-	286
SHF3D	396-7	SHF3	306	1BX67	422-12	1BX57	308	3-US-5	267-10	-	290
T1EH	478-15	T1JE	311	1BX78	422-12	1BX57	308	3-US-5A	267-10	3-US-5	290
T1EN	478-15	T1JE	311	1BX79	422-12	1BX57	308	3-X-521	226-7	-	287
T1JE	478-15	-	311	1MBT6	436-13	-	309	3-X-532	226-7	3-X-521	287
T2E	478-15	T1JE	311	1R81	156-10	-	279	3-X-533	226-7	3-X-521	287
T2J	478-15	T1JE	311	1T4E	499-11	1T4J	314	3-X-534	226-7	3-X-521	287
T2K	478-15	T1JE	311	1T4H	499-11	1T4J	314	3-X-535	226-7	3-X-521	287
TC1E	478-15	T1JE	311	1T4J	499-11	-	314	3-X-536	226-7	3-X-521	287
TC1FE	478-15	T1JE	311	1T5J	504-18	-	315	4-C-531	260-13	-	289
TC1JE	478-15	T1JE	311	1T5L	504-18	1T5J	315	4-C-532	260-13	4-C-531	289
TPM11	491-12	KS11	313	1X51	172-8	-	280	4-C-533	260-13	4-C-531	289
TPM12	491-12	KS11	313	1X52	172-8	1X51	280	4-C-534	260-13	4-C-531	289
TPM4	496-14	TPR-8	314	1X53	172-8	1X51	280	4-C-535	260-13	4-C-531	289
TPR8	496-14	TPR-8	314	1X54	172-8	1X51	280	4-C-541	273-11	-	292

BRAND/MODEL	PF SET	SIMILAR TO	PAGE	BRAND/MODEL	PF SET	SIMILAR TO	PAGE	BRAND/MODEL	PF SET	SIMILAR TO	PAGE
4-C-542	273-11	4-C-541	292	6-XD-5	309-14	-	295	75X11	33-21	-	262
4-C-543	273-11	4-C-541	292	6-XY-5A	333-11	-	298	75X12	33-21	75X11	262
4-C-544	273-11	4-C-541	292	6-XY-5B	333-11	6-XY-5A	298	77U	38-17	-	263
4-C-545	273-11	4-C-541	292	610V1	31-27	610V2	262	77V1	38-18	-	263
4-C-547	273-11	4-C-541	292	610V2	31-27	-	262	77V2	39-18	-	264
4-C-671	269-11	-	291	612V1	17-27	612V3	259	8-BT-10K	350-13	-	301
4-C-672	269-11	4-C-671	291	612V2	17-27	612V3	259	8-C-51	369-18	8-C-5E	304
4-X-551	271-10	-	291	612V3	17-27	-	259	8-C-5D	369-18	8-C-5E	304
4-X-552	271-10	4-X-551	291	64F1	4-16	64F3	254	8-C-5E	369-18	-	304
4-X-553	271-10	4-X-551	291	64F2	4-16	64F3	254	8-C-5F	369-18	8-C-5E	304
4-X-554	271-10	4-X-551	291	64F3	4-16	-	254	8-C-5L	369-18	8-C-5E	304
4-X-555	271-10	4-X-551	291	65AU	14-23	65U	257	8-C-6E	369-18	8-C-5E	304
4-X-661	265-9	-	290	65BR9	23-16	-	260	8-C-6J	369-18	8-C-5E	304
4-Y-511	261-12	-	289	65U	14-23	-	257	8-C-6L	369-18	8-C-5E	304
45-W-10	138-8	-	278	65X1	4-30	65X2	255	8-C-6M	369-18	8-C-5E	304
4X641	259-13	-	288	65X1	31-26	-	261	8-C-7EE	359-13	-	302
5-BX-41	278-10	-	292	65X2	31-26	65X1	261	8-C-7FE	359-13	8-C-7EE	302
5-C-581	284-11	-	293	65X2	4-30	-	255	8-C-7LE	359-13	8-C-7EE	302
5-C-591	285-12	-	294	66BX	14-24	-	258	8-C-8DE	359-13	8-C-7EE	302
5-C-592	285-12	5-C-591	294	66X1	7-23	-	256	8-C-8JJ	359-13	8-C-7EE	302
5-X-560	279-12	-	293	66X11	27-20	-	261	8-C-8ME	359-13	8-C-7EE	302
5-X-562	279-12	5-X-560	293	66X12	27-20	66X11	261	8-RF-13	390-9	-	305
5-X-564	279-12	5-X-560	293	66X13	27-20	66X11	261	8-X-5 Series	368-11	8-X-5D	303
54B1	7-22	-	256	66X14	27-20	66X11	261	8-X-51	368-11	8-X-5D	303
54B1-N	7-22	54B1	256	66X15	27-20	66X11	261	8-X-5D	368-11	-	303
54B2	7-22	54B1	256	66X2	7-23	66X1	256	8-X-6 Series	368-11	8-X-5D	303
54B3	7-22	54B1	256	66X3	7-23	66X1	256	8-X-8D	359-14	-	302
54B5	17-25	-	258	66X4	7-23	66X1	256	8-X-8J	359-14	8-X-8D	302
55F	4-6	-	254	66X9	7-23	66X1	256	8-X-8L	359-14	8-X-8D	302
59AV1	6-25	-	255	67AV1	9-27	-	257	8-X-8N	359-14	8-X-8D	302
59V1	6-25	59AV1	255	67V1	9-27	67AV1	257	8B41	76-16	8B42	270
6-X-5A	316-12	-	296	68R1	23-17	-	260	8B42	76-16	-	270
6-BX-5	299-7	-	294	68R2	23-17	68R1	260	8B43	76-16	8B42	270
6-BX-6A	299-7	6-BX-5	294	68R3	23-17	68R1	260	8BX5	46-20	-	265
6-BX-6B	299-7	6-BX-5	294	68R4	23-17	68R1	260	8F43	97-13	-	272
6-BX-6C	299-7	6-BX-5	294	6HF3	323-11	-	297	8R71	53-20	-	266
6-BY-4A	321-12	-	296	7-BT-10K	329-10	-	298	8R72	53-20	8R71	266
6-BY-4B	321-12	6-BY-4A	296	7-BT-9J	327-10	-	297	8R74	53-20	8R71	266
6-C-5A	340-15	-	299	7-BX-6	344-11	-	300	8R75	53-20	8R71	266
6-C-5B	340-15	6-C-5A	299	7-BX-7	344-11	7-BX-6	300	8R76	53-20	8R71	266
6-C-5C	340-15	6-C-5A	299	7-BX-8J	345-12	-	300	8V111	58-18	-	267
6-HF-1	334-8	-	299	7-BX-8L	345-12	7-BX-8J	300	8V112	58-18	8V111	267
6-HF-2	334-8	6-HF-1	299	7-BX-9H	349-10	-	301	8V90	56-20	-	267
6-X-5B	316-12	6-X-5A	296	7-C-6F	340-15	6-C-5A	299	8V91	56-20	8V90	267
6-X-5C	316-12	6-X-5A	296	7-C-6N	340-15	6-C-5A	299	8X521	52-17	-	266
6-X-7A	303-9	-	295	7-HFR-1	365-11	-	303	8X522	52-17	8X521	266
6-X-7B	303-9	6-X-7A	295	710V2	40-15	-	265	8X53	39-17	-	264
6-X-7C	303-9	6-X-7A	295	711V2	22-24	-	259	8X541	59-16	8X542	268

BRAND/MODEL	PF SET	SIMILAR TO	PAGE	BRAND/MODEL	PF SET	SIMILAR TO	PAGE	BRAND/MODEL	PF SET	SIMILAR TO	PAGE
RCA VICTOR *(cont.)*				9INT2	399-11	9INT1	307	9X571	107-7	-	275
8X542	59-16	-	268	9S103	73-10	9W101	269	9X572	107-7	9X571	275
8X547	59-16	8X542	268	9US5H	395-12	-	306	9X641	87-9	9X642	271
8X681	65-10	8X682	269	9US5KE	395-12	9US5H	306	9X642	87-9	-	271
8X682	65-10	-	269	9W101	73-10	-	269	9X651	104-9	9X652	275
8X71	63-15	-	268	9W105	73-10	9W101	269	9X652	104-9	-	275
8X72	63-15	8X71	268	9W106	97-12	-	272	9Y51	98-11	-	273
9-BT-9E	371-7	9-BT-9H	304	9X10FE	394-14	-	305	9Y510	131-13	-	277
9-BT-9H	371-7	-	304	9X10JE	394-14	9X10FE	305	9Y511	131-13	9Y510	277
9-BT-9J	371-7	9-BT-9H	304	9X10ME	394-14	9X10FE	305	9Y7	75-13	-	270
9BX56	79-13	-	271	9X561	101-9	-	273				
9INT1	399-11	-	307	9X562	101-9	9X561	273				

POPULAR HIT SONGS
1950-1960

1950   Autumn Leaves   *Johnny Mercer*

1951   Unforgettable   *Nat King Cole*

1952   Your Cheatin' Heart   *Hank Williams Sr.*

1953   Rock Around the Clock   *Bill Haley and the Comets*

1954   Shake, Rattle, and Roll   *Joe Turner*

1955   Maybelline   *Chuck Berry*

1956   Blueberry Hill   *Fats Domino*

1957   Jailhouse Rock   *Elvis Presley*

1958   Whole Lotta Shakin Goin On   *Jerry Lee Lewis*

1959   Put Your Head on My Shoulder   *Paul Anka*

1960   The Twist   *Chubby Checker*

Pictured Radios and Associated Tubes

BRAND/MODEL	PAGE	USES THESE TUBES
MOTOROLA		
A1W	80	12BE6, 12BA6, 12AV6, 50C5, 35W4
C3S-1	81	12BE6, 12BA6, 12AV6, 50C5, 35W4
C4B	81	12BE6, 12BA6, 12AV6, 50C5, 35W4
L12N	76	Transistorized
L14E	76	Transistorized
SK25MC	77	6BQ7A, 6BA6, 6BE6, 6BA6, 6AU6, 6AL5, EM81, 12AU7A, 6X4, 12AX7A, 12AX7A, 12AU7, EL84, EL84, 5Y3GT, 12AX7A, EL34, EL34, 5U4GB
SK32W	80	6BQ7A, 6BA6, 6BA6, 6BE6, 6BA6, 6AU6, 6AL5, EM81/6DA5, 12AX7A, 12AX7A
X11B	79	Transistorized
X11E	78	Transistorized
X12A	78	Transistorized
X12A-1	79	Transistorized
107F31	13	6SK7, 7Q7, 7F8, 6SG7, 7W7, 6SQ7, 6AL5, 6V6GT, 6V6GT, 5Y3GT
10KT12M	77	6BQ7A, 6AU6, 6BE6, 6BA6, 6AL5, 12AX7, 12AX7, EL84/6BQ5, EL84/6BQ5, EZ81/6CA4
10T28B	67	6BQ7A, 6BA6, 6BE6, 6BA6, 6AU6, 6AL5, 6AV6, 6AQ6, 6X4, 6DA5/EM81
13KT15M	71	6BQ7A, 6BA6, 6BA6, 6BE6, 6BA6, 6AU6, 6AL5, EM81/6DA5, 12AX7, 12AX7, EL84/6BQ5, EL84/6BQ5, EZ81/6CA4
15KT25MC-1	68	6BQ7A, 6BA6, 6BE6, 6BA6, 6AU6, 6AL5, 6DA5/EM81, 12AU7, 6X4, 12AU7, 12AU7, 12AX7, 6CA7/EL34, 5U4GB, 6CA7/EL34
42B1	40	1R5, 1U4, 1U5, 3S4
45B12	6	1A7GT, 1N5GT, 1H5GT, 3Q5GT
45P1	53	1R5, 1AH4, 1AJ5, 1AG4
47B11	9	1A7GT, 1N5GT, 1H5GT, 3Q5GT
48L11	15	1R5, 1U4, 1S5, 3S4
49L11Q	22	1R5, 1U4, 1U5, 3S4
51M1U	33	1R5, 1U4, 1U5, 3S4
52B1U	39	1R5, 1U4, 1U5, 3S4
52C1	40	12BE6, 12BD6, 12AT6, 50C5, 35W4
52C6	35	12BE6, 12BD6, 12AT6, 50C5, 35W4
52CW1	41	12BE6, 12BD6, 12AT6, 50C5, 35W4
52H11U	34	12BE6, 12BD6, 12AT6, 50C5, 35W4
52L1	39	1R5, 1U4, 1U5, 3S4
52M1U	37	1R5, 1U4, 1U5, 3S4
52R12A	36	12BE6, 12BD6, 12AT6, 50C5, 35W4
52R12U	36	12BE6, 12BD6, 12AT6, 50C5, 35W4
52R14	38	12BE6, 12BD6, 12AT6, 50C5, 35W4
53C1	44	12BE6, 12BA6, 12AT6, 50C5, 35W4
53C6	44	12BE6, 12BA6, 12AT6, 50C5, 35W4
53D1	48	12BE6, 12BA6, 12AT6, 50C5, 35W4
53F2	43	12BE6, 12BA6, 12AT6, 35L6GT, 50Y6GT
53H1	47	12BE6, 12BA6, 12AT6, 50C5, 35W4
53LC1	42	1V6, 1AH4, 1AJ5, 3S4
53R1	46	12BE6, 12BA6, 12AT6, 50C5, 35W4
53R1A	49	12BE6, 12BA6, 12AT6, 50C5, 35W4
53X1	45	12BE6, 12BA6, 12AT6, 50C5, 35W4
54L1	48	1V6, 1AH4, 1AJ5, 3S4
54X1	51	12BE6, 12BA6, 12AV6, 50C5, 35W4
55A1	51	12BE6, 12BA6, 12AV6, 50C5, 35W4
55B1	52	1R5, 1U4, 1U5, 3V4
55C1	50	12BE6, 12BA6, 12AT6, 50C5, 35W4
55F11	4	7Q7, 7A7, 7B6, 7B5, 5Y3GT
55J1	52	1R5, 1U4, 1U5, 3V4
56B1A	59	1R5, 1U4, 1U5, 3V4
56CC1	55	12BE6, 12BA6, 12AV6, 50C5, 35W4
56CE1	54	12BE6, 12BA6, 12AV6, 50C5, 35W4
56CJ1	57	12BE6, 12BA6, 12AV6, 50C5, 35W4
56R1	56	12BE6, 12BA6, 12AV6, 50C5, 35W4
56RF1	57	12BE6, 12BA6, 12AV6, 50C5, 35W4
56T1	59	Transistorized
56W1	54	12BE6, 12BA6, 12AV6, 50C5, 35W4
56X1	55	12BE6, 12BA6, 12AV6, 50C5, 35W4
56X11	8	12SA7GT, 12SK7GT, 12SQ7GT, 50L6GT, 35Z5GT
57A1	63	12BE6, 12BA6, 12AV6, 50C5, 35W4
57CD1	64	12BE6, 12BA6, 12CR6, 50C5, 35W4
57CE	62	12BE6, 12BA6, 12AV6, 50C5, 35W4
57CS1	65	12BE6, 12BA6, 12AV6, 50C5, 35W4
57H1	64	12BE6, 12BA6, 12AV6, 50C5, 35W4
57R1	60	12BE6, 12BA6, 12AV6, 50C5, 35W4
57RF1	61	12BE6, 12BA6, 12AV6, 50C5, 35W4
57W1	62	12BE6, 12BA6, 12CR6, 50C5, 35W4
57X1	60	12BE6, 12BA6, 12CR6, 50C5, 35W4
57X11	8	12SA7GT, 12SK7, 12SQ7GT, 50L6GT, 35Z5GT
57X12	8	12SA7GT, 12SK7, 12SQ7GT, 50L6GT, 35Z5GT
58A11	16	12SA7, 12SK7, 12SQ7, 50L6GT, 35Z5GT
58G11	20	12BE6, 12BA6, 12AT6, 50C5, 35W4
58L11	14	1R5, 1U4, 1S5, 3S4
58R11	15	12BE6, 12BA6, 12AT6, 50B5, 35W4
58R11A	21	12BE6, 12BA6, 12AT6, 50B5 or 50C5, 35W4

BRAND/MODEL	PAGE	USES THESE TUBES
MOTOROLA (cont.)		
58X11	16	12SA7, 12SK7, 12SQ7, 50L6GT, 35Z5GT
59F11	20	12BE6, 12BA6, 12AT6, 50C5, 35W4
59H11U	26	12BE6, 12BA6, 12AT6, 50C5, 35W4
59L12Q	22	1R5, 1U4, 1U5, 3S4
59R11	23	12BE6, 12BA6, 12AT6, 50C5, 35W4
59X11	24	12BE6, 12BA6, 12AT6, 50C5, 35W4
59X21U	26	12BE6, 12BA6, 12AT6, 50C5, 35W4
5A7	9	1R5, 1U4, 1S5, 3S4
5C1	31	12BE6, 12BA6, 12AT6, 50C5, 35W4
5C11E	72	12BE6, 12BA6, 12AV6, 50C5, 35W4
5C13M	74	12BE6, 12BA6, 12AV6, 50C5, 35W4
5C15BW	73	12BE6, 12BA6, 12AV6, 50C5, 35W4
5C22M	69	12BE6, 12BA6, 12AV6, 50C5, 35W4
5H11U	31	12BE6, 12BA6, 12AT6, 50C5, 35W4
5J1	27	1R5, 1U4, 1U5, 3S4
5L1	27	1R5, 1U4, 1U5, 3S4
5M1	27	1R5, 1U4, 1U5, 3S4
5P21N	70	1R5, 1U4, 1DN5, 3V4
5P31A	66	1R5, 1U4, 1DN5, 3V4
5R11U	30	12BE6, 12BA6, 12AT6, 50C5, 35W4
5R23G	68	12BE6, 12BA6, 12AV6, 50C5, 35W4
5T11M	70	12BE6, 12BA6, 12AV6, 50C5, 35W4
5T13P	74	12BE6, 12BA6, 12AV6, 50C5, 35W4
5T24GW-1	67	12BE6, 12BA6, 12AV6, 50C5, 35W4
5X11U	29	12BE6, 12BA6, 12AT6, 50C5, 35W4
5X21U	32	12BE6, 12BA6, 12AT6, 50C5, 35W4
62C1	38	12BD6, 12BE6, 12BD6, 12AT6, 35C5, 35W4
62CW1	41	12BD6, 12BE6, 12BD6, 12AT6, 35C5, 35W4
62L1U	37	1U4, 1R5, 1U4, 1U5, 3V4
62X11U	34	12BD6, 12BE6, 12BD6, 12AT6, 35C5, 35W4
62X21	43	12BA6, 12BE6, 12BA6, 12AT6, 35C5, 35W4
63C1	49	12BA6, 12BE6, 12BA6, 12AT6, 35C5, 35W4
63L1	42	1U4, 1R5, 1U4, 1U5, 3V4
63LSS	47	1U4, 1R5, 1U4, 1U5, 3V4
63X1	45	12BA6, 12BE6, 12BA6, 12AT6, 35C5, 35W4
63X21	46	12BA6, 12BE6, 12BA6, 12AT6, 35C5, 35W4
64X1	50	12BA6, 12BE6, 12BA6, 12AT6, 35C5, 35W4
65F11	5	7H7, 7A4, 7A7, 7B6, 7B5, 5Y3GT
65F21	3	6SG7, 6J5, 6SG7, 6SQ7, 6K6GT, 5Y3GT
65L1	53	1U4, 1R5, 1U5, 1U4, 3V4
65L11	6	1N5GT, 1A7GT, 1N5GT, 1H5GT, 3Q5GT, 117Z6GT
65X14A	3	12SJ7GT, 12SA7GT, 12SK7GT, 12SQ7GT, 35L6GT, 35Z5GT
66C1	58	12BA6, 12BE6, 12BA6, 12AT6, 35C5, 35W4
66L1	58	1T4, 1R5, 1U4, 1U5, 3V4
66T1	66	Transistorized
66X1	56	12BA6, 12BE6, 12BA6, 12AV6, 35C5, 35W4
67C1	63	12BA6, 12BE6, 12BA6, 12CR6, 35C5, 35W4
67F11	11	12AU6, 6BT6, 12BA6, 6AQ6, 50B5, 35W4
67F12	11	12AU6, 6BT6, 12BA6, 6AQ6, 50B5, 35W4
67F12B	11	12AU6, 6BT6, 12BA6, 6AQ6, 50B5, 35W4
67L11	11	1U4, 1R5, 1U4, 1S5, 3V4
67X1	61	12BA6, 12BE6, 12BA6, 12CR6, 35C5, 35W4
67X11	10	14C7, 14Q7, 14A7, 14B6, 35A5, 35Y4
67X12	10	14C7, 14Q7, 14A7, 14B6, 35A5, 35Y4
67X13	10	14C7, 14Q7, 14A7, 14B6, 35A5, 35Y4
67XM21	12	12AT7, 12BE6, 12BA6, 19T8, 50B5
68F11	19	12AU6, 6C4, 12BA6, 6AQ6, 50B5, 35W4
68F12	19	12AU6, 6C4, 12BA6, 6AQ6, 50B5, 35W4
68F14	19	12AU6, 6C4, 12BA6, 6AQ6, 50B5, 35W4
68L11	14	1U4, 1R5, 1U4, 1S5, 3V4
68T11	17	6SK7, 6SA7, 6SK7, 6SQ7 or 7B6, 7A5
68X11	18	14C7, 14Q7, 14A7/12B7, 14B6, 35A5, 35Y4
69L11	21	1U4, 1R5, 1U4, 1S5, 3V4
69X11	24	12BA6, 12BE6, 12BA6, 12AT6, 35C5, 35W4
6F11	32	6BA6, 6BE6, 6BA6, 6AV6, 6K6GT, 5Y3GT
6L1	28	1U4, 1R5, 1U4, 1U5, 3V4
6T15S	72	12BA6, 12BE6, 12BA6, 12AV6, 35C5, 35W4
6X11U	28	12BA6, 12BE6, 12BA6, 12AT6, 35C5, 35W4
6X28N	71	Transistorized
6X39A-2	69	Transistorized
72XM21	35	12BA6, 12BA7, 12BA6, 12BA6, 19T8, 50C5
75F21	7	7Q7, 6SG7, 6SQ7, 6SQ7, 6K6GT, 6K6GT, 5Y3GT
75F31	10	7F8, 7Q7, 6SG7, 6SG7 or 7W7, 6S8GT, 6K6GT or 6Y6GT, 5Y3GT
76T1	65	Transistorized
77FM21	12	12AT7, 12BE6, 12BA6, 12BA6, 19T8, 50B5

BRAND/MODEL	PAGE	USES THESE TUBES
77XM21	13	12AT7, 12BE6, 12BA6, 12BA6, 19T8, 50B5
77XM22	13	12AT7, 12BE6, 12BA6, 12BA6, 19T8, 50B5
77XM22B	13	12AT7, 12BE6, 12BA6, 12BA6, 19T8, 50B5
78F11	18	6SK7, 6SA7, 6SK7, 12SQ7, 50L6GT, 50L6GT
78F12	18	6SK7, 6SA7, 6SK7, 12SQ7, 50L6GT, 50L6GT
78FM21	19	12AT7, 12BE6, 12BA6, 12BA6, 19T8, 50B5
78FM22M	19	12AT7, 12BE6, 12BA6, 12BA6, 19T8, 50B5
79FM21R	25	12BA6, 12BA7, 12BA6, 12BA6, 19T8, 50C5
79XM21	25	12BA6, 12BA7, 12BA6, 12BA6, 19T8, 50C5
7F11	29	6BA6, 6BE6, 6BA6, 6AV6, 6V6GT, 6V6GT, 7Z4
7X24S	75	Transistorized
7X25P	75	Transistorized
85F21	5	6SG7, 6J5, 6SG7, 6SQ7, 6SQ7, 6K6GT, 6K6GT, 5Y3GT
85K21	4	6SG7, 6J5, 6SG7, 6SQ7, 6SQ7, 6K6GT, 5Y3GT, 6K6GT
88FM21	17	12AT7, 6BE6, 6BA6, 6BA6, 19T8, 50B5, 50B5
8FM21	33	6BA6, 6BA7, 6BA6, 6BA6, 6AL5, 6AV6, 6K6GT, 5Y3GT
8K26E	73	Transistorized
95F31	7	7F8, 7Q7, 6SG7, 6SG7, 6S8GT, 6SQ7, 6K6GT, 6K6GT, 5Y3GT
99FM21R	23	6AU6, 6BA7, 6BA6, 6AU6, 6AL5, 6V6GT, 6V6GT, 7Z4
9FM21	30	6AU6, 6BA7, 6BA6, 6AU6, 6AL5, 6AV6, 6V6GT, 6V6GT, 7Z4

MUNTZ

BRAND/MODEL	PAGE	USES THESE TUBES
R-10	82	12AU6, 12AV6, 50C5, 35W4
R-12	83	12AU6, 12AV6, 50C5, 50DC4
R-13	84	12BE6, 12BA6, 12AV6, 50C5, 35W4
804	82	12BE6, 12BA6, 12AV6, 12AV6, 35C5, 35C5
806A	83	6U8, 12AT7, 6BE6, 6BA6, 6BA6, 6AL5, 6AV6, 12AX7, 6973, 6973, 6CA4/EZ81

NANOLA (NANAO)

BRAND/MODEL	PAGE	USES THESE TUBES
6TP-106	84	Transistorized

NATIONAL

BRAND/MODEL	PAGE	USES THESE TUBES
HFS	89	6AK5, 9002, 6SG7, 6J5, 6J5, 6V6GT, 5Y3GT, 6SK7
HRO-50	89	6BA6, 6BA6, 6BE6, 6C4, 6K7, 6K7, 6J7, 6H6, 6SN7GT, 6H6, 6SJ7, 6V6GT, 6V6GT, 0B2, 5U4G, 6AK6, 6SK7, 6H6

BRAND/MODEL	PAGE	USES THESE TUBES
HRO-50R1	90	6BA6, 6BA6, 6BE6, 6C4, 6K7, 6SG7, 6SG7, 6H6, 6J7, 6H6, 6SJ7, 6SN7GT, 6V6GT, 6V6GT, 0B2, 5V4G
HRO-60	91	6BA6, 6BA6, 6C4, 6BE6, 6BE6, 6SG7, 6SG7, 6SG7, 6SJ7, 6H6, 6H6, 6SJ7, 6SN7GT, 6VGT, 6VGT, 5V4G, 0B2, 4H-4
HRO-7T	88	6K7, 6K7, 6J7, 6C4, 6J7 or 6K7, 0A2, 6K7, 6H6, 6J7, 6H6, 6SJ7, 6V6GT, 5Y3GT
NC-108T	87	6BA6, 6AG5, 6C4, 6SG7, 6SG7, 6SG7, 6H6, 6U5/6G5, 6SJ7, 6V6GT, 5Y3GT
NC-125	90	6SG7, 6SB7Y, 6SG7, 6SG7, 6H6, 6SL7GT, 6SL7GT, 6SL7GT, 6V6GT, 5Y3GT, 0D3/VR150
NC-173T	85	6SG7, 6SA7, 6J5, 6SG7, 6SG7, 6H6, 6H6, 6AC7, 6SJ7, 6SJ7, 6V6GT, 0D3/VR150, 5Y3GT
NC-183R	88	6SG7, 6SG7, 6SA7, 6J5, 6SG7, 6SG7, 6AC7, 6SJ7, 6H6, 6H6, 6SJ7, 6J5, 6V6GT, 6V6GT, 0D3/VR150, 5U4G
NC-2-40DR	86	6SK7, 6K8, 6J5, 6K7, 6SK7, 6SL7GT, 6SJ7, 6V6GT, 6SN7GT, 6V6GT, 6V6GT, 5Y3GT
NC-33	86	12SA7, 12SG7, 12H6, 12SL7GT, 35L6GT, 35Z5GT
NC-46	85	6K8, 6SG7, 6SG7, 6H6, 6SJ7, 6SF7, 6SC7, 25L6GT, 25L6GT, 25Z5GT
NC-57	87	6SG7, 6SB7Y, 6SG7, 6SG7, 6H6, 6SL7GT, 6V6GT, 0D3/VR150, 5Y3GT
NC-88	91	6BA6, 6C4, 6BE6, 6BD6, 6BD6, 6AL5, 12AX7, 6AQ5, 5Y3GT
NC-98	92	6BA6, 6BE6, 6C4, 6BD6, 6BD6, 6AL5, 12AX7, 6AQ5, 5Y3GT

NATIONAL UNION

BRAND/MODEL	PAGE	USES THESE TUBES
G-613	93	1A7GT, 1N5GT, 1N5GT, 1H5GT, 3Q5GT, 117Z6GT
571	92	12SA7GT/G, 12SK7GT/G, 12SQ7GT/G, 50L6GT, 35Z5GT

NEC

BRAND/MODEL	PAGE	USES THESE TUBES
NT-61	93	Transistorized
NT-620	94	Transistorized

NORELCO

BRAND/MODEL	PAGE	USES THESE TUBES
B2X98A/70R	98	ECC85, ECH81, EF89, UABC80, UL84, UY85
B3X88U/71	97	UABC80, UY85, UL84, UF89, UCH81, UCC85, UM80
B5X88A	94	ECC85, ECH81, EF89, EF85, EABC80, EL84, EL86, EM84, EZ80
L1X75T/64R	97	Transistorized
L2X97T	95	Transistorized
L3X86T	95	Transistorized
L3X88T	96	Transistorized
L4X95T	96	Transistorized

BRAND/MODEL	PAGE	USES THESE TUBES
OLSON		
RA-315	98	Transistorized
RA-323	99	12BE6, 12BA6, 12AV6, 50C5, 35W4
OLYMPIC		
HF500	109	12SA7, 12SK7, 12SQ7, 12AV6, 35L6GT, 35L6GT
402	111	12BA6, 12AU6, 12AT6, 50C5, 35W4
403	112	12BE6, 12BA6, 12AV6, 50C5, 35W4
412	122	12BE6, 12BA6, 12AV6, 50C5, 35W4
441	117	12BE6, 12BA6, 12AV6, 50C5, 35W4
442W-1	129	12SA7, 12BA6, 12AV6, 50C5, 35W4
445	111	1R5, 1U4, 1U5, 3S4
447	114	
450-V	114	1R5, 1T4, 1S5, 3S4
455	116	12AU6, 12AV6, 50C5, 35W4
461	122	(1AQ5)1R5-SF, (1AM4)1T4-SF, (1AS5)1U5-SF, (3W4)3S4-SF
465	124	12AU6, 12AV6, 50C5, 35W4
489	109	1R5, 1U4, 1U5, 3V4
505	110	6AG5, 6BE6, 6BA6, 6BA6, 6AL5, 6SQ7, 6SQ7, 6L6G, 6L6G, 5U4G
509	117	12BE6, 12BA6, 12AV6, 50C5, 35W4
51-421W	108	12SA7, 12SK7, 12SQ7, 50L6GT, 35Z5GT
544	116	12BE6, 12BA6, 12AV6, 50C5, 35W4
551	121	12BA6, 12AU6, 12AV6, 50C5, 35W4
552	126	12BE6, 12BA6, 12AV6, 50C5, 35W4
555	125	12BE6, 12BA6, 12AV6, 50C5, 35W4
557	129	12BA6, 12AU6, 12AV6, 50C5, 35W4
572B	110	6AG5, 6BE6, 6BA6, 6BA6, 6AL5, 6SQ7, 6SQ7, 6V6GT, 6V6GT, 5Y3GT
5720W	115	6SQ8/ECC85, 6AJ8/ECH81, EBF89, 6BM8/ECL82, 6V4/EZ80
574	113	12SA7GT, 12SK7GT, 12SQ7, 50L6GT, 35Z5GT
575	112	6BJ6, 12BE6, 6BJ6, 12AU6, 19T8, 35C5, 35W4
576	113	6AG5, 6BE6, 6BA6, 6BA6, 6AU6, 6AL5, 6SQ7, 6SQ7, 6V6GT, 6V6GT, 6SC7, 5Y3GT
5783W	115	6AQ8/ECC85, 6AJ8/ECH81, 6BY7/EF85, 6T8/EABC80, 6BR5/EM80, 6BQ5/EL84
593	118	6BE6, 6AG5, 6BA6, 6SC7, 6BA6, 5Y3GT, 6AL5, 6AU6, 12AX7, 6AV6, 6V6GT, 6V6GT
6-502P	100	12SA7GT, 12SK7GT, 12SQ7, 50L6GT, 35Z5GT
6-601W	101	6SG7, 6SA7, 6SK7, 6SQ7GT, 6V6GT, 5Y3GT
6-604W	101	12BA6, 12BE6, 6SS7, 6SZ7, 50B5, 35W4
6-606	100	1LN5, 1LA6, 1LN5, 1LH4, 1LB4, 35Z5GT
6-617	99	6SK7, 6SA7, 6SK7, 6SQ7, 6V6GT, 5Y3GT
666	118	Transistorized
683	119	6BC5, 6BE6, 6BA6, 6BA6, 6AU6, 6AL5, 6AV6, EM840, 12AX7, EL84/6BQ5, EL84/6BQ5, 5U4GB
688	121	12BE6, 12BA6, 12AV6, 50C5, 35W4
689M	123	6BJ6, 12BE6, 6BJ6, 12AU6, 19T8, 35C5, 35W4
694	123	6BC5, 6BE6, 6BA6, 6AU6A, 6T8, 6BQ5, ECL82, EZ81/6CA4
697	125	6BC5, 6BE6, 6BA6, 6BA6, 6AU6, 6AL5, EM840, 12AX7, EL84/6BQ5, 12AX7, EL84/6BQ5, 5Y3GT
7-421W	107	12SA7GT, 12SK7GT, 12SQ7GT, 50L6GT, 35Z5GT
7-435V	104	12SA7GT, 12SK7GT, 12SQ7, 50L6GT, 35Z5GT
7-526	102	1LN5, 1LA6, 1LN5, 1LH4, 3Q5GT
7-532W	103	12AT7 or 14F8, 12SA7, 12SK7, 12SQ7, 35L6GT, 35Z5GT
7-537	105	12AT7, 12SA7GT, 12SK7, 12SQ7, 35L6GT, 35Z5GT
7-622	104	12SA7GT, 12SG7, 12SQ7, 35L6GT, 35L6GT
7-724	102	6SG7, 6SA7, 6SK7, 6SQ7, 6SQ7, 6K6GT, 6V6GT, 5Y3GT
7-925	103	6BA6, 6BE6, 6BA6, 6BA6, 6AL5, 6SQ7GT, 6SQ7GT, 6K6GT, 6K6GT, 5Y3GT
730	126	12BE6, 12BA6, 12AX7, 35C5, 50C5
7511	128	6EZ8, 6BA6, 6BA6, 6AL5, EM840, 6BE6, 6BA6, 12AX7, 6BQ5, 6BQ5, EZ81/6CA4
766	119	Transistorized
768	120	Transistorized
770	124	Transistorized
771	120	Transistorized
777	128	Transistorized
8-451	106	1R5, 1U4, 1U5, 3S4
8-533W	107	14F8, 12SA7, 12SK7, 12SQ7, 35L6GT, 35Z5GT
8-618	105	6SG7, 6SA7, 6SK7, 6SQ7, 6V6GT, 5Y3GT
8-934	106	6BA6, 6BE6, 6BA6, 6BA6, 6AL5, 6SQ7, 6SQ7, 6K6GT, 6K6GT, 5Y3GT
808	127	Transistorized
859	127	Transistorized
9-435V	108	12SA7, 12SK7, 12SQ7, 50L6GT, 35Z5GT
OLYMPIC-CONTINENTAL		
GB374	130	ECC85, 20D4, 9D7, 6AL5, 12AX7, EL84, EL84, EZ80

BRAND/MODEL	PAGE	USES THESE TUBES
GB375	131	ECC85, 20D4, 9D7, EABC80, EM84, 12AX7, EL84, EL84, EZ80
GB376	130	ECC85, ECH81, 9D7, EABC80, EM84, 12AX7, EL84, EL84, EZ80
GBS388	132	6BR8, ECH81, 9D7, EABC80, 6AT6, 12AX7, EL84, EL84, 12AX7, EL84, EL84, EZ81
250	132	ECC85, ECH81, EF89, EABC80, EM84, EL84
300	131	ECC85, ECH81, EF89, EABC80, EM84, EL84

OLYMPIC-OPTA

BRAND/MODEL	PAGE	USES THESE TUBES
52804	134	ECC85/6AQ8, ECH81/6AJ8, EF89/6DA6, EABC80/6T8, EM80/6BR5, ECL82/6BM8, ECL82/6BM8
5711W	134	ECC85/6AQ8, ECH81/6AJ8, EF89/6DA6, EABC80/6T8, EM80/6BR5, EL84/6BQ5
5804T/W	133	6AQ8/ECC85, 6AJ8/ECH81, EF89, 6T8/EAB80, 6BR5/EM80, 6BQ5/EL84
5806T/W	133	6AQ8/ECC85, 6AJ8/ECH81, 6BY7/EF85, 6T8/EABC80, 6BR5/EM80, 6BQ5/EL84
5920	135	ECC85/6AQ8, ECH81/6AJ8, EF89, EABC80/6T8, EM80/6BR5, EC92/6AB4, EL84/6BQ5, EL84/6BQ5

PACKARD-BELL

BRAND/MODEL	PAGE	USES THESE TUBES
RPC-3	153	6BE6, 7025, 6AV6, 6E5, 7025, 6AU6A, 6BA6, 12AT7, 6BA6
100	143	12BE6 or 12SA7, 12SK7, 12SQ7, 50L6GT, 35Z5GT
1052A	136	6SK7, 6SA7, 6SK7, 6SF7, 6H6, 6SQ7, 6SF7, 6H6, 6V6GT, 5Y3GT
1054-B	137	6SK7, 6SA7, 6SK7, 6SF7, 6H6, 6SQ7, 6SF7, 6H6, 6V6GT, 5Y3GT
1063	139	6SK7, 6SA7, 6SK7, 6SF7, 6H6, 5Y3GT/G, 6SQ7, 6SF7, 6H6, 6V6GT/G
10RP1	148	6AW8, 12AT7, 6BE6, 6BA6, 6AU6, 6AL5, 6AV6, 12AU7, 6AQ5, 6AQ5, 5Y3GT
10RP2	150	6AU6, 12AT7, 6U8, 6BE6, 6AU6, 6AL5, 6AV6, 12AX7, 6AQ5, 6AQ5, 5Y3GT
1181	145	6BA6, 6BA6, 6AU6, 6BA6, 6BA6, 6AL5, 6H6, 6U5, 6SQ7, 6SK7, 6V6GT, 5Y3GT
11RP7S	152	6AU6, 12AT7, 6EA8, 6BE6, 6AU6, 6AL5, 6AU6, 7025, 6AQ5A, 6AQ5A, 6X4
1273	142	6BA6, 6BA6, 6AU6 or 6C4, 6BA6, 6BA6, 6AL5, 6H6, 6U5, 6SF7, 6SN7GT, 6V6GT, 6V6GT, 5Y3GT
1472	143	6BA6, 6BA6, 6C4, 6BA6, 6BA6, 6AL5, 6SF7, 6U5/6G5, 6SK7, 6H6, 6SN7GT, 6V6GT, 6V6GT, 5U4G
471	140	1R5, 1T4, 1S5, 3V4
4RB1	149	1R5, 1U4, 1U5, 3S4
4RC1	154	12AU6, 12AV6, 50DC4, 50EH5

BRAND/MODEL	PAGE	USES THESE TUBES
531	146	12BE6, 12BA6, 12AV6, 50C5, 35W4
532	146	12BE6, 12BA6, 12AV6, 50C5, 35W4
543	148	12BE6, 12BA6, 12AV6, 50C5, 35W4
568	139	12BE6, 12BA6, 12AT6, 50B5, 35W4
572	140	12BE6, 12BA6, 12AT6, 50B5, 35W4
5DA	138	12SA7, 12SK7, 12SQ7, 50L6GT, 35Z5GT
5R1	149	12BE6, 12BA6, 12AV6, 50C5, 35W4
5R5	151	12BE6, 12BA6, 12AV6, 50C5, 35W4
5R6	153	12BE6, 12BA6, 12AV6, 50C5, 35W4
5RC1	150	12BE6, 12BA6, 12AV6, 50C5, 35W4
621	145	6BJ6, 12BE6, 6BJ6, 12AV6, 50C5, 35W4
631	147	6BJ6, 12BE6, 6BJ6, 12AV6, 50C5, 35W4
632	147	6BJ6, 12BE6, 6BJ6, 12AV6, 50C5, 35W4
651	135	6SK7, 6SA7, 6SK7, 6SQ7, 6K6GT, 6X5GT
661	136	6SG7, 6SA7, 6SK7, 6SQ7, 6V6GT, 5Y3GT
662	137	6SK7, 6SA7, 6SK7, 6SQ7, 6V6GT, 5Y3GT
673A	141	6SK7, 6SA7, 6J5, 6SK7, 6J5, 6SK7, 6V6GT, 5Y3GT
682	144	6SS7, 12SA7, 6SS7, 12SQ7, 50L6GT, 35Z5GT
6RC1	151	12BA6, 12BE6, 12BA6, 12AV6, 35C5, 35W4
7R3	152	12BA6, 12AT7, 12AX7, 12BE6, 6BJ6, 12AU6, 50EH5
861	138	6SK7GT, 6SA7, 6SK7, 6H6, 6SF7, 6SQ7, 6V6GT/G, 5Y3GT/G
872	141	6BA6, 6BE6, 6BA6, 6BA6, 6BA6, 6AL5, 6C4, 5Y3
881-A	142	6SK7, 6SA7, 6SK7, 6H6, 6SQ7GT, 6SF7, 6V6GT, 5Y3GT
881-B	142	6SK7, 6SA7, 6SK7, 6H6, 6SQ7GT, 6SF7, 6V6GT, 5Y3GT
884	144	7F8, 6BE6, 6BA6, 6BA6, 6AL5, 6SF7, 6K6GT, 5Y3GT
892	144	7F8, 6BE6, 6BA6, 6BA6, 6AL5, 6SF7, 6K6GT, 5Y3GT

PHILCO

BRAND/MODEL	PAGE	USES THESE TUBES
B569	212	12BE6, 12BA6, 12AV6, 35C5, 35W4
B570	211	12BE6, 12BA6, 12AV6, 35C5, 35W4
B570	208	12BE6, 12BA6, 12AV6, 35C5, 35W4
B574	208	12BE6, 12BA6, 12AV6, 35C5, 35W4
B650	207	1R5, 1U4, 1U5, 3V4
B652	209	1R5, 1U4, 1U5, 3V4
B710	207	12BE6, 12BA6, 12AV6, 35C5, 35W4
B714	209	12BE6, 12BA6, 12AV6, 35C5, 35W4
C-570	214	12BE6, 12BA6, 12AV6, 35C5, 35W4
C-660	213	1R5, 1U4, 1U5, 3V4
C-663	213	1R5, 1U4, 1U5, 3V4

BRAND/MODEL	PAGE	USES THESE TUBES
PHILCO (cont.)		
C-666	215	1T4, 1R5, 1U5, 1U4, 3V4
C-667	214	1T4, 1R5, 1U4, 1U5, 3V4
D-1345	217	12BE6, 12BA6, 12AV6, 35C5, 35W4
D-579	216	12AU6, 12AV6, 50C5, 35W4
D-592	215	12BE6, 12BA6, 12AV6, 35C5, 35W4
D-664	216	1R5, 1U4, 1U5 or 1S5, 3V4
E-1370	219	12BE6, 12BA6, 12AV6, 50C5, 35W4
E-670	217	1R5, 1U4, 1U5, 3V4
E-740	218	12BE6, 12BA6, 12AV6, 50C5, 35W4
E-818	220	12BE6, 12BA6, 12AV6, 25F5, 25F5, 35W4
E-976	219	12AT7, 12BE6, 12BA6, 12BA6, 12AU6, 19T8, 35C5
F-1803	223	6BZ6, 12AT7, 6BA6, 6BA6, 6BE6, 6BA6, 6BY8, 6BJ7
F-809	221	12BE6, 12BA6, 12AV6, 50C5, 35W4
F-963	220	12BA6, 12BE6, 12BA6, 12AV6, 35C5, 35W4
F-974	222	12AT7, 12BA6, 12BE6, 12BA6, 12AU6, 19T8, 35C5
G-1707S	226	12AT7, 6BA6, 6BE6, 6BA6, 6AU6, 6BJ7
G-1907S	226	6BZ6, 6AB4, 6BA6, 6BA6, 6BE6, 6BA6, 6BY8, 6BJ7
G-681	224	1R5, 1U4, 1U5, 3V4
G-747	229	12AU6, 12AV6, 50EH5, 35W4
G-751	225	12BE6, 12BA6, 12AV6, 50C5, 35W4
G-822	225	12BE6, 12BA6, 12AV6, 50C5, 35W4
H-973	230	12BA6, 12BE6, 12BA6, 12AV6, 35C5, 35W4
H984AQ	231	12DT8, 12AU6, 12AU6, 12AL5, 12BE6, 12BA6, 12AV6, 35C5
T-1000	229	Transistorized
T-6	222	Transistorized
T-60	227	Transistorized
T-65	228	Transistorized
T-7	218	Transistorized
T-700	221	Transistorized
T-75	227	Transistorized
T-78	228	Transistorized
T-9	223	Transistorized
T4	224	Transistorized
TC-47	230	Transistorized
46-1201	155	7A8, 7B7, 7C6, 35L6GT, 50Y6GT
46-1201 Revised	160	7A8, 7B7, 7C6, 50A5 or 50B5, 35Y4 or 50Y6
46-1209	158	7F8, 7H7, 7H7, 7X7, 7AF7, 6V6GT, 6V6GT, 5Y3G/GT
46-1213	157	7W7, 7F8, 7H7, 7B7, 7H7, 6SQ7GT, 7AF7, FM1000, 6V6GT, 6V6GT, 5U4G
46-1226	158	7AF7, 7H7, 7H7, 7C6, 6J5, 6K6GT/G, 6K6GT/G, 5Y3GT/G

BRAND/MODEL	PAGE	USES THESE TUBES
46-131	156	1LA6, 1LN5, 1LH4, 1A5GT
46-131	161	1LA6, 1LN5, 1LH4, 1A5GT
46-132	154	1LA6, 1LN5, 1LH4, 1A5GT, 1A5GT
46-142	166	1LA6, 1LN5, 1LH4, 1A5GT, 1A5GT
46-350	157	1T4, 1R5, 1T4, 1U5, 3Q5, 117Z3
46-420	156	7C7, 7A8, 7B7, 7C6, 50L6GT, 35Z5GT
46-421	155	7C7, 7A8, 7B7, 7C6, 50L6GT, 35Z5GT
46-480	159	7F8, 7H7, 7H7, 6H6GT, 6SQ7GT, 6V6GT, 7Z4
46-906	182	6BJ6 or 12AU6, 117Z3, 14F8, 12AU7, 6BJ6, 6BJ6, 19T8, 50A5
47-204	163	7A8, 14A7, 14B6, 50L6GT, 35Y4
48-1253	166	7A8, 7B7, 7C6, 50A5, 50X6
48-1256	165	7C7, 7A8, 14A7, 7C6, 35L6GT, 50X6
48-1260	161	7A8, 7B7, 7C6, 50A5, 50X6
48-1262	165	14AF7, 7B7, 7B7, 7C6, 35L6GT, 50X6
48-1263	162	7AF7, 7A7, 7A7, 7C6, 6J5GT, 6K6GT, 6K6GT, 5Y3GT
48-1264	167	6AG5 or 6AU6, 7F8, 6BA6, 7R7, 7X7, 6J5GT, 6K6GT, 6K6GT, 5AZ4
48-1266	170	6AG5, 7F8, 6BA6, 7R7, 7X7, 7A4, 6V6GT, 6V6GT, 5AZ4
48-1270	172	7W7, 7F8, 7H7, 7B7, 7H7, FM1000, 6SQ7GT, 7E7, 7F7, 7AF7, 6V6GT, 6V6GT, 5U4G
48-1276	171	6AU6, 7F8, 7E5, 7H7, 7B7, 7H7, 7E6, FM1000, 6J5GT, 7F7, 7E7, 6J5GT, 6J5GT, 6L6GA, 6L6GA, 5U4G
48-1284	173	7AF7, 7A7, 7R7, 7F7, 6K6GT, 6K6GT, 7Z4
48-1286	177	6AU6, 7F8, 6BA6, 7R7, 7X7, 7E7, 7F7, 6J5GT, 6K6GT, 6K6GT, 5AZ4
48-1290	174	6AU6, 7F8, 6BA6, 7A7, 6BA6, FM1000, 6SQ7GT, 7AF7, 7F7, 7E7, 6V6GT, 6V6GT, 5U4G
48-150	164	1LG5, 1LA6, 1LN5, 1LH4, 3LF4
48-200-I	163	7A8, 14A7, 14B6, 50A5 or 50L6GT, 35Z5GT or 35Y4
48-214	163	7A8, 14A7, 14B6, 50A5 or 50L6GT, 35Z5GT or 35Y4
48-206	168	7A8, 14A7, 14B6, 50A5 or 50L6GT, 35Z5GT or 35Y4
48-225	167	7A8, 14A7, 14B6, 50A5, 35Z5GT
48-230	167	7A8, 14A7, 14B6, 50A5, 35Z5GT
48-250-I	162	7A8, 14A7, 14B6, 50A5 or 50B5, 35Z5GT or 35Y4
48-300	168	1R5, 1T4, 1U5, 3V4, 117Z3
48-360	169	1T4, 1R5, 1T4, 1U5, 3LF4, 117Z3
48-460	164	14AF7, 7B7, 7B7, 7C6, 50L6GT, 35Y4
48-461	169	14AF7, 7B7, 7B7, 7C6, 50A5, 35Y4
48-464	159	14AF7, 7B7, 7B7, 7C6, 50A5, 35Y4
48-472	175	12AU6, 12AU7, 14F8, 6BJ6, 6BJ6, 19T8, 50A5, 117Z3
48-472-I	173	12AW6, 14F8, 14H7, 14X7, 50A5, 117Z3

BRAND/MODEL	PAGE	USES THESE TUBES
48-475	170	6AG5, 7F8, 6BA6, 6BA6, 6H6GT, 7C6, 7B5, 7Z4
48-482	160	7W7, 7F8, 7H7, 7B7, 7H7, FM1000, 7B6 or 6SQ7GT, 6V6GT, 5Y3GT
49-101	187	1R5, 1T4, 1U5, 3V4
49-1100	176	14AF7, 7B7, 7B7, 7C6, 35L6GT, 50X6
49-1401	174	12BE6, 12BA6, 6AQ6, 35L6GT, 50Y6GT
49-1405	180	12BE6, 12BA6, 6AQ6, 35L6GT, 50Y7GT
49-1600	177	14Q7, 12BA6, 7C6, 35L6GT, 50X6
49-1613	188	6AU6, 7F8, 6BA6, 7R7, 7X7, 6J5GT, 6K6GT or 6V6GT, 6K6GT or 6V6GT, 7F7, 7E7, 5AZ4
49-1615	184	6AU6, 7F8, 6BJ6 or 6BA6, 6BJ6 or 6BA6, 6T8, 7F7, 7E7, 7A4, 6V6GT, 6V6GT, 5U4G
49-500	175	7A8, 14A7, 14B6, 50A5, 35Z5
49-501	181	7A8, 14A7, 14B6, 50A5, 35Y4
49-503	178	7A8, 14A7, 14B6, 50A5, 35Z5GT
49-504	180	7A8, 14A7, 14B6, 50A5, 35Z5GT
49-505	179	7A8, 14A7, 14B6, 50A5, 35Y4
49-601	172	1R5, 1T4, 1U5, 3V4
49-602	171	1R5, 1T4, 1U5, 3V4
49-603	183	12BE6, 12BA6, 12AT6, 50B5, 35W4
49-607	182	1T4, 1R5, 1T4, 1U5, 3LF4, 117Z3
49-900-E	176	14AF7, 7B7, 7B7, 7C6, 50L6GT, 35Y4
49-901	181	7A8, 14A7, 14B6, 50A5, 35Y4
49-902	178	14AF7, 12BA6, 12AU6, 14B6, 35A5, 35Y4
49-904	183	14AF7, 12BA6, 12AU6, 14B6, 35A5, 35Y4
49-905	179	12AT7, 6BA6, 12AU6, 19T8, 35C5, 35W4
50-1421	190	12BE6, 12BA6, 6AQ6, 35L6GT, 50Y7GT
50-1720	189	12AU6, 14F8, 12AU7, 6BJ6, 6BJ6, 19T8, 50C6G
50-1724	190	6BA6, 7F8, 6BA6, 6AU6, 6BC7, 6AV6, 6Y6G, 5AZ4
50-1727	186	6AU6, 7F8, 6BJ6, 6BJ6, 6T8, 7A4, 6V6GT, 6V6GT, 7F7, 7E7, 5U4G
50-520	184	7A8, 12BA6, 14B6, 50L6GT, 35Z5GT
50-522	185	7A8, 12BA6, 14B6, 50L6GT, 35Z5GT
50-526	189	7A8, 12BA6, 14B6, 50L6GT, 35Z5GT
50-527	185	7A8, 14A7, 50L6GT, 35Y4, 14B6
50-620	186	1R5, 1T4, 1U5, 3V4
50-621	188	1T4, 1R5, 1U4, 1U5, 3V4
50-921	187	7A8, 7B7, 14B6, 50C5 or 50L6GT, 35Z5GT, 7B7
50-925	191	12BA6, 12AT7, 12BA6, 12BA6, 19C8, 50C5
51-1330	194	7A8, 7B7, 7C6, 50L6GT, 35Z5GT
51-1730	195	7B7, 7A8, 7B7, 7B6, 6W6GT, 7X6
51-1732	193	6AU6, 7F8, 6BA6, 6AU6, 6V8, 6Y6GT or 6W6GT, 5AZ4
51-1733	195	6BA6, 7F8, 6BA6, 6AU6, 6BC7, 6AV6, 6Y6G, 5AZ4
51-530	192	7A8, 12BA6, 14B6, 50L6GT, 35Z5GT
51-537	193	7A8, 14A7, 14B6, 50L6GT, 35Y4
51-629	194	1R5, 1U4, 1U5, 3V4
51-631	192	1R5, 1U4, 1U5, 3V4
51-930	196	7B7, 7A8, 7B7, 14B6, 35L6GT, 35Z5GT or 35Y4
51-934	191	12AU6, 12AT7, 12BA6, 12AU6, 19C8, 50C5
52-1340	197	7A8, 7B7, 7C6, 35L6GT, 50Y7GT
52-544-I	198	7A8, 12BA6, 12AV6, 50L6GT, 35Z5GT
52-640	196	1R5, 1U4, 1U5, 3V4
52-643	198	1T4, 1R5, 1U4, 1U5, 3V4
52-940	197	7B7, 7A8, 7B7, 14B6, 35L6, 35Z5GT
52-944	199	12AU6, 12AT7, 12BA6, 12AU6, 19V8, 35C5
53-1750	204	7A8, 7B7, 7C6, 35L6GT, 50Y7GT
53-1754	206	6BJ6, 6BE6, 6BJ6, 6AV6, 6AQ5, 6X4
53-559	205	12BE6, 12BA6, 12AV6, 35C5, 35W4
53-560	201	12BE6, 12BA6, 12AV6, 35C5, 35W4
53-562	200	12BE6, 12BA6, 12AV6, 35C5, 35W4
53-563	202	12BE6, 12BA6, 12AV6, 35C5, 35W4
53-566	199	7A8, 7B7, 7C6, 50C5, 35W4
53-656	200	1T4, 1R5, 1U4, 1U5, 3V4
53-701	201	12BE6, 12BA6, 12AV6, 35C5, 35W4
53-702	204	12BE6, 12BA6, 12AV6, 35C5, 35W4
53-804	205	6BJ6, 12BE6, 6BJ6, 6AQ6, 35C5, 35W4
53-950	203	6BJ6, 12BE6, 6BJ6, 6AQ6, 35C5, 35W4
53-958	203	12BA6, 12AT7, 12BA6, 12AU6, 19V8, 35C5
53-960	202	6BA6, 7S7, 7E7, 7B6, 12AU7, 7B5, 7B5, 5AZ4
B-1352	210	7A8, 7B7, 7C6, 35L6GT, 50Y7GT
B-1752	210	7A8, 7B7, 7C6, 35L6GT, 50Y7GT
B-956	206	12AU6, 12AT7, 12BA6, 12AU6, 19V8, 35C5
B1349	212	7A8, 7B7, 7C6, 35L6GT, 50Y7GT
B1756	211	6AU6, 6AU6, 12AT7, 6BA6, 6AU6, 6AL5, 6AL5, 6C4A, 12AU7, 6C4, 12AX7, 6V6GT, 6V6GT, 5U4G

PHILHARMONIC

100C	232	6SA7GT, 6SF7, 6SL7GT, 25L6GT, 25L6GT, 25Z6GT, 25Z6GT
349C	232	12BA6, 12BA6, 12BE6, 12BA6, 12H6, 6AQ6, 50B5
8712	231	6SK7GT, 6SA7, 6SK7, 6SQ7, 6J5, 6V6GT, 6V6GT, 5Y3GT

PHILLIPS

R-100	233	12BE6, 12BA6, 12AV6, 50C5, 35W4

BRAND/MODEL	PAGE	USES THESE TUBES
PHILLIPS 66		
3-81A	233	7H7, 6SB7Y, 7A4, 7Q7, 7H7, 7H7, 7H7, 7C7, 7A6, 1273, 7A4, 7C5, 7C5, 5U4G
PILOT		
PT-1031	238	6BA6, 6U8, 6AB4, 6BA6, 6AU4, 12AX7, 12AX7, EL84/6BQ5, EL84/6BQ5, 6X4, EZ81/6CA4
PT-1036	237	6BA6, 6U8, 6AB4, 6BA6, 6AU6, 6AL5, EM81, 12AX7, 12AX7, 12AX7, EC84, EL84, 6X4, EZ81
T-411-U	235	12SK7, 12SA7, 12SK7, 12SQ7, 35L6GT, 35Z5GT
T-500U	234	12SA7GT, 12SK7GT, 12SQ7GT, 50L6GT, 35Z5GT
T-511	234	12SK7, 12SA7, 12SK7, 12SQ7, 35L6GT or 50L6G 35Z5GT
T-521	236	6AG5, 6SB7Y, 6SG7, 6SG7, 6H6, 6SJ7, 25L6GT, 25Z6GT
T-530AB	235	6SB7Y, 6SG7, 6SG7, 6SJ7, 6H6, 6SJ7, 25L6GT, 25Z6GT
T-601	236	6BA6, 6BE6, 6BA6, 6BA6, 6AL5
T-741	237	6SB7Y, 25Z6GT, 6SG7, 6SG7, 6H6, 6SJ7, 25L6GT
POLICALARM		
PR-31	238	12AT7, 6BJ6, 6BJ6, 19T8, 35B5, 35W4
PR-31A	239	12AT7, 6BJ6, 6BJ6, 19T8, 35C5, 35W4
PR-9	239	12AT7, 6BJ6, 6BJ6, 19T8, 35C5, 35W4
PORTO BARADIO		
PB-520	240	12SA7, 12SK7, 12SQ7, 50L6GT, 35Z5GT
PB-520 Revised	240	12SA7GT, 12SK7GT, 12SQ7GT, 50L6GT, 35Z5GT
PREMIER		
15LW	241	14Q7, 7B7, 14B6, 50A5, 35Y4
PURITAN		
501	241	12SA7, 12SK7, 12SQ7, 50L6GT, 35Z5GT
502X	242	12SA7, 12SK7, 12SQ7, 50L6GT, 35Z5GT
503	243	12SA7, 12SK7, 12SQ7, 50L6GT, 35Z5GT
504	243	6SA7, 6SK7, 6SF7, 6SJ7, 6V6GT, 6X5GT
508	242	6SK7, 6SA7, 6SK7, 6SQ7, 6SJ7, 6K6GT/G, 5Y3GT/G
509	244	1A7GT, 117Z6GT, 3Q5GT, 1N5GT, 1N5GT, 1H5GT
RADIOETTE		
PR-2	244	1R5, 1T4, 1S5, 3Q4

BRAND/MODEL	PAGE	USES THESE TUBES
RADIOLA		
61-1	245	12SG7, 12J5GT, 12SK7, 12SQ7, 35L6GT, 35Z5GT
61-5	245	12SG7, 12SA7, 12SK7, 12SQ7, 35L6GT, 35Z5GT
61-8	246	12SA7, 12SK7, 12SQ7, 50L6GT, 35Z5GT
61-9	246	12SA7, 12SK7, 12SQ7, 50L6GT, 35Z5GT
75ZU	246	12SA7, 12SK7, 12SQ7, 50L6GT, 35Z5GT
76ZX11	247	12SG7, 12J5GT, 12SK7, 12SQ7, 35L6GT, 35Z5GT
RADIONIC		
Y62W	247	7B7, 14Q7, 7B7, 14B6, 50A5, 35Y4
Y728	247	7B7, 14Q7, 7B7, 14B6, 50A5, 35Y4
RANGER		
118	248	1T4, 1R5, 1T4, 1S5, 3Q4
RAYENERGY		
AD	248	12SA7, 12SK7GT, 12SQ7GT, 50L6GT, 35Z5GT
AD4	249	12SA7GT, 12SK7GT, 12SQ7GT, 50L6GT, 35Z5GT
SRB-1X	249	12SA7 or 14Q7, 12SK7 or 14A7, 12SQ7 or 14B6, 50L6GT or 50A5, 35Z5GT or 35Y4
RAYTHEON		
CR-41	250	12BE6, 12BA6, 12AV6 or 12AT6, 50C5
FR81A	251	6AU6, 12AT7, 6BE6, 6BA6, 6AU6, 6AL5, 6AV6, 6V6GT, 5Y3GT
PR51	250	1R5, 1U4, 1U5, 3V4
T-100-1	253	Transistorized
T-150-1	253	Transistorized
T2500	252	Transistorized
5R-10B	252	12BE6, 12BA6, 12AV6 or 12AT6, 50C5, 35W4
8TP-1	251	Transistorized
RCA VICTOR		
A-82	278	6BA6, 6BE6, 6BA6, 6AV6, 6C4, 6V6GT, 6V6GT, 5Y3GT
A55	276	12SA7, 12BA6, 12SQ7, 50L6GT, 35Z5GT
BX55	274	1R5, 1T4 or 1U4, 3V4, 1U5
BX6	274	1R5, 1T4, 1U5, 3V4, 1T4
C-4EM	312	12BE6, 12BA6, 12AV6, 50C5, 35W4
C1E	311	12BE6, 12BA6, 12AV6, 50C5, 35W4
PX600	280	1T4, 1R5, 1T4, 1U5, 3V4

BRAND/MODEL	PAGE	USES THESE TUBES
SHC4	309	6CB6, 6X8, 6BA6, 6AU6, 6AU6, 6AL5, 6AV6, 6AL7GT, 6CG7, 6CG7, 6V6GT, 6V6GT, 5AS4A
SHF-1	308	6CB6, 6J6, 6BA6, 6AU6, 6AU6, 6AL5, 6AV6, 6AL7GT, 6CG7, 6CG7
SHF3	306	6CB6, 6X8, 6BA6, 6AU6, 6AU6, 6AL5, 6AV6, 6CG7, 6CG7, 6V6GT, 6V6GT, 6AL7GT, 5Y3GT
T1JE	311	Transistorized
TPM11	313	12DT8, 12BA6, 12BA6, 12AU6, 19T8, 12BE6, 6BA6, 35W4
KS11	313	12DT8, 12BA6, 12BA6, 12AU6, 19T8, 12BE6, 6BA6, 35W4
TPR-8	314	6DT8, 6BA6, 6AU6, 6AU6, 6AU6, 6AL5, 6BA6, 6BE6, 6BA6, 6AV6, EM84/6FG6, 6CG7, 6CG7, 6CG7, 6BQ5, 6BQ5, 6CG7, 6BQ5, 6BQ5, 5AS4A
TX1JE	313	Transistorized
X552	276	12BE6, 12BA6, 12AV6, 50L6GT, 35W4
X711	277	19J6, 6BJ6, 12AU6, 12AL5, 6AQ6, 35C5, 35W4
XF3EM	312	12DT8, 12BA6, 12BE6, 12BA6, 12AU6, 19T8, 35C5
1BT21	310	Transistorized
1BT32	310	Transistorized
1BT46	307	Transistorized
1BX57	308	1R5, 1U4, 1U5, 3V4
1MBT6	309	Transistorized
1R81	279	6AU6, 6X8, 6BA6, 6AU6, 6AL5, 6AU6, 6V6GT, 5Y3GT
1T4J	314	Transistorized
1T5J	315	Transistorized
1X51	280	12SA7, 12BA6, 12SQ7, 50L6GT, 35Z5GT
1X592	279	12SA7, 12SK7, 12SQ7, 50L6GT, 35Z5GT
2-R-51	283	12BE6, 6BJ6, 12AV6, 6AK6
2-S-10	285	6CB6, 6J6, 6BA6, 6AU6, 6AL5, 6AV6, 6C4, 6V6GT, 6V6GT, 5Y3GT
2-S-7	286	12BE6, 12BA6, 6AQ6, 6AQ6, 35C5, 35C5
2-X-621	284	12BA6, 12BE6, 12BA6, 12SQ7, 35L6GT, 35Z5GT
2-XF-91	284	6BJ6, 19X8, 12BA6, 12AU6, 12AU6, 12AL5, 12AV6, 35C5
2-XF-931	285	6BJ6, 19X8, 12BA6, 12AU6, 12AU6, 12AL5, 12AV6, 35C5
2B400	281	1R5, 1U4, 1U5, 3V4
2BX63	282	1T4, 1R5, 1T4, 1U5, 3V4
2C521	282	12BE6, 12BA6, 12AV6, 50C5, 35W4
2US7	281	12BE6, 12BA6, 6AQ6, 6AQ6, 35C5, 35C5
2X61	283	12SK7, 12SA7, 12SK7, 12SQ7, 35L6GT, 35Z5GT
3-BX-51	287	1R5, 1T4, 1U5, 3V4
3-BX-671	288	1U4, 1L6, 1U4, 1U5, 3V4
3-RF-91	286	6CB6, 6X8, 6BA6, 6AU6, 6AU6, 6AL5, 6AV6, 6V6GT, 5Y3GT
3-US-5	290	12BE6, 12BA6, 12AV6, 50C5, 35W4
3-X-521	287	12BE6, 12BA6, 12AV6, 50C5, 35W4
4-C-531	289	12BE6, 12BA6, 12AV6, 50C5, 35W4
4-C-541	292	12BE6, 12BA6, 12AV6, 50C5, 35W4
4-C-671	291	12BA6, 12BE6, 12BA6, 12AV6, 35C5, 35W4
4-X-551	291	12BE6, 12BA6, 12AV6, 50C5, 35W4
4-X-661	290	12BA6, 12BE6, 12BA6, 12AV6, 35C5, 35W4
4-Y-511	289	12BE6, 12BA6, 12AV6, 50C5, 35W4
45-W-10	278	6CB6, 6J6, 6BA6, 6AU6, 6AL5, 6AV6, 6C4, 6V6GT, 6V6GT, 5Y3GT
4X641	288	12BA6, 12BE6, 12BA6, 12AV6, 35C5, 35W4
5-BX-41	292	1R5, 1U4, 1U5, 3V4
5-C-581	293	12BE6, 12BA6, 12AV6, 50C5, 35W4
5-C-591	294	12BE6, 12BA6, 12AV6, 50C5, 35W4
5-X-560	293	12BE6, 12BA6, 12AV6, 50C5, 35W4
54B1	256	1R5, 1T4, 1S5, 3S4
54B5	258	1R5, 1T4, 1S5, 3S4
55F	254	1N5GT, 1A7GT, 1N5GT, 1H5GT, 3Q5GT
59AV1	255	6SG7, 6SA7, 6SK7, 6R7, 6SQ7, 6SQ7, 6F6G, 6F6G, 5U4G
6-BX-5	294	1R5, 1U4, 1U5, 3V4
6-BY-4A	296	1R5, 1U4, 1U5, 3V4
6-C-5A	299	12BE6, 12BA6, 12AV6, 50C5, 35W4
6-HF-1	299	6CB6, 6J6, 6BA6, 6AU6, 6AU6, 6AL5, 6AV6, 6AL7GT, 6C4, 12AU7, 6AU6, 6AU6, 6V6GT, 6V6GT, 6V6GT, 6V6GT, 5U4G
6-X-5A	296	12BE6, 12BA6, 12AV6, 50C5, 35W4
6-X-7A	295	12BE6, 12BA6, 12AV6, 50C5, 35W4
6-XD-5	295	12BE6, 12BA6, 12AV6, 50C5, 35W4
6-XY-5A	298	12BE6, 12BA6, 12AV6, 50C5, 35W4
610V2	262	6BE6, 6BE6, 6BA6, 6AU6, 6AL5, 6SQ7, 6SQ7, 6K6GT, 6K6GT, 5Y3GT
612V3	259	6BA6, 6BA6, 6BE6, 6BA6, 6AU6, 6AU6, 6AL5, 6AT6, 6J5, 6F6G, 6F6G, 5AU4G
64F3	254	1A7GT, 1N5GT, 1H5GT, 3Q5GT
65BR9	260	1T4, 1T4, 1R5, 1S5, 3V4
65U	257	12SA7, 12SK7, 12SQ7, 50L6GT, 35Z5GT
65X1	261	12SA7, 12SK7, 12SQ7, 50L6GT, 35Z5GT
65X2	255	12SA7, 12SK7, 12SQ7, 50L6GT, 35Z5GT
66BX	258	1T4, 1R5, 1T4, 1S5, 3Q4 or 3V4, 117Z3
66X1	256	12SG7, 12SA7, 12SK7, 12SQ7, 35L6GT, 35Z5GT

RCA VICTOR (cont.)

BRAND/MODEL	PAGE	USES THESE TUBES
66X11	261	12SG7, 12J5GT, 12SK7, 12SQ7, 35L6GT, 35Z5GT
67AV1	257	6SA7, 6SG7, 6SQ7, 6SQ7, 6K6GT, 6K6GT, 5Y3GT
68R1	260	6BE6, 6BE6, 6BA6, 6AU6, 6AL5, 6SQ7, 6K6GT, 5Y3GT
6HF3	297	12AU6, 12AU6, 6BJ6, 12AL5, 12AV6, 12BA6, 19X8, 35W4GT, 6CG7, 6CG7, 6V6GT, 6V6GT, 5Y3GT
7-BT-10K	298	Transistorized
7-BT-9J	297	Transistorized
7-BX-6	300	1R5, 1U4, 1U5, 3V4
7-BX-8J	300	1T4, 1R5, 1T4, 1U5, 3V4
7-BX-9H	301	1T4, 1L6, 1T4, 1U5, 3V4
7-HFR-1	303	6CB6, 6X8, 6BA6, 6AU6, 6AU6, 6AL5, 6AV6, 6AL7GT, 6CG7, 6CG7, 6V6GT, 6V6GT, 5Y3GT, 12AX7, 12AU7, 6CG7, 6E5, 6X4
710V2	265	6BE6, 6BE6, 6BA6, 6AU6, 6AL5, 6SQ7, 6SQ7, 6K6GT, 6K6GT, 5Y3GT
711V2	259	6BA6, 6BA6, 6BE6, 6BA6, 6AU6, 6AL5, 6SQ7, 6J5, 6F6G, 6F6G, 5U4G
75X11	262	12SA7, 12SK7, 12SQ7, 50L6GT, 35Z5GT
77U	263	12SA7, 12SK7, 6AQ6, 6C4, 35L6GT, 35L6GT
77V1	263	6SA7, 6SK7, 6SQ7, 6SQ7, 6V6GT, 6V6GT, 6X5GT
77V2	264	6SA7, 6SK7, 6SQ7, 6SQ7, 6V6GT, 6V6GT, 6X5GT
8-BT-10K	301	Transistorized
8-C-5E	304	12BE6, 12BA6, 12AV6, 50C5, 35W4
8-C-7EE	302	12BE6, 12BA6, 12AV6, 50C5, 35W4
8-RF-13	305	6CB6, 6X8, 6BA6, 6AU6, 6AU6, 6AL5, 6AV6, 6AL7GT, 6CG7, 6CG7, 6V6GT, 6V6GT, 5Y3GT
8-X-5D	303	12BE6, 12BA6, 12AV6, 50C5, 35W4
8-X-8D	302	12BE6, 12BA6, 12AV6, 50C5, 35W4
8B42	270	1R5, 1U4, 1U5, 3S4
8BX5	265	1R5, 1T4, 1U5, 3V4, 117Z3
8F43	272	1A7GT, 1N5GT, 1H5GT, 3Q5GT
8R71	266	6J6, 6BA6, 6AU6, 6AL5, 6AV6, 6V6GT, 6X5GT
8V111	267	6J6, 6BA6, 6AU6, 6AL5, 6AV6, 6AV6, 6BA6, 6AU6, 6V6GT, 6V6GT, 6X5GT
8V90	267	6J6, 6BA6, 6AU6, 6AL5, 6AU6, 6AV6, 6V6GT, 6V6GT, 6X5GT
8X521	266	12BE6, 12BA6, 12AT6, 50C5, 35W4
8X53	264	12SA7, 12SK7, 12SQ7, 50L6GT, 35Z5GT
8X542	268	12SA7, 12SK7, 12SQ7, 50L6GT, 35Z5GT
8X682	269	12BA6, 12BE6, 12BA6, 12AT6, 35C5, 35W4
8X71	268	19J6, 6BJ6, 12AU6, 12AL5, 6AQ6, 35C5, 35W4
9-BT-9H	304	
9BX56	271	1R5, 1U4, 1U5, 3V4
9INT1	307	12AT7, 6AJ8/ECH81, 6DA6/EF89, 6DA6/EF89, 6AK8/EABC80, 6CD7/EM34, 6BQ5/EL84
9US5H	306	12BE6, 12BA6, 12AV6, 50C5, 35W4
9W101	269	6J6, 6BA6, 6AU6, 6AL5, 6AV6, 6BF6, 6AV6, 6V6GT, 6V6GT, 6X5GT
9W106	272	6BJ6, 6J6, 6BA6, 6AU6, 6AL5, 6AV6, 6AV6, 6V6GT, 6V6GT, 6X5GT
9X10FE	305	12BE6, 12BA6, 12AV6, 50C5, 35W4
9X561	273	12SA7, 12SK7, 12SQ7, 35Z5GT
9X571	275	12SA7, 12SK7, 12SQ7, 50L6GT, 35Z5GT
9X642	271	12SK7, 12SA7, 12SK7, 12SQ7, 35L6GT, 35Z5GT
9X652	275	12BA6, 12BE6, 12BA6, 12SQ7, 35L6GT, 35Z5GT
9Y51	273	12BE6, 12BA6, 12AV6, 50L6GT, 35W4
9Y510	277	12BE6, 12BA6, 12AV6, 50L6GT, 35W4
9Y7	270	12SA7, 12SK7, 6AQ6, 6AQ6, 35L6GT, 35L6GT

Tube Substitutions

TUBE	MAY BE REPLACED BY
0A2	0A2WA, 6073, 6626, 150C2, 150C4, M8223, STV150/30
0B2	0B2WA, 6074, 6627, 108C1, M8224, STV108/30
0D3/VR150	0D3A, W, VR150, W, WT294, 150C3, KD25
1A5GT	1A5G, 1A5GT/G, 1T5GT
1A7GT	1A7G, 1B7G*, GT*, DK32, X14
1AG4	-
1AH4	-
1AJ5	-
1AM4	1AF4, 1AJ4, 1T4SF,1AE4*, 1T4*, 1F1, DF96, W25
1AQ5	1R5SF, 1R5*, 1H33
1AS5	1U5SF, 1DN5*, 1U5*, DAF92*
1DN5	1U5, 1AS5*, 1U5SF*, DAF92
1H5GT	1H5, G, GT/G, DAC32, HD14
1L6	1U6*
1LA6	1LC6
1LB4	-
1LG5	-
1LH4	-
1LN5	1LC5
1N5GT	1N5G, 1N5GT/G, DF33, Z14, 1P5G, 1P5GT
1R5	1R5SF*, 1R5WA, 1AQ5*, 1C1, DK91, X17, 1H33*
1R5-SF	1AQ5, 1R5*, 1H33
1S5	1AF5*, 1AH5*, 1AR5*, 1FD9, DAF91, ZD17
1T4	1AE4*, 1AF4*, 1AJ4*, 1AM4*, 1T4SF*, 1T4WA, 1F3, DF91, W17
1T4-SF	1AE4*, 1AF4, 1AJ4* 1AM4, 1T4*, 1F1, DF96, W25
1U4	1AF4*, 1U4WA, 5910
1U5	1DN5, 1AS5*, 1U5SF*, 1U5WA, DAF92
1U5-SF	1AS5, 1DN5*, 1U5*
1V6	DCF60
3LF4	3LE4#, 3D6*
3Q4	3S4#, 3S4SF*#, 3W4*#, 3Z4*#, DL95, N18
3Q5	3Q5G, GT, GT/G, 3B5, G5, 3C5GT
3Q5GT	3Q5, G, GT/G, 3B5, GT, 3C5GT, WT389, DL33, N15, N16
3S4	3S4SF*, 3Q4#, 3W4*, 3Z4*, 1P10, DL92, N17
3S4-SF	3W4, 3Z4, 3S4*, 3Q4*#
3V4	3C4*, 3E5*, 3V4WA, 1P11, DL94, N19

TUBE	MAY BE REPLACED BY
3W4	3S4SF, 3Z4, 3S4*, 3Q4*#
4H-4	-
5AS4A	5AS4, 5DB4, 5U4GB, 5V3, A
5AU4G	-
5AZ4	-
5U4G	5U4GA, GB, WG, WGB, 5AR4, 5AS4, A, 5AU4, 5DB4, 5R4G, GTY, GY, GYA, GYB, 5T4, 5V3, 5V3A, 5931, WTT135, 5Z10, GZ32, U52
5U4GB	5AS4, A, 5AU4, 5DB4, 5V3, A
5V4G	5V4GA, GY, 5AR4, 274, 52KU, 53KU, 54KU, GZ30, GZ32, GZ33, GZ34, OSW3107, R52, U54
5Y3	-
5Y3G/GT	5Y3G, GA, GT, 5AR4, 5AX4GT, 5CG4, 5R4G, GTY, GY, GYA, GYB, 5T4, 5V4G, GA, GY, 5Z4, G, GT, GT/G, MG
5Y3GT	5Y3G, GA, GT/G, WTG, WGTA, WGTB, 5AR4, 5AX4GT, 5CG4, 5R4G, GTY, GY GYA, GYB, 5T4, 5V4, GA, GY, 5Z4, G, GT, GT/G, MG, 6087, 6106, 6853, WTT202, U50, U52
5Y3GT/G	5Y3G, GA, GT, 5AR4, 5AX4GT, 5CG4, 5R4G, GTY, GY, GYA, GYB, 5T4, 5V4G, GA, GY, 5Z4, G, GT, GT/G, MG
6AB4	6664, EC92
6AC7	6AC7A, W, WA, Y, 6AB7, Y, 6AJ7, 6SG7*#, GT*#, Y*#, 6SH7*#, GT*#, 1852, 6134, 6F10, OSW2190, OSW2600
6AG5	6AG5WA, 6BC5, 6CE5, 6AK5*, 6AU6#, A#, 6AW6#, 6CB6#, A#, 6CF6#, 6DC6#, 6DE6#, 6CY5*#, 6EA5*#, 6EV5*#, 6186, EF96
6AJ8/ECH81	6AJ8, ECH81, 6C12, 20D4, X719
6AK5	6AK5W, WA, WB, 403A, 6AG5*, 6BC5*, 6CE5*, 6CY5*#, 6EA5*#, 6EV5*#, 1220, 5654, 5591*, 6096, 6968, 6F32, DP61, E95F, E95#, EF905, M8100#, PM05#
6AK6	-
6AK8/EABC80	6AK8, EABC80, 6T8A, 6LD12, DH719
6AL5	6AL5W, 6EB5, 5726, 6058, 6097, 6663, 7631, D27, 6B32, 6D2, D2M9, D77, D152, D717, DD6, EAA91, EB91, EAA901, EAA901S
6AL7GT	6AL7
6AQ5	6AQ5A, W, 6BM5, 6HG5, 6005, 6095, 6669, 6L31, BPM04, EL90, M8245, N727
6AQ5A	6HG5, 6AG5*!, GBM5*!

TUBE	MAY BE REPLACED BY
6AQ6	6AT6*, 6AV6*, 6BK6*, 6BT6*6AQ7GT
6AQ8/ECC85	6AQ8, ECC85, 6L12, B719
6AT6	6AV6, 6BK6, 6BT6, 6AQ6*, 6066, 6BC32, DH77, EBC90, EBC91
6AU6	6AU6A, WA, WB, 6BA6, 7543, 6AW6#, 6AG5#, 6BC5#, 6CB6#, A#, 6CE5#, 6CF6#, 6DE6#, 6DK6#, 6136, EF94
6AU6A	6AU6*!, 7543*!, 6BA6*!, 6CB6A#, 6CE5#
6AV6	6AT6, 6BK6, 6BT6, 6AQ6*, 6066, 6BC32, DH77, EBC90, EBC91
6AW8	6AW8A, 6AU8, A, 6BA8, A, 6JV8, 6KS8, 6LF8, 6EB8*, 6EH8*#, 6X8*#, 6X8A*#
6BA6	6BA6W, WA, 6AU6, A, 6BD6, 6CG6, 7543, 6BZ6#, 5749, 6660, 7496, 6F31, EF93, M8101, PM04, W727
6BA7	-
6BC5	6AG5, 6CE5, 6AK5*, 6AU6#, A#, 6AW6#, 6CB6#, A#, 6CF6#, 6DC6#, 6DE6#, 6DK6#, 6CY5*#, 6EA5*#, 6EV5*#, EF96
6BC7	-
6BD6	6BA6, 6CG6
6BE6	6BE6W, 6BY6, 6CS6, 5750, 7502, 6H31, EK90, HM04, X77, X727, 6H31, EK90, HM04
6BF6	6BU6
6BJ6	6BJ6A, 6662, 7694, N78, E99F
6BJ7	-
6BM8/ECL82	6BM8, ECL82, 6PL12
6BQ5	7189, 7189A, 7320, 6P15, EL84, EL84L, N709
6BQ5/EL84	6BQ5, EL84, 7189, 7189A, 7320, 6P15, EL84L, N709
6BQ7A	6BQ7, 6BC8, 6BS8, 6BX8, 6BZ7, 6BZ8, 6HK8, X155, 6BK7*, A*, B*, ECC180
6BR5/EM80	6BR5, EM80, 6DA5, 6DA5
6BR8	6BRA8, 6VF8, A, 6JN8, 6BE8#, A#, 6CL8#, A#
6BT6	6AT6, 6AV6, 6BK6, 6AQ6*, 6BC32, DH77, EBC90, EBC91
6BY7/EF85	6BY7, EF85, 6BX6, 6BY8, 6F19, 6F26, W719
6BY8	-
6BZ6	6DC6, 6HQ6, 6JH6, 6BA6#
6C4	6C4W, WA, 6100, 6135*, EC90, L77, M8080, QA2401, QL77
6C4A	6C4WA, 6100, 6135*
6CA4/EZ81	6CA4, EZ81, U709, UU12

TUBE	MAY BE REPLACED BY	TUBE	MAY BE REPLACED BY	TUBE	MAY BE REPLACED BY
6CA7/EL34	6CA7, EL34, 7D11, 12E13, KT77, KT88	6SJ7	6SJ7GT, GTX, GTY, W, WGT, WGTY, Y, 5693, WTT122	7C5	7C5LT, 7B5*, 7B5LT*, EL22, KT81, N148
6CB6	6CB6A, 6AW6, 6CF6, 6DC6, 6DE6, 6DK6, 6HQ6, 6AG5#, 6AU6#, 6AU6A#, 6BC5#, 6CE5#, 6BH6#, 6HS6*#, 6676, 7732, EF190	6SK7	6SG7, 6SK7GT, GT/G, GTX, GTY, W, WA, WGT, Y, 6SS7*, GT*, 6137, OSW3111	7C6	7B6*, 7B6LM*, DH149
		6SK7GT	6SG7, 6SH7, 6SK7, G, GT/G, GTX, GTY, Y, 6SS7*, GT*	7C7	7AJ7*, 7G7*, 7L7*
6CD7/EM34	6CD7, EM34, 64ME	6SL7GT	6SL7A, GTY, L	7E5	1201
6CG7	6FQ7	6SN7GT	6SN7A, GTA, GTB, GTY, L, 13D2, B65, ECC32*, QA2408, QB65	7E6	-
6DA5/EM81	6DA5, EM81, 6BR5, 65ME, EM80			7E7	7R7
6DA6/EF89	6DA6, EF89	6SQ7	6SQ7G, GT, GT/G, W, 6SZ7*, OSW3105	7F7	-
6DT8	6AT7N	6SQ7GT	6SQ7, G, GT/G, 6SZ7*	7F8	7F8W
6E5	6G5#, 6S5G, OSW3110	6SS7	6SS7GT, 6SK7*, G*, GT*, GT/G*, GTX*, GTY*, Y*	7H7	7A7, LM, 7AH7*, 7B7*, EF22*, W143*, W148
6EA8	6CQ8#, 6GH8#, 6KD8#, 6MQ8*, 6U8A*	6SZ7	6SQ7*, G, GT*, GT/G*	7Q7	-
6EZ8	-	6T8	6T8A, 6AK8, 6LD12, DH719, EABC80	7R7	7E7
6F6G	6F6, GT, GT/G, MG, KT63	6T8/EABC80	6T8, EABC80, 6T8A, 6AK8, 6LD12, DH719	7S7	7J7, XB1, X148
6H6	6H6G, GT, GT/G, WGT, MG, WTT103, WT261, D63, EB34*, OSW3109	6U5	6G5, 6G5/6H5, 6H5, 6T5, 6U5/6G5	7W7	-
		6U5/6G5	6U5, 6G5, 6G5/6H5, 6H5, 6T5	7X6	-
6H6GT	6H6, G, GT/G,MG	6U8	6EA8, 6LN8, 6CQ8#, 6KD8*, 6MQ8*, 6U8A, 1252, 6678, 7731, ECF80, ECF82	7X7	XXFM
6J5	6J5G, GT, GT/G, GTX, GX, MG, WGT, 6C5, G, GT/G, MG, 6L5G, WTT129, L63, L63B, OSW3112			7Z4	-
		6V4/EZ80	6V4, EZ80	9D7	-
6J5GT	6J5, G, GT/G, GTX, GX, MG, 6C5, G, GT, GT/G, MG, 6L5G	6V6GT	6V6, G, GTA, GT/G, GTX, GTY, GX, Y, 7408, OSW3106*	12AL5	10D2, HAA91, UB91
6J6	6J6A, W, WA, 5964, 6030, 6045, 6099, 6101, 6927*, 6CC31, 6M-HH3, ECC91, M8081, T2M05	6V6GT/G	6V6, G, GT, GTA, GTX, GTY, GX, Y, 7408	12AT6	12AT6A, 12AV6, A, 12BK6, 12BT6, 12BC32, HBC90, HBC91
		6V8	-	12AT7	12AZ7*, A*, 12AT7WA, WB, 6060, 6201, 6671, 6679, 7492, 7728, A2900, B152, B309, B739, ECC81, ECC801, ECC801S, E81CC, M8162, QA2406, QB309
6J7	6J7G, GT, GTX, MG, 6W7G*, 1223, 1620, 7000, A863, EF37*, Z63	6W6GT	6W6, 6DG6GT, 6EF6*, 6EY6*, 6EZ5*		
6K6GT	6K6, G, GT/G, MG	6X4	6X4W, WA, 6BX4, 6AV4*, 6063, 6202, WTT100, 6Z31, E90Z, EZ90, EZ900, U707, V2M70, QA2407*, U78*	12AU6	12AU6A, 12BA6, A, 12AW6#, 12F31, HF93, HF94
6K6GT/G	6K6, G, GT, MG			12AU7	12AU7A, W, WA, 12AX7, A#, 5814*, 5814A, 5814WA*, 5963, 6067, 6189, 6680, 7316, 7489, 7730, B329, B749, E82CC, ECC82, ECC186, ECC802, ECC802S, M8136
6K7	6K7G, GT, GTX, MG, 6U7G, 5732, PF9, W61, W63, EF39*, OM6*, W147*	6X5GT	6X5, G, GT/G, L, MG, 6AX5GT*, 6W5G*, GT*		
6K8	6K8G, GT, GTX, WTT128	6X8	6X8A, 6AU8*#, A*#, 6AW8*#, A*#, 6EH8#	12AU7A	12AU7, ECC82, 12AX7#, A#
6L6G	6L6, A, GA, GAY, GB, GC, GT, GX, Y, 5881, 7581, 7581A	6Y6G	6Y6GA, GT, 6U6GT*	12AV6	12AV6A, 12AT6, A, 12BK6, 12BT6, 12BC32, HBC90, HBC91
6L6GA	6L6, A, G, GAY, GB, GC, GT, GX, Y, 5881, 7581, 7581A	6Y6GT	6Y6G, GA, 6U6GT*	12AW6	12AU6#, 12AU6A#
6R7	6R7G, GT, GT/G, MG	7A4	XXL	12AX7	12AX7A, WA, 12DF7, 12DT7, 7025, A, 12AD7*, 12AU7#, A#, 12BZ7*, 12DM7*, 5751*, 5751WA*, 6057, 6681, 7494, 7729, 6L13, B339, B759, E83CC, ECC83, ECC803, M8137
6S8GT	-	7A5	-		
6SA7	6SA7G, GT, GT/G, GTX, GTY, WGT, Y, 6SB7, Y, GTY, 5961, OSW3104	7A6	5679		
		7A7	7A7LM, 7H7, 7B7*, W81, W148, EF22*, W143*	12AX7A	12AX7, 12DF7, 12DT7, 7025, A, 12AD7*, 12AU7#, A#, 12BZ7*, 12DM7*
6SA7GT	6SA7, G, GT/G, GTX, GTY, Y, 6SB7, Y, GTY	7A8	7B8*, 7B8LM*	12BA6	12AU6, A, 12BA6A, 12BZ6#, 12F31, HF93, HF94
6SB7Y	6SB7, GTY	7AF7	-	12BA7	-
6SC7	6SC7GT, GTY, 1655	7B5	7B5LT, 7C5*, 7C5LT*, EL22, KT81, N148	12BD6	12BA6, 12BA6A
6SF7	6SF7GT			12BE6	12BE6A, 12CS6, 12H31, HK90
6SG7	6SG7Y, 6SH7, GT, 6AB7*#, Y*#, 6AC7#, A*#, Y*#, 6AJ7*#, 6006	7B6	7B6LM, 7C6*, DH81, DL82	12CR6	-
		7B7	7AH7, 7A7*, 7A7LM*, 7H7*, W149	12DT8	-

* OK for parallel-filament circuits   # May not work in all circuits   ! OK for series circuits not requiring controlled warm-up

TUBE	MAY BE REPLACED BY	TUBE	MAY BE REPLACED BY	TUBE	MAY BE REPLACED BY
12H6	-	25Z6GT	25Z6, G, GT/G, MG	ECH81	6AJ8, 6C12, 20D4, X719
12J5GT	12J5	35A5	35A5LT	ECH81/6AJ8	6AJ8, ECH81, 6C12, 20D4, X719
12SA7	12SA7G, GT, GT/G, GTY, Y, 12SY7, GT	35B5	-	ECL82	6BM8, 6PL12, ECL82
		35C5	35C5A, 30A5, HL94	ECL82/6BM8	6BM8, ECL82, 6PL12
12SA7GT	12SA7, G, GT/G, GTY, Y, 12SY7, GT	35L6	35L6G, GT, GT/G	EF85	6BY7, 6BX6, 6BY8, 6F19, 6F26, EF85, W719
12SA7GT/G	12SA7, G, GT, GTY, Y, 12SY7, GT	35L6GT	35L6G, GT/G		
		35W4	35W4A, HY90	EF89	6DA6
12SG7	12SG7GT, Y	35W4GT	-	EF89/6DA6	6DA6, EF89
12SJ7GT	12SJ7	35Y4	-	EL34	6CA7, 7D11, 12E13, KT77, KT88
12SK7	12SK7G, GT, GT/G, GTY, Y, 5661	35Z5	35Z5G, GT, GT/G	EL84	6BQ5, 7189, 7189A, 7320, 6P15, EL84L, N709
12SK7GT	12SK7, G, GT/G, GTY, Y	35Z5GT	35Z5, G, GT/G		
12SK7GT/G	12SK7, G, GT, GTY, Y	50A5	-	EL84/6BQ5	6BQ5, EL84, 7189, 7189A, 7320, 6P15, EL84L, N709
12SL7GT	2C52*	50B5	WTT126		
12SQ7	12SQ7G, GT, GT/G, OBC3	50C5	50C5A, HL92	EL86	6CW5
12SQ7GT	12SQ7, G, GT/G	50C6G	50C6GA	EM80/6BR5	6BR5, EM80, 6DA5, 6DA5
12SQ7GT/G	12SQ7, G, GT	50DC4	-	EM81	6DA5, 6BR5, 65ME, EM80
14A7	12B7, ML, 14H7, 14A7/12B7, 14A7ML, 14A7ML/12B7ML	50EH5	50EH5A, 50CA5	EM81/6DA5	6DA5, EM81, 6BR5, 65ME, EM80
		50L6GT	50L6G, KT71	EM84	6FG6, EM840
14A7/12B7	12B7, ML, 14A7, 14H7, 14A7ML, 14A7ML/12B7ML	50X6	-	EM84/6FG6	6FG6, EM84, EM840
		50Y6	-	EM840	6FG6
14AF7	XXD, 14N7*	50Y6GT	50Y6G, GT/G	EZ80	6V4
14B6	-	50Y7GT	50Z7G, GT	EZ81	6CA4, U709, UU12
14C7	1280	117Z3	-	EZ81/6CA4	6CA4, EZ81, U709, UU12
14F8	-	117Z6GT	117Z6G, GT/G, WT377	FM1000	-
14H7	12B7, ML, 14A7, 14A7/12B7, 14A7ML, 14A7ML/12B7ML	1273	7AJ7	UABC80	-
		6973	-	UCC85	26AQ8
14Q7	-	7025	7025A, 12AX7, A, 12DF7, 12DT7, 12AD7*, 12BZ7*, 12DM7*, B339, B759, ECC83, M8137	UCH81	19D8
14X7	-			UF89	12DA6
19C8	19T8, 19T8A, HABC80			UL84	45B5, 10P18, N119
19J6	-	9002	-	UM80	19BR5
19T8	19T8A, 19C8, HABC80	EABC80	6T8A, 6AK8, 6LD12, DH719	UY85	38A3
19V8	-	EABC80/6T8	6T8, EABC80, 6T8A, 6AK8, 6LD12, DH719		
19X8	-				
20D4	6AJ8	EBF89	6DC8, 6AD8, 6N8, 6FD12, EBF85		
25F5	25F5A	EC84	6AJ4		
25L6GT	25L6, G, GT/G, 25W6GT, 6046	EC92/6AB4	6AB4, EC92, 6664		
25Z5GT	-	ECC85	6AQ8, 6L12, B719		
		ECC85/6AQ8	6AQ8, ECC85, 6L12, B719		

Manufacturers and Their Brands

MANUFACTURER/SUPPLIER	CITY	BRANDS
A.E. Dufenhorst Co.	Rockford, IL	Saba
Admiral	Chicago, IL	Admiral
Aermotive Equipment Corp.	Kansas City, MO	Aeromotive
Affiliated Retailers	New York, NY	Artone
Air King Products Co.	Brooklyn, NY	Air King
Airadio Inc.	Stanford, CT	Airadio
Alamo Electronics	San Antonio, TX	Radioette
Algene Radio Corp.	Brooklyn, NY-	Algene
Allen B. Dumont Laboratories	Passaic, NJ	Dumont
Allied Radio Corp.	Chicago, IL	Knight
American Communications Co.	-	Liberty
American Elite	New York, NY	Telefunken
American Geloso Electric	New York, NY	Geloso
Andrea Radio Corp.	Long Island City, NY	Andrea
Ansley Radio	Trenton, NJ	Ansley
Apex Electric Mfg. Co.	Chicago, IL	Apex
Arthur Ansley Mfg.	Doylestown, PA	Arthur Ansley
Arvin Industries	Columbus, IN	Arvin
Associated Merchandising Corp.	-	AMC
Atomic Heater/Radio	New York, NY	Atlas
Audar, Inc.	Argos, IN	Audar
Audio Industries	Michigan City, IN	Ultratone
Auto Musical Instruments	Grand Rapids MI	AMI
Automatic Radio Mfg. Co.	Boston, MA	Automatic
Aviola Radio Corp.	-	Aviola
B.F. Goodrich Tire & Rubber Co.	Akron, OH	Mantola
Bell Sound Div.	Columbus, OH	Pacemaker
Belmont Radio Corp.	Chicago, IL	Belmont
Bendix Radio Div., Bendix Corp.	Baltimore, MD	Bendix
Benrus Watch Co.	New York, NY	Benrus
Blonder-Tongue Laboratories	-	Blonder-Tongue
Burstein-Applebee Co.	-	Calrad
Butler Brothers	Chicago, IL	Air Knight, Sky Rover
Capehart-Farnsworth Corp.	Ft. Wayne, IN	Capehart
Capital Appliance Distributor Div.	-	Robin
CBS Columbia Div.	Long Island City, NY	CBS
CBS Electronics	-	Columbia
CBS-Hytron	-	Columbia
Channel Master Corp.	Ellenville, NY	Channel Master
Cities Service Oil Co.	New York, NY	Cisco
Collins Radio Co.	Cedar Rapids, IA	Collins
Columbia Phonographs	New York, NY	Columbia
Columbia Records	New York, NY	Columbia
Concord Radio Corp.	Chicago, IL	Concord
Continental	-	Sharp
Continental Electronics Ltd.	New York, NY	Continental Electronics

MANUFACTURER/SUPPLIER	CITY	BRANDS
Continental Merchandise Co.	New York, NY	Continental
Coronado	Minneapolis, MN; Los Angeles, CA	Coronado
Crescent Industries	Chicago, IL	Crescent
Crosley Corp.	Cincinnati, OH	Crosley
Crystal Products Co.	-	Coronet
Curtis Mathes Manufacturing Co.	-	Curtis Mathes
Dalbar Manufacturing Co.	Dallas, TX	Dalbar
David Bogen Co.	Paramus, NJ	David Bogen
Dearborn	Chicago, IL	Dearborn
Delco Radio, Div. of GM Corp.	Kokomo, IN	Delco
Delmonico International	Long Island City, NY	Delmonico, Emud, Sony
DeWald Radio Mfg. Co.	Long Island City, NY	DeWald
Dynamic Electronics	Richmond Hill, NY	Dynamic
Dynavox	Long Island City, NY	Dynavox
Eckstein Radio & Television Co.	-	Karadio
Electro Appliances Mfg. Co.	New York, NY	Electro
Electro-Tone Corp.	Hoboken, NJ	Electro-Tone
Electro-Voice	Buchanan, MI	Electro-Voice
Electromatic	New York, NY	Electromatic
Electronic Corp. of America	Brooklyn, NY	ECA
Electronic Devices Co.	New York, NY	20th Century
Electronic Laboratories	Indianapolis, IN	Electronic Labs
Electronic Specialty	Los Angeles, CA	Ranger
Electronic Utilities	Chicago, IL	Hitachi, Braun
Electronics Guild	Long Island, NY	Bulova
Emerson Radio & Phonograph Corp.	Jersey City, NJ; New York, NY	Emerson
Empire Designing Corp.	New York, NY	Empress
Espey Mfg. Co.	New York, NY	Espey, Philharmonic
Esquire Radio Corp.	New York, NY	Esquire
Excel Corp. of America	New York, NY	Excel, Toshiba
Fada Radio & Electric Co.	Long Island City, NY; Belleville, NJ	Fada
Fanon Electronic Sales Corp.	-	Fanfare
Farnsworth Television & Radio Corp.	Ft. Wayne, IN	Capehart, Farnsworth
Federal Telephone & Radio	Newark, NJ	Federal
Ferrar Radio & Television	New York, NY	Ferrar
Firestone Tire & Rubber Co.	Akron, OH	Air Chief, Firestone
Fisher Radio Corp.	Long Island, NY	Fisher
Flush-Wall Radio Mfg.	Newark, NY	Flush-Wall
Galvin Mfg. Corp.	Chicago, IL	Motorola
Gamble-Skogmo	Minneapolis, MN	Coronado
Garod Radio Corp.	Brooklyn, NY	Garod

MANUFACTURER/SUPPLIER	CITY	BRANDS
General Electric (GE)	Utica, NY; Syracuse, NY; Bridgeport, CN	General Electric
General Implement	Cleveland, OH	General Implement
General Television & Radio Corp.	Chicago, IL	General Television
Gilfillan Brothers	Los Angeles, CA	Gilfillan
Globe Electronics	New York, NY	Globe
Gonset	Burbank, CA	Gonset
Granco Products	Long Island City, NY	Granco
Grossman Music Co.	-	Stratovox
Guild Radio & Television Co.	Inglewood, CA	Guild
Hallicrafters Co.	Chicago, IL	Echophone, Hallicrafters
Hammarlund Mfg.	New York, NY	Hammarlund
Harman-Kardon	Westbury, NY; Long Island, NY	Harman-Kardon
Harvey-Wells Electronics	Southbridge, MA	Harvey-Wells
Heath Co.	Benton Harbor, MI	Heath
H.H. Scott	Maynard, MA; Cambridge MA	H.H. Scott
Hinners-Galanek Radio Corp.	Long Island City, NY	Cavalier
Hoffman Radio Corp.	Los Angeles, CA	Hoffman
Howard Radio	Chicago, IL	Howard
I.D.E.A.	Indianapolis, IN	Regency, Monitoradio, Policalarm
Industrial Electronic	New York, NY	Simplon
Intercontinental Industries, Inc.	-	Minute Man
International Detrola	Detroit, MI	Aria, Detrola
Jackson Industries	Chicago, IL	Jackson
Jefferson-Travis	New York, NY	Jefferson-Travis
Jewel Radio Corp.	Newark, NJ; Long Island City, NY	Belltone
John Meck Industries	Plymouth, IN	Meck, Trail Blazer, Mirror-Tone
J.W. Davis Co.	-	Watterson
J.W. Miller	Los Angeles, CA	Miller
Kanematsu	New York, NY	NEC
Kappler Co.	Los Angeles, CA	Kappler
La Magna Mfg. Co.	E. Rutherford, NJ	Lamco
Lafayette Radio	Jamaica, NY	Kowa, Lafayette
Lawrence Co., Radio Div.	Cincinnati, OH	Bagpiper
Lear Inc.	Grand Rapids, MI	Learadio
Leopold Sales Corp.	-	Nanola (Nanao)
Lewyt Corp.	New York, NY	Lewyt
Lindex	New York, NY	Swank
Maco Electric Corp.	-	Maco
Madison Fielding	New York, NY	Madison Fielding
Magnavox Co.	Ft. Wayne, IN; Oakland, CA	Magnavox, Spartan
Maguire Industries Inc.	Greenwich, CT	Maguire
Majestic International Sales Corp.	Chicago, IL	Grundig Majestic
Majestic Radio & Television	Elgin, IL; St Charles, IL	Majestic

MANUFACTURER/SUPPLIER	CITY	BRANDS
Mark Simpson Co.	Long Island, NY	Masco
Mason Radio Products	Kingston, NY	Mason
Meissner Mfg. Co.	Mt. Carmel, IL	Meissner
Midwest Radio & Television	Cincinnati, OH	Midwest
Minerva Corp. of America	New York, NY	Minerva
Mitchell Mfg. Co.	Chicago, IL	Mitchell
Molded Insulation	Philadelphia, PA	VIZ
Monitor Equipment Co.	New York, NY	Monitor
Monitoradio Div. I.D.E.A.	Indianapolis, IN	Monitoradio, Policalarm
Monitro	New York, NY	Monitor
Montgomery Ward & Co.	Chicago, IL	Airline
Motorola Inc.	Chicago, IL	Motorola
Muntz TV	Evanston, IL	Muntz
N. Pickens Import Co.	-	Blaupunkt
National Co.	Malden, MA	National
National Co-op	Chicago, IL	Co-op
National Union Radio Corp.	Newark, NJ	National Union
Newcomb Audio Products	Hollywood, CA	Newcomb
Noblitt Sparks Industries	Columbus, IN	Arvin
North American Philips Co. Inc.	-	Noreloo
Northeastern Engineering Inc.	-	Electone
Olson Radio Corp.	-	Olson
Olympic, Div. of Siegler Corp.	-	Olympic-Continental
Olympic-Opta	Long Island City, NY	Cremona, Magnet
Olympic Radio & Television	Long Island City, NY	Olympic
Packard-Bell Electronics Corp.	Los Angeles, CA	Packard-Bell
Pedersen Electronics	Lafayette, CA	Pederson
Pentron Corp.	Chicago, IL	Astra-Sonic, Pentron
Philco Corp.	Philadelphia, PA	Philco
Philharmonic Radio Corp.	New York, NY	Philharmonic
Phillips Petroleum Co.	Bartlesville, OK	Woolaroc
Phillips Radio	Kokomo, IN	Phillips
Pilot Radio	Long Island City, NY	Pilot
Polyrad	Cincinnati, OH	Standard
Porto Products	Chicago, IL	Porto Baradio
Precision Electronics Inc.	Franklin Park, IL	Grommes
Premier Crystal Laboratories Inc.	New York, NY	Premier
Pure Oil Co.	Chicago, IL	Puritan
Radiaphone	Los Angeles, CA	Mayfair
Radio & Television Inc.	New York, NY	Brunswick
Radio Apparatus Co.	Indianapolis, IN	Monitoradio, Policalarm
Radio Corp. of America	Camden, NJ	Radiola, RCA Victor
Radio Craftsmen	Chicago, IL,	Radio Craftsmen, Kitchenaire
Radio Development & Research Corp.	New York, NY	Magic Tone

MANUFACTURER/SUPPLIER	CITY	BRANDS
Radio Manufacturing Engineers, Inc.	Peoria, IL	RME
Radio Wire & Television Inc.	New York, NY	Lafayette
Radionic	New York, NY	Chancellor
Rauland-Borg	Chicago, IL	Rauland
Rauland Corp.	-	Lyric
Rayenergy Radio & Television Corp.	New York, NY	Rayenergy
Raytheon Television & Radio Corp.	Chicago, IL	Raytheon
RCA Engineering Products	Camden, NJ	RCA Victor
RCA Home Instrument Division	Camden, NJ	Radiola
RCA Victor	Camden, NJ	RCA Victor
Realtone Electronics	-	Realtone
Regal Electronics Corp.	New York, NY	Regal
Regency Division I.D.E.A. Inc.	Indianapolis, IN	Regency
Remler Co.	San Francisco, CA	Remler
Renard Radio Mfg. Co.	-	Renard
Revere Camera Co.	Chicago, IL	Revere
Roland Radio Corp.	Mt. Vernon, NY	Roland
Royal	New York, NY	Royal
Sarkes Tarzian, Inc.	-	Sarkes Tarzian
SC Ryan	Minneapolis, MN	Darb
Scott Radio Laboratories Inc.	Chicago, IL	Scott
Sears & Roebuck & Co.	Chicago, IL	Silvertone
Sentinel Radio Corp.	Evanston, IL; Ft. Wayne, IN	Sentinel
Setchell-Carlson Inc.	New Brighton, MN; St. Paul, MN	Setchell-Carlson
Sheridan Electronics	Chicago, IL	Vogue
Sherwood Electronic Laboratories	Chicago, IL	Sherwood
Shriro, Inc.	New York, NY	Crown, Linmark
Signal Electronics, Inc.	New York, NY	Signal
Sonic Industries Inc.	Lynbrook, NY	Sonic
Sonora Radio & Television Corp.	Chicago, IL	Sonora
Sound	Chicago, IL	Sound
Sparks-Withington Co.	Jackson, MI	Sparton
Spartan	Ft. Wayne, IN	Spartan
Speigel Inc.	Chicago, IL	Aircastle, Continental
Stark	Ft. Wayne, IN	Cromwell, Plymouth, Stark
Steelman Phonograph & Radio Co.	Mt. Vernon, NY	Steelman
Stewart-Warner Electric Corp.	Chicago, IL	Stewart-Warner
Stewart-Warner Radio Corp.	Chicago, IL	Stewart-Warner
Stromberg-Carlson Co.	Rochester, NY	Stromberg-Carlson
Superex Electronics Corp.	-	Superex
Sylvania Electric Products	Buffalo, NY	Sylvania
Sylvania Home Electronics	Batavia, NY	Sylvania
Symphonic Radio & Electronics Corp.	-	Symphonic
Tech-Master Corp.	-	Tech-Master
Tele King Corp.	New York, NY	Tele King
Tele-Tone Radio Corp.	New York, NY	Tele-Tone
Telesonic Corp. of America	-	Telesonic (Medco)
Televox, Inc.	Mt. Vernon, NY	Televox
Templetone Radio Mfg.	New London, CT	Temple
Transistor World Corp.	-	Toshiba
Trav-ler Karenola Radio & Television Corp.	Chicago, IL	Trav-ler
Union Electric	Long Island City, NY	Unitone
United Motors, GM Building	Detroit, MI	Delco
US Televison Mfg. Co.	New York, NY	Clearsonic
V-M Co.	Benton Harbor, MI	V-M
Van Camp Hardware & Iron Co.	Indianapolis, IN	Van Camp
Videola-Erie	Brooklyn, NY	Fonovox, Tonfunk
W.T. Grant Co.	New York, NY	Grantline
Warwick Mfg. Corp.	Chicago, IL	Clarion
Waterproof Electric Co.	Burbank, CA	Gon-set
Watterson Radio Mfg. Corp.	Dallas, TX	Watterson
Webcor Inc.	-	Webcor
Webster	Chicago, IL	Webcor
Wells-Gardner & Co.	-	Wells-Gardner
Westinghouse Electric Co.	Sunbury, PA	Westinghouse
Westinghouse Home Radio Div.	Sunbury, PA	Westinghouse
Western Auto Supply Co.	Kansas City, KS	Truetone
Whitney & Co.	-	Arcadia
Wilcox-Gay Corp.	Brooklyn, NY	Majestic
Wilcox-Gay Corp.	Charlotte, MI	Recordio
Wilmak Corp.	-	Wilmak
Zenith Radio Corp.	Chicago, IL	Zenith

FIXING UP YOUR OLD RADIO

Here are some basic troubleshooting hints to help you identify and solve typical problems you might encounter when working on your radio.

First Step: DON'T PLUG IT IN!

For your own safety and to avoid circuit damage, make a visual inspection of the radio before powering up:

- Check the power cord for dried-out, cracked insulation and bare spots.

- Check the circuit wire insulation for bare spots. If the radio has been stored in an attic or shed, mice and other animals may have nested in it and chewed the wires.

- Check the dial cord, if your set has one, for frayed spots.

- Check the speaker for cracks and tears. Because of the age of the radio you're working on, the speaker has probably deteriorated over time. You'll likely need to replace or repair it.

- Check to make sure the correct tube type is in each socket. If you need a schematic to verify placement, order the appropriate PHOTOFACT.

- Check all fuses for obvious breaks.

Then use an AC-DC variable isolation transformer to power the set up *gradually*. This gradual power-up avoids a sudden current surge which could damage defective circuits. It can also allow old, dried-out electrolytic capacitors to regenerate themselves, making replacement unnecessary.

Troubleshooting the Symptoms

Once you've identified and corrected any of the above-mentioned problems, you may encounter less obvious problems to bringing the radio up to full working order. On the next few pages are nine common complaints, the areas to check, the nature of the problem, and several possible causes listed in order of probability.

	CHECK	PROBLEM	POSSIBLE CAUSE
Set doesn't play at all	Power supply	Not all tubes glow or warm up	Defective tube
			Tubes in wrong sockets
			Defective line cord
			Defective line switch
		Abnormal or no hum level	Shorts or opens in filter capacitors
		Abnormal voltage on B plus line	Defective rectifier tube
			Open filter resistor
	Audio frequency section	Abnormal or no test tone response	Defective tube
			Defective first AF tube
			Shorted plate bypass capacitor
			Open coupling capacitor
			Open cathode resistor
			Open first AF plate resistor
			Open output transformer
			Shorted AF grid
			Shorted first AF plate
			Shorted first AF grid
			Defective speaker
	Intermediate frequency and detector section	No signal note heard	Defective IF tube
			Defective detector tube
			Short or open in volume control
			Defective IF transformer
			Shorted IF grid circuit
			Antenna/oscillator misaligned

	CHECK	PROBLEM	POSSIBLE CAUSE
Set doesn't play at all (cont'd)	Converter	No signal note heard	Defective converter tube
			Open converter cathode circuit
			Short in oscillator tuning capacitor
			Open input IF transformer
			Open oscillator coil
			Short in oscillator-grid circuit
			Shorted converter-grid circuit

	CHECK	PROBLEM	POSSIBLE CAUSE
Set doesn't pick up enough stations	Power supply	Low sensitivity due to set location or insufficient gain	Poor set location
	Audio frequency section		No antenna
	Intermediate frequency and detector section		Weak tube
	Converter		Defective loop antenna
			Misalignment
			Shorted or open AVC bypass capacitor
			Conductive dust in the gang tuning capacitor

	CHECK	PROBLEM	POSSIBLE CAUSE
Set fades	Connections	Radio operates intermittently	Thermal tube
	Tubes		House voltage change
	Wiring		Loose connection anywhere in radio

	CHECK	PROBLEM	POSSIBLE CAUSE
Set doesn't play loudly enough	Power supply Audio frequency section Intermediate frequency and detector section Converter	Low volume due to weak response or defective components	Weak rectifier tube Weak tube(s) in audio stages Defective filter capacitors Defective coupling capacitor Resistors changing in value Defective output transformer Defective speaker

	CHECK	PROBLEM	POSSIBLE CAUSE
Poor tone quality	Speaker Filter system Tubes Capacitors	Poor tone due to hum or distortion in audio circuits	Defective tubes Incorrect cathode voltage Defective filter capacitors Leaking coupling capacitor AVC bypass capacitor leakage Open volume control Defective cathode-bias resistor Lower-than-normal plate voltage Positive voltage on control grid Rubbing voice coil Rattling cone

	CHECK	PROBLEM	POSSIBLE CAUSE
Hum at all points of tuning range	Filter capacitors Tubes Grid circuits Capacitor blocks	Hum interferes with tone quality	Tube cathode-heater leakage Old or defective filter capacitors Leakage between capacitor block sections Open grid circuit

	CHECK	PROBLEM	POSSIBLE CAUSE
Hum only at specific points of tuning range	Capacitors	Hum interferes with tone quality	Overload on strong signal due to defective AVC capacitor Defective capacitor

	CHECK	PROBLEM	POSSIBLE CAUSE
Squealing and motorboating	Output filter capacitor Grid circuits Bypass capacitor Shield cans Wire dress	Interference with audio output across the entire tuning range	Noisy defective tubes Old output filter capacitor Open bypass capacitor Open grid circuit Ungrounded shield can Incorrectly dressed wiring
		Interference on a specific station only	Image-frequency interference

	CHECK	PROBLEM	POSSIBLE CAUSE
Set makes crackling noises	Set location Connections Tubes Wiring	Noise interferes with audio quality	Set may receive outside electrical noise Defective tube(s) Corrosion in transformer windings Loose or poorly soldered connections Wires may be touching

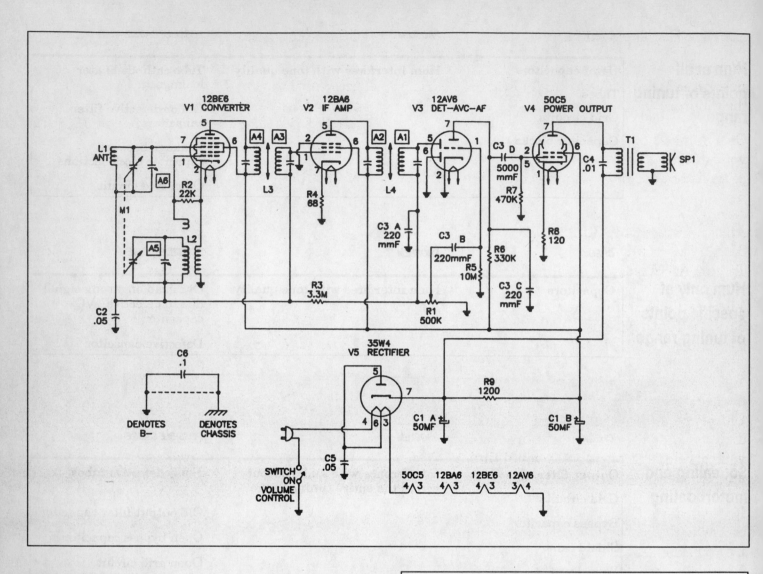

Typical Schematic

This 5-tube schematic represents the basic superheterodyne receiver. Tubes added to this basic configuration enhance the radio's performance by providing power regulation, audio gain, other band frequencies, better sensitivity, and other uses.

If you want a complete schematic for your specific radio, order the appropriate PHOTOFACT.

Tubes		C5	Line Filter
V1	Converter	C6	Line Isolation
V2	IF Amp		
V3	DET. -AVC-AF	**Controls**	
V4	Power Output	R1A	Volume Control
V5	Rectifier	R1B	Power Switch

Capacitors		Resistors	
C1A	Filter-Red	R2	Oscillator Grid
C1B	Filter-Green	R3	AVC Network
C2	AVC Filter	R4	IF Cathode
C3A	Diode RF Filter	R5	AF Amplifier
C3B	Audio Coupling		Grid
C3C	AF Amp Plate Bypass	R6	AF Amplifier Plate
C3D	Audio Coupling	R7	Output Grid
C4	Output Plate Bypass	R8	Output Cathode
		R9	Filter

Locating Replacement Parts

Worn or defective parts can be replaced with new pieces available through electronic parts distributors.

Call Sams at 800-428-7267 for the name of your nearest distributor. Or order unusual pieces from specialty supply companies such as:

Antique Electronic Supply
6221 S. Maple Avenue.
Tempe, AZ 85283

Vintage TV & Radio
3498 W. 105th Street
Cleveland, OH 44111

You can also find a lot of still-functioning pieces at your local radio club swap meet. Using the indexes in this book, you'll discover models similar to yours from which you can salvage parts, and you'll find appropriate tube substitutions.

If you order a PHOTOFACT for your radio, it comes with a complete parts list to help you in your search.

Other Resources

A wealth of technical and collecting information is available from other radio enthusiasts through special-interest clubs and publications. Here are a few to get you started.

Magazines   Radio Age
636 Cambridge Road
Augusta, GA 30909

Antique Radio Classified
498 Cross Street
P.O. Box 2
Carlisle, MA 01741

National Clubs   Antique Wireless Assn.
Main Street
Holcomb, NY 14469
716-657-7489

Antique Radio Club
of America
81 Steeplechase Road
Devon, PA 19333
215-688-2976

If you enjoyed this volume and want to collect all six in the Radios of the Baby Boom Era series, look for them at your authorized Sams distributor or your favorite bookstore.

TO ORDER SAMS PHOTOFACT®
SERVICE DATA

If you'd like to work on a radio but need more information than we provide here, simply order a Sams PHOTOFACT for that model.

PHOTOFACT service data includes schematics, parts lists, alignments/adjustments, testing procedures, and troubleshooting guidelines. A PHOTOFACT is available for every radio listed in this book.

To identify the PHOTOFACT for your particular radio, find the radio among the pictures, or locate the model number in the index entitled "Pictured Radios and Similar Models." A PHOTOFACT set number is provided there for each model. This is the number to order.

Your local authorized Sams distributor carries a large stock of PHOTOFACT service data. Call Sams customer service at

800-428-7267

for the name of your local distributor or to place a phone order. When you call, ask for Operator RB-4.

Howard W. Sams & Company is a recognized leader in technical publishing. Nearing our first half-century in the service documentation business, we have covered more than 150,000 models of products varying from radios and TVs to ovens and helicopters. When Sams provides technical information, you can count on it being accurate and complete.